D1309358

Simio and Simulation:
Modeling, Analysis, Applications
Third Edition

W. David Kelton (University of Cincinnati)
Jeffrey S. Smith (Auburn University)
David T. Sturrock (Simio LLC)

Published by Simio LLC
www.simio.com

Simio and Simulation:
Modeling, Analysis, Applications

Third Edition
Revision: February 4, 2014

Copyright ©2014 by W. David Kelton, Jeffrey S. Smith, and David T. Sturrock. All rights reserved. Except as permitted under the United States Copyright Act of 1976, no parts of this publication may be reproduced or distributed in any form or by any means, or stored in a data base retrieval system, without prior written permission of the publisher.

We welcome feedback and other contributions from instructors and students to textbook@simio.com. This textbook was written for Simio 5.91 or later.

- Evaluation software is available for download at www.simio.com.

- Professors desiring a grant for using the full Simio Academic Version at no charge can apply at www.simio.com/academics.

- Registered students may use university-supplied software at no charge or may obtain a personal copy for a nominal fee through their professor.

Simio is a registered trademark of Simio LLC. Microsoft, Windows, Excel, Word, PowerPoint, and Visual Studio are either trademarks or registered trademarks of Microsoft Corporation. OptQuest is a registered trademark of OptTek Systems, Inc. Stat::Fit is a registered trademark of Geer Mountain Software Corporation. @RISK and StatTools are registered trademarks of Palisade Corporation. Minitab is a registered trademark of Minitab, Inc. Flexsim is a trademark of Flexsim software products. Arena is a trademark of Rockwell Automation. ILOG is a trademark of IBM. APO-PP/DS is a trademark of SAP. All other trademarks and registered trademarks are acknowledged as being the property of their respective owners. The instructions in this book do not cover all the software aspects or details of the Simio software. Consult the product documentation for the latest and most complete descriptions.

Preface

This third edition explains how to use simulation to make better business decisions in application domains from healthcare to mining, heavy manufacturing to supply chains, and everything in between. It is written to help both technical and non-technical users better understand the concepts and usefulness of simulation. It can be used in a classroom environment or in support of independent study. Modern software makes simulation more useful and accessible than ever and this book illustrates simulation concepts with Simio®, a leader in simulation software.

In this edition we added sections on Randomness in Simulation, Model Debugging, and Monte Carlo simulation, and a new appendix on simulation-based scheduling. In addition, the coverage of animation, input analysis and output analysis has been significantly expanded. End-of-chapter problems have been improved and expanded, and we have incorporated many reader suggestions. We have reorganized the material for an improved flow, and have updates throughout the book for many of the new Simio features recently added. A new format better supports our e-book users, and a new publisher supports significant cost reduction for our readers.

This book can serve as the primary text in first and second courses in simulation at both the undergraduate and beginning-graduate levels. It is written in an accessible tutorial-style writing approach centered on specific examples rather than general concepts, and covers a variety of applications including an international flavor. Our experience has shown that these characteristics make the text easier to read and absorb, as well as appealing to students from many different cultural and applications backgrounds.

A first simulation course would probably cover Chapter 1 through 8 thoroughly, and likely Chapters 9 and 10, particularly for upper class or graduate-level students. For a second simulation course, it might work to skip or quickly review Chapters 1-3 and 6, thoroughly cover all other chapters up to Chapter 10, and use Chapter 11 as reinforcing assignments.

The text or components of it could also support a simulation module of a few weeks within a larger survey course in programs without a stand-alone simulation course (e.g., MBA). For a simulation module that's part of a larger survey course, we recommend concentrating on Chapters 1, 4, and 5, and then perhaps lightly touch on Chapters 7 and 8.

The extensibility introduced in Chapter 10 could provide some interesting project work for a graduate student with some programming background, as it could be easily linked to other research topics. Likewise Appendix A could be used as the lead-in to some advanced study or research in the latest techniques in simulation-based planning and scheduling.

We assume basic familiarity with the Microsoft® Windows® operating system and common applications like Microsoft Excel® and Microsoft Word®. This book also assumes prior coursework in, and comfort with, probability and statistics. Readers don't need to be experts, but do need command of the basics of probability and statistics; more specific topics are outlined at the beginning of Chapters 2 and 6.

This textbook was written for use with the Simio simulation software. The following Simio products are available for academic use:

- The *Simio Evaluation Edition* permits full modeling capability but supports saving and experimentation on only small models (up to 18 objects and 16 steps). It can be downloaded without cost from `www.simio.com/evaluate.php`. While this is useful for personal learning and short classes, the small-model limitation generally makes it inadequate for a classroom environment where larger problems and projects will be involved.

- The *Simio Academic Edition* is full featured software equivalent to the commercially available Simio Design Edition. In many regions (including the USA) it has no model-size limitations; in other areas it is limited to models up to 200 objects (fairly large). In all cases it is limited to non-commercial use (see full details at `www.simio.com/academics/simio-academic-simulation-products.htm`) and limited to be used only on university and instructor's computers. *Instructors desiring a grant for using Simio at no charge for their department and labs can apply at* `www.simio.com/academics`.

- The *Simio Student Edition* is identical to the Simio Academic Edition, but licensed for use by students on their own computers. A one-year license is available for a nominal fee to students who are registered in an accredited course. Note that *students may use a university-supplied Simio Academic Edition at no charge.* Instructors who have obtained a software grant should look to their activation letters for instructions on how to arrange software availability for their students, or instructors may contact `academic@simio.com`. Students should contact their instructors for availability.

Simio follows an agile development process — there are minor releases about every three weeks, and major releases about every eight months. This is good from the standpoint of having new features and bug fixes available as soon as they're created. It's bad from the standpoint of "keeping up" — downloading, learning, and documenting. This textbook edition was written for use with Simio Version 5.91 (the last major release) or later[1]. While new features will continue to be added, the concepts presented in this edition should be accurate for any version 5.xx and beyond. The examples and figures were created with

---

[1] If you are using a newer version of Simio, look to the student area of the textbook web site where supplemental on-line content will be posted as it becomes available.

Simio Version 5.91 — using a different version may produce slightly different results (see the explanation in Chapters 1 and 4).

Supplemental course material is also available on-line. On-line resources are available in three categories. A web site containing general textbook information and resources available to the public can be found at `www.simio.com/publications/SASMAA`. Information and resources available only to students are available via links on that page: here you'll find the model files and other files used in the examples and end-of-chapter problems, additional problems, and other useful resources. The username is `student` and the password is `Reg!stered5tudent`. This student area of the web site will also contain post-publication updates, such as Version 6-specific information. There are special restricted-access links also on that page that are available to instructors, which contain slides and other helpful teaching resources. An instructor of record must contact Simio (`academic@simio.com`) for the login information.

Many people helped us get to this point. First, as co-author on the first edition, Dr. Alexander Verbraeck has provided immeasurable contributions to the structure, quality, and content. Dr. C. Dennis Pegden provided important contributions to the scheduling appendix. The Simio LLC technical staff — Cory Crooks, Glenn Drake, Glen Wirth, Dan Spice, Dave Takus, Renee Thiesing, and Christine Watson — were great in helping us understand the features, find the best way to describe and illustrate them, and even provided proofreading and help with the case studies. Jan Burket helped us with proofreading. Molly Arthur, Christie Miller, and Rich Ritchie of Simio LLC provided great support in helping get the word out and working with early adopters. Hazal Karaman, Cagatay Mekiker, Karun Alaganan, and Daniel Marquez generously shared their University of Pittsburgh class projects for our case studies. From Auburn University, Chris Bevelle and Josh Kentrick worked on the new introductory case studies and James Christakos and Yingde Li worked on material for the first edition. While we appreciate the participation of all of the early adopters, we'd like to give special thanks to Jim Grayson, Josh Kendrick, Gary Kochenberger, Deb Medeiros, Barry Nelson, Leonard Perry, and Laurel Travis (and her students at Virginia Tech) for providing feedback to help us improve.

W. David Kelton
University of Cincinnati
david.kelton@uc.edu

Jeffrey S. Smith
Auburn University
jsmith@auburn.edu

David T. Sturrock
Simio LLC and the University of Pittsburgh
dsturrock@simio.com

Please send feedback to any of the above authors or to textbook@simio.com

To those most important to us:

Albert, Anna, Anne, Christie, and Molly
Drew, Katy, and Kristi
Diana, Kathy, Melanie, and Victoria

About the Authors

W. David Kelton is a Professor in the Department of Operations and Business Analytics at the University of Cincinnati, where he also serves as Director of the MS Program in Quantitative Analysis. He received a BA in mathematics from the University of Wisconsin-Madison, an MS in mathematics from Ohio University, and MS and PhD degrees in industrial engineering from Wisconsin. He was formerly on the faculty at Penn State, the University of Minnesota, The University of Michigan, and Kent State. Visiting posts have included the Naval Postgraduate School, the University of Wisconsin-Madison, the Institute for Advanced Studies in Vienna, and the Warsaw School of Economics. He is a Fellow of both INFORMS and IIE.

Dr. Kelton's research interests and publications are in the probabilistic and statistical aspects of simulation, applications of simulation, statistical quality control, and stochastic models. His papers have appeared in *Operations Research*, *Management Science*, the *INFORMS Journal on Computing*, *IIE Transactions*, *Naval Research Logistics*, the *European Journal of Operational Research*, and the *Journal of the American Statistical Association*, among others. He is co-author of *Simulation with Arena*, which received McGraw-Hill's award for Most Successful New Title in 1998. He was also coauthor, with Averill M. Law, of the first three editions of *Simulation Modeling and Analysis*. He was Editor-in-Chief of the *INFORMS Journal on Computing* from 2000 to mid-2007; he has also served as Simulation Area Editor for *Operations Research*, the *INFORMS Journal on Computing*, and *IIE Transactions*.

He has received the TIMS College on Simulation award for best simulation paper in *Management Science*, the IIE Operations Research Division Award, the INFORMS College on Simulation Distinguished Service Award, and the INFORMS College on Simulation Outstanding Simulation Publication Award. He was President of the TIMS College on Simulation, and was the INFORMS co-representative to the Winter Simulation Conference Board of Directors from 1991 through 1999, where he served as Board Chair for 1998. In 1987 he was Program Chair for the WSC, and in 1991 was General Chair.

Jeffrey S. Smith is the Joe W. Forehand Professor of Industrial and Systems Engineering at Auburn University and a founding partner of Conflexion, LLC. Prior to his position at Auburn, he was an Associate Professor of Industrial Engineering at Texas A&M University. In addition to his academic positions, Dr. Smith has held professional engineering positions with Electronic Data Systems (EDS) and Philip Morris USA. Dr. Smith has a BS in Industrial Engineering from Auburn University and MS and PhD degrees in Industrial Engineering from The Pennsylvania State University. His primary research interests are in manufacturing systems design and analysis, and discrete-event simulation.

Dr. Smith is on the editorial boards of the *Journal of Manufacturing Systems* and *Simulation*, and his research work has been funded by the Defense Advanced Research Projects Agency (DARPA), NASA, the National Science Foundation (NSF), Sandia National Laboratories, SEMATECH, the USDA, and the FHWA. His industrial partners on sponsored research include Alcoa, DaimlerChrysler, Siemens VDO, Continental, Rockwell Software, Systems Modeling Corporation, JC Penney, Fairchild Semiconductor, IBM, Nacom Industries, and the United States Tennis Association (USTA). Dr. Smith has served as principal investigator on over $4 million of sponsored research and won the annual Senior Research Award of the College of Engineering at Auburn University in 2004. In addition, he has been selected as the Outstanding Faculty Member in the Industrial and Systems Engineering Department at Auburn three times. Dr. Smith has over 50 scholarly publications in journals and conference proceedings, numerous book chapters, and invited presentations. He has served on several national conference committees, was the General Chair for the 2004 Winter Simulation Conference, and is currently on the Winter Simulation Conference board of directors. Dr. Smith is a Fellow of the Institute of Industrial Engineers (IIE) and senior member of INFORMS.

David T. Sturrock is Member and Vice-President of Operations for Simio LLC. He is responsible for development, support, and services for Simio LLC simulation and scheduling products. In that role he not only manages new product development, but also teaches frequent commercial courses and manages a variety of consulting projects. He also teaches simulation classes as a Field Faculty member at the University of Pittsburgh. With over 30 years of experience, he has applied simulation techniques in the areas of manufacturing, transportation systems, scheduling, high-speed processing, plant layout, business processes, call centers, capacity analysis, process design, health care, plant commissioning, and real-time control. He received his bachelor's degree in industrial engineering from The Pennsylvania State University with concentrations in manufacturing and automation.

David began his career at Inland Steel Company where he worked as a plant industrial engineer, and later formed and led a simulation/scheduling group that solved a wide array of modeling problems for the company, its suppliers and its customers. He subsequently joined Systems Modeling where, as a Senior Consultant, then Development Lead, and ultimately Vice-President of Development, he was instrumental in building SIMAN and Arena into a market-leading position. When Systems Modeling was acquired by Rockwell Automation, he became the Product Manager for Rockwell's suite of simulation and emulation products. He is an ardent promoter of simulation, having had speaking engagements in over 40 countries across six continents. He was the General Chair for the international 1999 Winter Simulation Conference (WSC). He co-authored the 3rd and 4th editions of *Simulation with Arena* (Kelton et al. 2004, 2007). He has participated in several funded research projects, written a fist full of papers, and has been an active member of the Institute of Industrial Engineers (IIE), INFORMS, PDMA, SME, AMA, and other professional groups.

Contents

Chapter 1

Introduction to Simulation

Simulation has been in use for over 40 years, but it's just moving into its prime. Gartner (`www.gartner.com`) is a leading provider of technical research and advice for business. In 2010, Gartner [12] identified *Advanced Analytics*, including simulation, as number two of the top ten strategic technologies. In 2012 [48] and 2013 [13] Gartner reemphasized the value of analytics and simulation:

> "Because analytics is the 'combustion engine of business,' organizations invest in business intelligence even when times are tough. Gartner predicts the next big phase for business intelligence will be a move toward more simulation and extrapolation to provide more informed decisions."

> "With the improvement of performance and costs, IT leaders can afford to perform analytics and simulation for every action taken in the business. The mobile client linked to cloud-based analytic engines and big data repositories potentially enables use of optimization and simulation everywhere and every time. This new step provides simulation, prediction, optimization and other analytics, to empower even more decision flexibility at the time and place of every business process action."

Advancements in simulation-related hardware and software over the last decade have been dramatic. Computers now provide processing power unheard of even a few years ago. Improved user interfaces and product design have made software significantly easier to use, lowering the expertise required to use simulation effectively. Breakthroughs in object-oriented technology continue to improve modeling flexibility and allow accurate modeling of highly complex systems. Hardware, software, and publicly available symbols make it possible for even novices to produce simulations with compelling 3D animation to support communication between people of all backgrounds. These innovations and others are working together to propel simulation into a new position as a critical technology.

This book opens up the world of simulation to you by providing the basics of general simulation technology, identifying the skills needed for successful simulation projects, and introducing a state-of-the-art simulation package.

1.1   About the Book

We will start by introducing some general simulation concepts, to help understand the underlying technology without yet getting into any software-specific
concepts. Chapter 1, *Introduction to Simulation*, covers typical simulation applications, how to identify an appropriate simulation application, and how to
carry out a simulation project. Chapter 2, *Basics of Queueing Theory*, introduces the concepts of queueing theory, its strengths and limitations, and in
particular how it can be used to help validate components of later simulation
modeling. Chapter 3, *Kinds of Simulation*, introduces some of the technical
aspects and terminology of simulation, classifies the different types of simulations along several dimensions, then illustrates this by working through several
specific examples.

Next we introduce more detailed simulation concepts illustrated with numerous examples implemented in Simio. Rather than breaking up the technical
components (like validation, and output analysis) into separate chapters, we
look at each example as a mini project and introduce successively more concepts with each project. This approach provides the opportunity to learn the
best overall practices and skills at an early stage, and then reinforce those skills
with each successive project.

Chapter 4, *First Simio Models*, starts with a brief overview of Simio itself,
and then directly launches into building a single-server queueing model in Simio.
The primary goal of this chapter is to introduce the simulation model-building
process using Simio. While the basic model-building and analysis processes
themselves aren't specific to Simio, we'll focus on Simio as an implementation
vehicle. This process not only introduces modeling skills, but also covers the
statistical analysis of simulation output results, experimentation, and model
verification. That same model is then reproduced using lower-level tools to
illustrate another possible modeling approach, as well as to provide greater
insight into what's happening "behind the curtain." The chapter continues
with a third, more interesting model of an ATM machine, introduces additional
output analysis using Simio's innovative SMORE plots, and discusses output
analysis outside of Simio. The chapter closes with a discussion of how to discover
and track down those nasty "bugs" that often infest models.

The goal of Chapter 5, *Intermediate Modeling With Simio*, is to build on
the basic Simio modeling-and-analysis concepts presented earlier so that we can
start developing and experimenting with models of more realistic systems. We'll
start by discussing a bit more about how Simio works and its general framework. Then we'll build an electronics-assembly model and successively add
additional features, including modeling multiple processes, conditional branching and merging, etc. As we develop these models, we'll continue to introduce
and use new Simio features. We'll also resume our investigation of how to set
up and analyze sound statistical simulation experiments, this time by considering the common goal of comparing multiple alternative scenarios. By the end
of this chapter, you should have a good understanding of how to model and
analyze systems of intermediate complexity with Simio.

At this point we will have covered some interesting simulation applications,
so we'll then discuss issues regarding the input distributions and processes that

drive the models. Chapter 6, *Input Analysis*, discusses different types of inputs to simulations, methods for converting observed real-world data into something useful to a simulation project, and generating the appropriate input random quantities needed in most simulations.

Chapter 7, *Working With Model Data*, takes a wider view and examines the many types of data that are often required to represent a real system. We'll start by building a simple emergency-department (ED) model, and will show how to meet its input-data requirements using Simio's data-table construct. We'll successively add more detail to the model to illustrate the concepts of sequence tables, relational data tables, arrival tables, and importing and exporting data tables. We'll continue enhancing the ED model to illustrate work schedules, rate tables, and function tables. The chapter ends with a brief introduction to lists, arrays, and changeover matrices. After completing this chapter you should have a good command of the types of data frequently encountered in models, and the Simio choices for representing those data.

Animation and Entity Movement, Chapter 8, discusses the enhanced validation, communication, and credibility that 2D and 3D animation can bring to a simulation project. Then we explore the various animation tools available, including background animation, custom symbols, and status objects. We'll revisit our previous electronics-assembly model to practice some new animation skills, as well as to explore the different types of links available, and add conveyors to handle the work flow. Finally, we'll introduce the Simio Vehicle and Worker objects for assisted entity movement, and revisit our earlier ED model to consider staffing and improve the animation.

Chapter 9 is *Advanced Modeling With Simio*. We start with a simpler version of our ED model, with the goal of demonstrating the use of models for decision-making, and in particular simulation-based optimization. Then we'll introduce a new pizza-shop example to illustrate a few new modeling constructs, as well as bring together concepts that were previously introduced. A third and final model, an assembly line, allows study of buffer-space allocation to maximize throughput.

Chapter 10, *Customizing and Extending Simio* starts with some slightly more advanced material — it builds on the prior experience using add-on processes to provide guidance in building your own custom objects and libraries. It includes examples of building objects hierarchically from base objects, and sub-classing standard library objects. This chapter ends with an introduction to Simio's extendability through programming your own rules, components, and add-ons to Simio.

Chapter 11, *Case Studies Using Simio* includes four introductory and two advanced case studies involving the development and use of Simio models to analyze systems. These problems are larger in scope and are not as well-defined as the homework problems in previous chapters and provide an opportunity to use your skills on more realistic problems. Finally, in Appendix A we explore the use of simulation as a planning and scheduling tool. While simulation-based planning and scheduling has been discussed and used for many years, recent advances in simulation software tools have made these applications significantly easier to implement and use. We conclude the appendix with a description of Simio's Risk-based Planning and Scheduling (RPS) technology.

1.2   Systems and Models

A *system* is any set of related components that together work toward some purpose. A system might be as simple as a waiting line at an automated teller machine (ATM), or as complex as a complete airport or a worldwide distribution network. For any system, whether existing or merely contemplated, it's necessary and sometimes even essential to understand how it will behave and perform under various configurations and circumstances.

In an existing system, you can sometimes gain the necessary understanding by careful observation. One drawback of this approach is that you may need to watch the real system a long time in order to observe the particular conditions of interest even once, let alone making enough observations to reach reliable conclusions. And of course, for some systems (such as a worldwide distribution network), it may not be possible to find a vantage point from which you can observe the entire system at an adequate level of detail.

Additional problems arise when you want to study changes to the system. In some cases it may be easy to make a change in the real system — for example, add a temporary second person to a work shift to observe the impact. But in many cases, this is just not practical: consider the investment required to evaluate whether you should use a standard machine that costs \$300,000 or a high-performance machine that costs \$400,000. Finally, if the real system doesn't yet exist, no observation is possible at all.

For these reasons among others, we use *models* to gain understanding. There are many types of models, each with its own advantages and limitations. *Physical models*, such as a model of a car or airplane, can provide both a sense of reality as well as interaction with the physical environment, as in wind-tunnel testing. *Analytical models* use mathematical representations which can be quite useful in specific problem domains, but applicable domains are often limited. Simulation is a modeling approach with much broader applicability.

Computer simulation imitates the operation of a system and its internal processes, usually over time, and in appropriate detail to draw conclusions about the system's behavior. Simulation models are created using software designed to represent common system components, and record how they behave over time. Simulation is used for predicting both the effect of changes to existing systems, and the performance of new systems. Simulations are frequently used in the design, emulation, and operation of systems.

Simulations may be stochastic or deterministic. In a *stochastic* simulation (the most common), randomness is introduced to represent the variation found in most systems. Activities involving people always vary (for example in time taken to complete a task or quality of performance); external inputs (such as customers and materials) vary; and exceptions (failures) occur. *Deterministic* models have no variation. These are rare in design applications, but more common in model-based decision support such as scheduling and emulation applications. Section 3.1.3 discusses this further.

There are two main types of simulation, *discrete* and *continuous*. The terms discrete and continuous refer to the changing nature of the states within the system. Some states (e.g., the length of a queue, status of a worker) can change only at discrete points in time (called *event times*). Other states (e.g., pressure in a tank or temperature in an oven) can change continuously over time. Some systems are purely discrete or continuous, while others have both types of states present. Section 3.1.2 discusses this further, and gives an example of a continuous simulation.

Continuous systems are defined by *differential equations* that specify the rate of change. Simulation software uses numerical integration to generate a solution for the differential equations over time. *System dynamics* is a graphical approach for creating simple models using the same underlying concept, and is often used to model population dynamics, market growth/decay, and other relationships based on equations.

Four discrete modeling paradigms have evolved in simulation. *Events* model the points in time when the system state changes (a customer arrives or departs). *Processes* model a sequence of actions that take place over time (a part in a manufacturing system seizes a worker, delays by a service time, then releases the worker). *Objects* allow more intuitive modeling by representing complete objects found in the facility. *Agent-based modeling* (ABM) is a special case of the object paradigm in which the system behavior emerges from the interaction of a large number of autonomous intelligent objects (such as soldiers, firms in a market, or infected individuals in an epidemic). The distinction between these paradigms is somewhat blurred because modern packages incorporate multiple paradigms. Simio is a multi-paradigm modeling tool that incorporates all these paradigms into a single framework. You can use a single paradigm, or combine multiple paradigms in the same model. Simio combines the ease and rapid modeling of objects with the flexibility of processes.

Simulation has been applied to a huge variety of settings. The following are just a few samples of areas where simulation has been used to understand and improve the system effectiveness:

Airports: Parking-lot shuttles, ticketing, security, terminal transportation, food court traffic, baggage handling, gate assignments, airplane de-icing.

Hospitals: Emergency department operation, disaster planning, ambulance dispatching, regional service strategies, resource allocation.

Ports: Truck and train traffic, vessel traffic, port management, container storage, capital investments, crane operations.

Mining: Material transfer, labor transportation, equipment allocation, bulk material mixing.

Amusement parks: Guest transportation, ride design/startup, waiting line management, ride staffing, crowd management.

Call centers: Staffing, skill-level assessment, service improvement, training plans, scheduling algorithms.

Supply chains: Risk reduction, reorder points, production allocation, inventory positioning, transportation, growth management, contingency planning.

Manufacturing: Capital-investment analysis, line optimization, product-mix changes, productivity improvement, transportation, labor reduction.

Military: Logistics, maintenance, combat, counterinsurgency, search and detection, humanitarian relief.

Telecommunications: Message transfer, routing, reliability, network security against outages or attacks.

Criminal-justice system: Probation and parole operations, prison utilization and capacity.

Emergency-response system: Response time, station location, equipment levels, staffing.

Public sector: Allocation of voting machines to precincts.

Customer service: Direct-service improvement, back-office operations, resource allocation, capacity planning.

Far from being a tool for manufacturing only, the domains and applications of simulation are wide-ranging and virtually limitless.

1.3   Randomness and the Simulation Process

In this section we discuss the typical steps involved in the simulation process. We also describe the important roles that uncertainty and randomness play in both the inputs to and outputs from simulation models.

1.3.1   Randomness in Simulation and Random Variables

Although some examples of simulation modeling use only deterministic values, the vast majority of simulation models incorporate some form of randomness because it is inherent in the systems being modeled. Typical random components include processing times, service times, customer or entity arrival times, transportation times, machine/resource failures and repairs, and similar occurrences. For example, if you head to the drive-through window at a local fast-food restaurant for a late-night snack, you cannot know exactly how long it will take you to get there, how many other customers may be in front of you when you arrive, or how long it will take to be served, to name just a few variables. We may be able to *estimate* these values based on prior experience or other knowledge, but we cannot predict them with certainty. Using deterministic estimates of these stochastic values in models can result in invalid (generally overly optimistic) performance predictions. However, incorporating these random components in standard analytical models can be difficult or impossible. Using simulation, on the other hand, makes inclusion of random components quite easy and, in fact, it is precisely its ability to easily incorporate stochastic behavior that makes simulation such a popular modeling and analysis tool. This will be a fundamental theme throughout this book.

Because randomness in simulation models is expressed using *random variables*, understanding and using random variables is fundamental to simulation

Table 1.1: Probability mass (PMF) and density (PDF) functions for random variables.

Discrete Random Variables	**Continuous Random Variables**
$p(x_i) = Pr(X = x_i)$	$f(x)$ has the following properties:
$F(x) = \sum_{\substack{\forall i \ni \\ x_i \leq x}} p(x_i)$	1. $f(x) \geq 0 \; \forall$ real values, x
	2. $\int_{-\infty}^{\infty} f(x)\mathrm{d}x = 1$
	3. $Pr(a \leq x \leq b) = \int_a^b f(x)\mathrm{d}x$

modeling and analysis (see [49], [41] to review). At its most basic, a random variable is a function whose value is determined by the outcome of an experiment; that is, we do not know the value until after we perform the experiment. In the simulation context, an experiment involves running the simulation model with a given set of inputs. The probabilistic behavior of a random variable, X, is described by its distribution function (or *cumulative distribution function*, CDF), $F(x) = Pr(X \leq x)$, where the right hand side represents the probability that the random variable X takes on a value less than or equal to the value x. For discrete random variables, the probability mass function, $p(x_i)$ must be considered, and for continuous random variables, we evaluate the probability density function, $f(x)$ (see Table 1.1). Once we've characterized a random variable X, we measure metrics such as the expected value ($E[X]$), the variance ($\mathrm{Var}[X]$), and various other characteristics of the distribution such as quantiles, symmetry/skewness, etc. In many cases, we must rely on the sample statistics such as the sample mean, $\overline{X}$, and sample variance, $S^2(X)$, because we cannot feasibly characterize the corresponding population parameters. Determining the appropriate sample sizes for these estimates is important. Unlike many other experimental methods, in simulation analysis, we can often control the sample sizes to meet our needs.

Simulation requires inputs and outputs to evaluate a system. From the simulation *input* side, we characterize random variables and generate samples from the corresponding distributions; from the *output* side we analyze the characteristics of the distributions (i.e., mean, variance, percentiles, etc.) based on observations generated by the simulation. Consider a model of a small walk-in healthcare clinic. System inputs include the patient arrival times and the care-giver diagnosis and treatment times, all of which are random variables (see Figure 1.1 for an example). In order to simulate the system, we need to understand and generate observations of these random variables as inputs to the model. Often, but not always, we have data from the "real" system that we use to characterize the input random variables. Typical outputs may include the patient waiting time, time in the system, and the care-giver and space utilizations. The simulation model will generate observations of these random variables. By controlling the execution of the simulation model, we can use the generated observations to characterize the outputs of interest. In the following

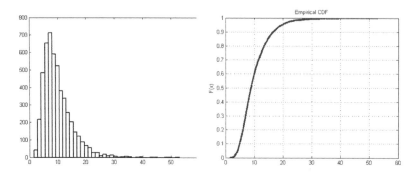

Figure 1.1: Sample patient treatment times and the corresponding empirical CDF.

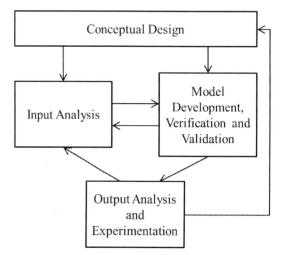

Figure 1.2: The simulation process.

section, we will discuss input and output analysis in the context of the general simulation process.

1.3.2 The Simulation Process

The basic simulation process is shown in Figure 1.2. Note that the process is not strictly sequential and will often turn out to be iterative. We will briefly discuss each of these components in the following sections and will develop the topics in detail throughout the book.

1.3.3 Conceptual Design

Conceptual design requires a detailed understanding of the system being modeled as well as a basic modeling approach for creating the simulation model(s). Conceptual design can be done with pen and paper or on a whiteboard or similar collaboration space that promotes free thinking. It helps to be outside of the constraints of the simulation software package that you happen to be using. Although a well-defined process or methodology for conceptual design would be ideal, we do not know of one. Instead, planning the project is an informal process involving "thinking about" and discussing the details of the problem and the potential modeling approaches. Then the modelers can outline a systematic detailing of the modeling approach and decide on the application of software-specific details. Note that simulation models are developed for *specific objectives* and an important aspect of conceptual design is ensuring that the model will answer the questions being asked. In general, new simulationists (as well as new model builders in other domains) spend far too little time in the conceptual design phase. Instead, they tend to jump in and start the model development process. Allocating too little time for conceptual design almost always increases the overall time required to complete the project.

1.3.4 Input Analysis

Input analysis (which is covered in detail in Chapter 6) involves characterizing the system inputs, and then developing the algorithms and computer code to generate observations on the input random variables and processes. Virtually all commercial simulation software (including Simio) has built-in features for generating the input observations. So the primary input-analysis task involves characterizing the input random variables and specifying corresponding distributions and processes to the simulation software. Often we have sample observations of the real-world data, and our task is to "fit" standard or empirical distributions to these data that can then be used to generate the samples during the simulation (as show in Figure 1.1). If we don't have real-world data on inputs, we can use general rules-of-thumb and sensitivity analysis to help with the input-analysis task.

1.3.5 Model Development, Verification, and Validation

Model development is the "coding" process by which the conceptual model is converted into an "executable" simulation model. We don't want to scare anybody off with the term "coding" — most modern simulation packages provide sophisticated graphical user interfaces to support modeling building/maintenance so the "coding" generally involves dragging and dropping model components and filling in dialog boxes and property windows. However, effective model development does require a detailed understanding of simulation methodology in general and how the specific software being used works in particular. The verification and validation steps ensure that the model is correct. Verification is the process that ensures that the model behaves as the developer intended, and the validation component ensures that the model is accurate relative to the actual system being modeled. Note that proving correctness in any but the simplest models will not be possible. Instead, we focus on collecting evidence until we (or our customers) are satisfied. Although this may disturb early simulationists, it is reality. Model development, verification, and validation topics are covered starting in Chapter 4 and throughout the remainder of the book.

1.3.6   Output Analysis and Experimentation

Once a model has been verified and validated, you then exercise the model to glean information about the underlying system. In the above examples, you may be interested in assessing performance metrics like the average time a patient waits before seeing a care-giver, the 90th percentile of the number of patients in the waiting room, the average number of vehicles waiting in the drive-through lane, etc. You may also be interested in making design decisions such as the number of care-givers required to ensure that the average patient waits no more than 30 minutes, the number of kitchen personnel to ensure that the average order is ready in 5 minutes, etc. Assessing performance metrics and making design decisions using a simulation model involves output analysis and experimentation. Output analysis takes the individual observations generated by the simulation, characterizes the underlying random variables (in a statistically valid way), and draws inferences about the system being modeled. Experimentation involves systematically varying the model inputs and model structure to investigate alternative system configurations. Output analysis topics are spread throughout the modeling chapters (4, 5, and 9).

1.4   When to Simulate (and When Not To)

Simulation of complicated systems has become quite popular. One of the main reasons for this is embodied in that word "complicated." If the system of interest were actually simple enough to be validly represented by an exact analytical model, simulation wouldn't be needed, and indeed shouldn't be used. Instead, exact analytical methods like queueing theory, probability, or simple algebra or calculus could do the job. Simulating a simple system for which we can find an exact analytical solution only adds uncertainty to the results, making them less precise.

However, the world tends to be a complicated place, so we quickly get out of the realm of such very "simple" models. A *valid* model of a complicated system is likely be fairly complicated itself, and not amenable to a simple analytical analysis. If we go ahead and build a simple model of a complicated system with the goal of preserving our ability to get an exact analytical solution, the resulting model might be *overly* simple (simplistic, even), and we'd be uncertain whether it validly represents the system. We may be able to obtain a nice, clean, exact, closed-form analytical solution to our simple model, but we may have made a lot of simplifying assumptions (some of which might be quite questionable in reality) to get to our analytically-tractable model. We may end up with a solution to the *model*, but that model might not bear much resemblance to reality so we may not have a solution to the *problem*.

It's difficult to measure *how* unrealistic a model is; it's not even clear whether that's a reasonable question. On the other hand, if we don't concern ourselves with building a model that will have an analytical solution in the end, we're freed up to allow things in the model to become as complicated and messy as they need to be in order to mimic the system in a valid way. When a simple analytically tractable model is not available, we turn to simulation, where we simply mimic the complicated system, via its complicated (but realistic) model, and study what happens to the results. This allows some model inputs to be *stochastic* — that is, random and represented by "draws" from probability distributions rather than by fixed constant input values — to represent the way things are in reality. The results from our simulation model will likewise be stochastic, and thus uncertain.

Clearly, this uncertainty or imprecision in simulation output is problematic. But, as we'll see, it's *not hard* to measure the degree of this imprecision. If the results are too imprecise we have a remedy. Unlike most statistical sampling experiments, we're in complete control of the "randomness" and numbers of replications, and can use this control to gain any level of precision desired. Computer time used to be a real barrier to simulation's utility. But with modern software running on readily available fast, multi-processor computers and even cloud computing, we can do enough simulating to get results with imprecision that's measurable, acceptably low, and perceptively valid.

In years gone by, simulation was sometimes dismissed as "the method of last resort," or an approach to be taken only "when all else fails" ([59], pp. 887, 890). As noted above, simulation should not be used if a *valid* analytically-tractable model is available. But in many (perhaps most) cases, the actual system is just too complicated, or does not obey the rules, to allow for an analytically tractable model of any credible validity to be built and analyzed. In our opinion, it's better to simulate the right model and get an approximate answer whose imprecision can be objectively measured and reduced, than to do an exact analytical analysis of the wrong model and get an answer whose error cannot be even be quantified, a situation that's worse than imprecision.

While we're talking about precise answers, the examples and figures in this text edition were created with Simio Version 5.91[1]. Because each version of Simio may contain changes that could affect low-level behavior (like the processing order of simultaneous events), different versions could produce different numerical output results for an interactive run. You may wonder "Which results are correct?" Each one is as correct (or as incorrect) as the others! In this book you'll learn how to create *statistically* valid results, and how to recognize when you have (or don't have) them. With the possible exception of a rare bug fix between versions, every version should generate *statistically* equivalent (and valid) results for the same model, even though they may differ numerically across single interactive runs.

1.5 Simulation Success Skills

Learning to use a simulation tool and understanding the underlying technology will not guarantee your success. Conducting successful simulation projects requires much more than that. Newcomers to simulation often ask how they can be successful in simulation. The answer is easy: "Work hard and do everything right." But perhaps you want a bit more detail. Let's identify some of the more important issues that should be considered.

---

[1] If you are using a newer version of Simio, look to the student area of the textbook web site where supplemental on-line content will be posted as it becomes available.

1.5.1   Project Objectives

Many projects start with a fixed deliverable date, but often only a rough idea of what will be delivered and a vague idea of how it will be done. The first question that comes to mind when presented with such a challenge is "What are the project objectives?" Although it may seem like an obvious question with a simple answer, it often happens that stakeholders don't know the answer.

Before you can help with objectives, you need to get to know the stakeholders. A *stakeholder* is someone who commissions, funds, uses, or is affected by the project. Some stakeholders are obvious — your boss is likely to be stakeholder (if you're a student, your instructor is most certainly a stakeholder). But sometimes you have to work a bit to identify all the key stakeholders. Why should you care? In part because stakeholders usually have differing (and conflicting) objectives.

Let's say that you're asked to model a specific manufacturing facility at a large corporation, and evaluate whether a new $4 million crane will provide the desired results (increases in product throughput, decreases in waiting time, reductions in maintenance, etc.). Here are some possible stakeholders and what their objectives might be in a typical situation:

- Manager of industrial engineering (IE) (your boss): She wants to prove that IE adds value to the corporation, so she wants you to demonstrate dramatic cost savings or productivity improvement. She also wants a nice 3D animation she can use to market your services elsewhere in the corporation.

- Production Manager: He's convinced that buying a new crane is the only way he can meet his production targets, and has instructed his key people to provide you the information to help you prove that.

- VP-Production: He's been around a long time and is not convinced that this "simulation" thing offers any real benefit. He's marginally supporting this effort due to political pressure, but fully expects (and secretly hopes) the project will fail.

- VP-Finance: She's very concerned about spending the money for the crane, but is also concerned about inadequate productivity. She's actually the one who, in the last executive meeting, insisted on commissioning a simulation study to get an objective analysis.

- Line Supervisor: She's worked there 15 years and is responsible for material movement. She knows that there are less-expensive and equally effective ways to increase productivity, and would be happy to share that information if anyone bothered to ask her.

- Materials Laborer: Much of his time is currently spent moving materials, and he's afraid of getting laid off if a new crane is purchased. So he'll do his best to convince you that a new crane is a bad idea.

- Engineering Manager: His staff is already overwhelmed, so he doesn't want to be involved unless absolutely necessary. But if a new crane is going to be purchased, he has some very specific ideas of how it should be configured and used.

This scenario is actually a composite of some real cases. Smaller projects and smaller companies might have fewer stakeholders, but the underlying principles remain the same. Conflicting objectives and motivations are not at all unusual. Each of the stakeholders has valuable project input, but it's important to take their biases and motivations into account when evaluating their input.

So now that we've gotten to know the stakeholders a bit, we need to determine how each one views or contributes to the project objectives and attempt to prioritize them appropriately. In order to identify key objectives, you must ask questions like these:

- What do you want to evaluate, or hope to prove?

- What's the model scope? How much detail is anticipated for each component of the system?

- What components are critical? Which less-important components might be approximated?

- What input information can be made available, how good is it, who will provide it, and when?

- How much experimentation will be required? Will optimum-seeking be required?

- How will any animation be used (animation for validation is quite different from animation presented to a board of directors)?

- In what form do you want results (verbal presentation, detailed numbers, summaries, graphs, text reports)?

One good way to help identify clear objectives is to design a mock-up of the final report. You can say, *"If I generate a report with the following information in a format like this, will that address your needs?"* Once you can get general agreement on the form and content of the final report, you can often work backwards to determine the appropriate level of detail and address other modeling concerns. This process can also help bring out unrecognized modeling objectives.

Sometimes the necessary project clarity is not there. If so, and you go ahead anyway to plan the entire project including deliverables, resources, and date, you're setting yourself up for failure. Lack of project clarity is a clear call to do the project in phases. Starting with a small prototype will often help clarify the big issues. Based on those prototype experiences, you might find that you can do a detailed plan for subsequent phases. We'll talk more about that next.

1.5.2 Functional Specification

"If you don't know where you're going,
how will you know when you get there?"

Carpenter's advice: "Measure twice. Cut once."

If you've followed the advice from Section 1.5.1, you now have at least some basic project objectives. You're ready to start building the model, right? Wrong! In most cases your stakeholders will be looking for some commitments.

- When will you get it done (is yesterday too soon)?

- How much will it cost (or how many resources will it require)?

- How comprehensive will the model be (or what specific system aspects will be included)?

- What will be the quality (or how will it be verified and validated)?

Are you ready to give reliable answers to those questions? Probably not.

Of course the worst possible, but quite common, situation is that the *stakeholder* will supply answers to all of those questions and leave it to you to deliver. Picture a statement like "I'll pay you $5000 to provide a thorough, validated analysis of ... to be delivered five days from now." If accepted, such a statement often results in a lot of overtime and produces an incomplete, unvalidated model a week or two late. And as for the promised money ... well, the customer didn't get what he asked for, now, did he?

It's OK for the customer to specify answers to *two* of those questions, and in rare cases maybe even *three*. But you must reserve the right to adjust at least one or two of those parameters. You might cut the scope to meet a deadline. Or you might extend the deadline to achieve the scope. Or, you might double both the resources and the cost to achieve the scope and meet the date (adjusting the quality is seldom a good idea).

If you're fortunate, the stakeholder will allow you to answer all four questions (of course, reserving the right to reject your proposal). But how do you come up with good answers? By creating a *functional specification*, which is a document describing exactly what will be delivered, when, how, and by whom. While the details required in a functional specification vary by application and project size, typical components may include the following:

1. Introduction

   a) Simulation objectives: Discussion of high-level objectives. What's the desired outcome of this project?

   b) Identification of stakeholders: Who are the primary people concerned with the results from this model? Which other people are also concerned? How will the model be used and by whom? How will they learn it?

2. System description and modeling approach: Overview of system components and approaches for modeling them including, but not limited to, the following components:

   a) Equipment: Each piece of equipment should be described in detail, including its behavior, setups, schedules, reliability, and other aspects that might affect the model. Include data tables and diagrams as needed. Where data do not yet exist, they should be identified as such.

b) Product types: What products are involved? How do they differ? How do they relate to each other? What level of detail is required for each product or product group?

c) Operations: Each operation should be described in detail including its behavior, setups, schedules, reliability, and other aspects that might affect the model. Include data tables and diagrams as needed. Where data do not yet exist, they should be identified as such.

d) Transportation: Internal and external transportation should be described in adequate detail.

3. Input data: What data should be considered for model input? Who will provide this information? When? In what format?

4. Output data: What data should be produced by the model? In this section, a mock-up of the final report will help clarify expectations for all parties.

5. Project deliverables: Discuss all agreed-upon project deliverables. When this list is fulfilled, the project is deemed complete.

a) Documentation: What model documentation, instructions, or user manual will be provided? At what level of detail?

b) Software and training: If it's intended that the user will interact directly with the model, discuss the software that's required; what software, if any, will be included in the project price quote; and what, if any, custom interface will be provided. Also discuss what project or product training is recommended or will be supplied.

c) Animation: What are the animation deliverables and for what purposes will the animations be used (model validation, stakeholder buy-in, marketing)? 2D or 3D? Are existing layouts and symbols available, and in what form? What will be provided, by whom, and when?

6. Project phases: Describe each project phase (if more than one) and the estimated effort, delivery date, and charge for each phase.

7. Signoffs: Signature section for primary stakeholders.

At the beginning of a project there's a natural inclination just to start modeling. There's time pressure. Ideas are flowing. There's excitement. It's very hard to stop and do a functional specification. But trust us on this — *doing a functional specification is worth the effort.* Look back at those quotations at the beginning of this section. Pausing to determine where you're going and how you're going to get there can save misdirected effort and wasted time.

We recommend that approximately the first 10% of the total estimated project time be spent on creating a prototype and a functional specification. Do not consider this to be extra time. Rather, like in report design, you are just shifting some specific tasks to early in the project — when they can have the most planning benefit. Yes, that means if you expect the project may take 20 days, you should spend about two days on this. As a result, you may well find that the project will require 40 days to finish — certainly bad news, but much better to find out up front while you still have time to consider alternatives (reprioritize the objectives, reduce the scope, add resources, etc.).

1.5.3   Project Iterations

Simulation projects are best done as an iterative process, even from the first steps. You might think you could just define your objectives, create a functional specification, and then create a prototype. But while you're writing the functional specification, you're likely to discover new objectives. And while you're doing the prototype, you'll discover important new things to add to the functional specification.

As you get further into the project, an iterative approach becomes even more important. A simulation novice will often get an idea and start modeling it, then keep adding to the model until it's complete — and *only then* run the model. But even the best modeler, using the best tools, will make mistakes. But when all you know is that your mistake is "somewhere in the model," it's very hard to find it and fix it. Based on our collective experience in teaching simulation, this is a huge problem for students new to the topic.

More experienced modelers will typically build a small piece of the model, then run it, test it, debug it, and verify that it does what the modeler expected it to do. Then repeat that process with another small piece of the model. As soon as enough of the model exists to compare to the real world, then validate, as much as possible, that the entire section of the model matches the intended system behavior. Keep repeating this iterative process until the model is complete. At each step in the process, finding and fixing problems is much easier because it's very likely a problem in the small piece that was most recently added. And at each step you can save under a different name (like `MyModelV1`, `MyModelV2`, or with full dates and even times appended to the file names), to allow reverting to an earlier version if necessary.

Another benefit of this iterative approach, especially for novices, is that potentially-major problems can be eliminated early. Let's say that you built an entire model based on a faulty assumption of how entity grouping worked, and only at the very end did you discover your misunderstanding. At that point it might require extensive rework to change the basis of your model. However, if you were building your model iteratively, you probably would have discovered your misunderstanding the very first time you used the grouping construct, at which time it would be relatively easy to take a better strategy.

A final, and extremely important benefit of the iterative approach is the ability to prioritize. *For each iteration, work on the most important small section of the model that's remaining.* The one predictable thing about software development of all types is that it almost always takes much longer than expected. Building simulation models often shares that same problem. If you run out of project time when following a non-iterative approach and your model is not yet even working, let alone verified or validated, you essentially have nothing useful to show for your efforts. But if you run out of time when following an iterative approach, you have a portion of the model that's completed, verified, validated, and ready for use. And if you've been working on the highest-priority task at each iteration, you may find that the portion completed is actually enough to fulfill most of the project goals (look up the 80-20 rule or the Pareto principle to see why).

Although it may vary somewhat by project and application, the general steps in a simulation study are:

1. Define high-level objectives and identify stakeholders.

2. Define the functional specification, including detailed goals, model boundaries, level of detail, modeling approach, and output measures. Design the final report.

3. Build a prototype. Update steps 1 and 2 as necessary.

4. Model or enhance a high-priority piece of the system. Document and verify it. Iterate.

5. Collect and incorporate model input data.

6. Verify and validate the model. Involve stakeholders. Return to step 4 as necessary.

7. Design experiments. Make production runs. Involve stakeholders. Return to step 4 as necessary.

8. Document the results and the model.

9. Present the results and collect your kudos.

As you're iterating, don't waste the opportunity to *communicate regularly with the stakeholders*. Stakeholders don't like surprises. If the project is producing results that differ from what was expected, learn together why that's happening. If the project is behind schedule, let stakeholders know early so that serious problems can be avoided. Don't think of stakeholders as just clients, and certainly not as adversaries. Think of stakeholders as partners — you can help each other to obtain the best possible results from this project. And those results often come from the detailed system exploration that's necessary to uncover the actual processes being modeled. In fact, in many projects a large portion of the value occurs before any simulation "results" are even generated — due to the knowledge gained from the early exploration by modelers, and frequent collaboration with stakeholders.

1.5.4 Project Management and Agility

There are many aspects to a successful project, but one of the most obvious is meeting the completion deadline. A project that produces results after the decision is made has little value. Other, often-related, aspects are the cost, resources, and time consumed. A project that runs over budget may be canceled before it gets close to completion. You must pay appropriate attention to completion dates and project costs. But both of those are outcomes of how you manage the day-to-day project details.

A well-managed project starts by having clear goals and a solid functional specification to guide your decisions. Throughout the project, you'll be making large and small decisions, like the following:

- How much detail should be modeled in a particular section?

- How much input data do I need to collect?

- To which output data should I pay most attention?

- When is the model considered to be valid?

- How much time should I spend on animation? On analysis?

- What should I do next?

In almost every case, the functional specification should directly or indirectly provide the answers. You've already captured and prioritized the objectives of your key stakeholders. That information should become the basis of most decisions.

One of the things you'll have to prioritize is "evolving specifications" or new stakeholder requests, sometimes called "scope creep." One extreme is to take a hard line and say "If it's not in the functional specification, it's not in the model." While in some rare cases this response may be appropriate and necessary, in most cases it's not. Simulation is an exploratory and learning process. As you explore new areas and learn more about the target system, it's only natural that new issues, approaches, and areas of study will evolve. Refusing to deal with these severely limits the potential value of the simulation (and your value as a solution provider).

Another extreme is to take the approach that the stakeholders are always right, and if they ask you to work on something new, it *must* be the right thing to do. While this response makes the stakeholder happy in the short-term, the most likely longer-term outcome is a late or even unfinished project — and a *very unhappy* stakeholder! If you're always chasing the latest idea, you may never have the time to finish the high-priority work necessary to produce any value at all.

The key is to manage these opportunities — that management starts with open communication with the stakeholders and revisiting the items in the functional specification and their relative priorities. When something is added to the project, something else needs to change. Perhaps addressing the new item is important enough to postpone the project deadline a bit. If not, perhaps this new item is more important than some other task that can be dropped (or moved to the "wish list" that's developed for when things go better than expected). Or perhaps this new item itself should be moved to the "wish list."

Our definition of *agility* is the ability to react quickly and appropriately to change. Your ability to be agile will be a significant contributor to your success in simulation.

Simulation Stakeholder Bill of Rights

The people who request, pay for, consume, or are affected by a simulation project and its results are often referred to as its stakeholders. For any simulation project the stakeholders should have reasonable expectations from the people actually doing the work. Here are some basic stakeholder rights that should be assured.

1 **Partnership** – The modeler will do more than provide information on request. The modeler will assume some ownership of helping stakeholders determine the right problems and identify and evaluate proposed solutions.

2 **Functional Specification** – A specification will be created at the beginning of the project to help define clear project objectives, deadlines, data, responsibilities, reporting needs, and other project aspects. This specification will be used as a guide throughout the project, especially when tradeoffs must be considered.

3 **Prototype** – All but the simplest projects will have a prototype to help stakeholders and the modeler communicate and visualize the project scope, approach, and outcomes. The prototype is often done as part of the functional specification.

4 **Level of Detail** – The model will be created at an appropriate level of detail to address the stated objectives. Too much or too little detail could lead to an incomplete, misunderstood, or even useless model.

5 **Phased Approach** – The project will be divided into phases and the interim results should be shared with stakeholders. This allows problems in approach, detail, data, timeliness, or other areas to be discovered and addressed early and reduces the chance of an unfortunate surprise at the end of a project.

6 **Timeliness** – If a decision-making date has been clearly identified, usable results will be provided by that date. If project completion has been delayed, regardless of reason or fault, the model will be re-scoped so that the existing work can provide value and contribute to effective decision-making.

7 **Agility** – Modeling is a discovery process and often new directions will evolve over the course of the project. While observing the limitations of level of detail, timeliness, and other aspects of the functional specification, a modeler will attempt to adjust project direction appropriately to meet evolving needs.

8 **Validated and Verified** – The modeler will certify that the model conforms to the design in the functional specification and that the model appropriately represents the actual operation. If there is inadequate time for accuracy, there is inadequate time for the modeling effort.

9 **Animation** – Every model deserves at least simple animation to aid in verification and communication with stakeholders.

10 **Clear Accurate Results** – The project results will be summarized and expressed in a form and terminology useful to stakeholders. Since simulation results are an estimate, proper analysis will be done so that the stakeholders are informed of the accuracy of the results.

11 **Documentation** – The model will be adequately documented both internally and externally to support both immediate objectives and long term model viability.

12 **Integrity** – The results and recommendations are based only on facts and analysis and are not influenced by politics, effort, or other inappropriate factors.

Note: This is the companion piece to Simulationist Bill of Rights, which outlines reasonable expectations a modeler should have in a simulation project. To read that and more, visit our website.

Simio LLC -- Forward Thinking -- www.simio.com -- © 2010

Figure 1.3: Simulation Stakeholder Bill of Rights.

Simulationist Bill of Rights

The companion *Simulation Stakeholder Bill of Rights* proposed some reasonable expectations that a consumer of a simulation project might have. But this is not a one-way street. The modeler or simulationist should have some reasonable expectations as well.

1 Clear Objectives – A simulationist can help stakeholders discover and refine their objectives, but clearly the stakeholders must agree on project objectives. The primary objectives must remain solid throughout the project.

2 Stakeholder Participation – Adequate access and cooperation must be provided by the people who know the system both in the early phases and throughout the project. Stakeholders will need to be involved periodically to assess progress and resolve outstanding issues.

3 Timely Data – The functional specification should describe what data will be required, when it will be delivered and by whom. Late, missing, or poor quality data can have a dramatic impact on a project.

4 Management Support – The simulationist's manager should support the project as needed not only in issues like tools and training discussed below, but also in shielding the simulationist from energy sapping politics and bureaucracy.

5 Cost of Agility – If stakeholders ask for project changes, they should be flexible in other aspects such as delivery date, level of detail, scope, or project cost.

6 Timely Review/Feedback – Interim updates should be reviewed promptly and thoughtfully by the appropriate people so that meaningful feedback can be provided and any necessary course corrections can be immediately made.

7 Reasonable Expectations – Stakeholders must recognize the limitations of the technology and project constraints and not have unrealistic expectations. A project based on the assumption of long work hours is a project that has been poorly managed.

8 "Don't shoot the messenger" – The modeler should not be criticized if the results promote an unexpected or undesirable conclusion.

9 Proper Tools – A simulationist should be provided the right hardware and software appropriate to the project. While "the best and latest" is not always required, a simulationist should not have to waste time on outdated or inappropriate software and inefficient hardware.

10 Training and Support – A simulationist should not be expected to "plunge ahead" into unfamiliar software and applications without training. Proper training and support should be provided.

11 Integrity – A simulationist should be free from coercion. If a stakeholder "knows" the right answer before the project starts, then there is no point to starting the project. If not, then the objectivity of the analysis should be respected with no coercion to change the model to produce the desired results.

12 Respect – A good simulationist may sometimes make the job look easy, but don't take them for granted. A project often "looks" easy only because the simulationist did everything right, a feat that in itself is very difficult. And sometimes a project looks easy only because others have not seen the nights and weekends involved.

Simio LLC -- Forward Thinking -- www.simio.com -- © 2010

Figure 1.4: Simulationist Bill of Rights.

1.5.5   Stakeholder and Simulationist Bills of Rights

We'll end this chapter with an acknowledgement that stakeholders have reasonable expectations of what you will do for them (Figure 1.3). Give these expectations careful consideration to improve the effectiveness and success of your next project. But along with those expectations, stakeholders have some responsibilities to you as well (Figure 1.4). Discussing both sets of these expectations ahead of time can enhance communications and help ensure that your project is successful — a win-win situation that meets everyone's needs. These "rights" were excerpted from the Success in Simulation [56] blog at `www.simio.com/blog` and used with permission. We urge you to peruse the early topics of this non-commercial blog for its many success tips and short interesting topics.

Chapter 2

Basics of Queueing Theory

Many (not all) simulation models are of *queueing* systems representing a wide variety of real operations. For instance, patients arrive to an urgent-care clinic (i.e., they just show up randomly without appointments), and they all must first sign in, possibly after waiting in a line (or a *queue*) for a bit; see Figure 2.1. After signing in, patients go either to registration or, if they're seriously ill, go to a trauma room, and could have to wait in queue at either of those places too before being seen. Patients going to the Exam room then either exit the system, or go to a treatment room (maybe queueing there first) and then exit. The seriously ill patients that went to a trauma room then all go to a treatment room (possibly after queueing there too), and then they exit. Questions for designing and operating such a facility might include how many staff of which type to have on duty during which time periods, how big the waiting room should be, how patient waiting-room stays would be affected if the doctors and nurses decreased or increased the time they tend to spend with patients, what would happen if 10% more patients arrived, and what might be the impact of serving patients in an order according to some measure of acuity of their

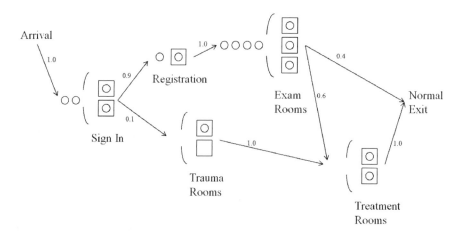

Figure 2.1: A queueing system representing an urgent-care clinic.

presented condition instead of first-come, first served.

This short chapter will cover just the basics of queueing *theory* (not queueing *simulation*), since familiarity with this material and the terminology is important for developing many simulation models. The relatively simple mathematical formulas from elementary queueing theory can also prove valuable for *verification* of simulation models of queueing systems (helping to determine that the simulations are correct). Queueing-theory models can do that by providing a benchmark against which simulation results can be compared, *if* the simulation model is simplified (probably unrealistically) to match the more stringent assumptions of queueing theory (e.g., assuming exponential probability distributions even though such is not true for the real system we're simulating). If the (simplified) simulation model approximately agrees with the queueing-theoretic results, then we have better confidence that the simulation model's logic, at least, is correct. As we develop our Simio simulation models in later chapters, we'll do this repeatedly, which is helpful since simulation "code" can be quite complex and thus quite difficult to verify.

Queueing theory is an enormous subject, and has been studied mathematically since at least 1909, at first by A.K. Erlang in connection with how the recently-invented "telephone" systems in Copenhagen, Denmark might be designed and perform ([9], [10]). There are many entire books written on it (e.g., [16], [28]), as well as extensive chapters in more general books on probability and stochastic processes (e.g. [49]) or on other application-oriented topics like manufacturing (e.g. [2]); a web search on "queueing theory" returned more than 1.4 million results. Thus, we make *no* attempt here to give any sort of complete treatment of the subject; rather, we're just trying to introduce some terminology, give some basic results, contrast queueing theory with simulation via their relative strengths and weaknesses, and provide some specific formulas that we'll use in later chapters to help verify our Simio simulation models.

In this chapter (and really, in the whole book), we assume that you're already familiar with basic probability, including:

- The foundational ideas of a probabilistic *experiment*, the *sample space* for the experiment, and *events*.

- *Random variables* (RVs), both the *discrete* and *continuous* flavors.

- *Distributions* of RVs — *probability mass functions* (PMFs) for discrete, *probability density functions* (PDFs) for continuous, and *cumulative distribution functions* (CDFs) for both flavors.

- *Expected values* of RVs and their distributions (a.k.a. expectations or just means), *variances*, and how to use PMFs, PDFs, and CDFs to compute probabilities that RVs will land in intervals or sets.

- *Independence* (or lack thereof) between RVs.

If not, then you should first go back and review those topics before reading on.

In this chapter, and elsewhere in the book, we'll often refer to specific probability distributions, like the exponential, uniform, triangular, etc. In days gone by books using probability notions contained lots of pages of distribution facts like definitions of PMFs, PDFs, and giving CDFs, expected values, variances, and so on, for many distributions. But this book does not have such a compendium since that material is readily available elsewhere, including the Simio documentation itself (described in Section 6.1.3), and on-line, such as en.wikipedia.org/wiki/List_of_probability_distributions, which in turn has links to web pages on well over 100 specific univariate distributions. Encyclopedic books, such as [21], [22], [20], and [11], have been compiled. References to further material on distributions are in Section 6.1.3, along with some discussion of their properties, such as ranges.

In Section 2.1 we'll describe the general structure, terminology, and notation for queueing systems. Section 2.2 states some important relationships among different output performance metrics of many queueing systems. We'll quote a few specific queueing-system output results in Section 2.3, and briefly describe how to deal with networks of queues in Section 2.4. Finally, in Section 2.5 we'll compare and contrast queueing theory vs. simulation as analysis tools, and show that each has pros and cons that dovetail nicely with each other (though on balance, we still like simulation better most of the time).

2.1 Queueing-System Structure and Terminology

A *queueing*[1] *system* is one in which entities (like customers, patients, jobs, or messages) arrive, get served either at a single station or at several stations in turn, might have to wait in one or more queues for service, and then may leave (if they do leave the system is called *open*, but if they never leave and just keep circulating around within the system it's called *closed*).

The urgent-care clinic described earlier, and in Figure 2.1, is modeled as an open queueing system. There are five separate service stations (Sign In, Registration, Trauma Rooms, Exam Rooms, and Treatment Rooms), each of which could be called a *node* in a network of *multiserver queueing nodes* (Registration has just a single server, but that's a special case of multiserver). If there are multiple individual parallel servers at a queueing node (e.g., three for Exam Rooms), a single queue "feeds" them all, rather than having a separate queue for each single server, and we usually assume that the individual servers are identical in terms of their capabilities and service rates. The numbers by the arcs in Figure 2.1 give the probabilities that patients will follow those arcs. When coming out of a station where there's a choice about where to go next (out of Sign In and Exam Rooms) we need to know these probabilities; when coming out of a station where all patients go to the same next station (out of Arrival, Registration, Trauma Rooms, Treatment Rooms) the 1.0 probabilities noted in the arcs are obvious, but we show them anyway for completeness. Though terminology varies, we'll say that an entity is *in queue* if it's waiting in the line but not in service, so right now at the Exam Rooms in Figure 2.1 there are four patients in queue and seven patients in the Exam-Room *system*; there are three patients *in service* at the Exam Rooms.

---

[1]There's disagreement about whether the spelling should be "queueing" or "queuing," but the former seems more prevalent in the technical literature, and besides, we like it since we get to use five vowels in a row.

When a server finishes service on an entity and there are other entities in queue for that queueing node, we need to decide which specific entity in queue will be chosen to move into service next — this is called the *queue discipline*. You're no doubt familiar with *first-come, first-served* (better known in queueing as *first-in, first-out*, or FIFO), which is common and might seem the "fairest." Other queue disciplines are possible, though, like *last-in, first-out* (LIFO), which might represent the experience of chilled plates in a stack waiting to be used at a salad bar; clean plates are loaded onto the top, and taken from the top as well by customers. Some queue disciplines use *priorities* to pay attention to differences among entities in the queue, like *shortest-job-first* (SJF), also called *shortest processing time* (SPT). With an SJF queue discipline, the next entity chosen from the queue is one whose processing time will be lowest among all those then in queue (you'd have to know the processing times beforehand and assign them to individual entities), in an attempt to serve quickly those jobs needing only a little service, rather than making them wait behind jobs requiring long service, thereby hopefully improving (reducing) the average time in queue across all entities. Pity the poor big job, though, as it could be stuck near the back of the queue for a very long time, so whether SJF is "better" than FIFO might depend on whether you care more about average time in system or maximum (worst) time in system. A kind of opposite of SJF would be to have values assigned to each entity (maybe profit upon exiting service) and you'd like to choose the next job from the queue as the one with the highest value (time is money, you know). In a health-care system like that in Figure 2.1, patients might be initially triaged into several acuity levels, and the next patient taken from a queue would one in the most serious acuity-level queue that's non-empty; within each acuity-level queue a tie-breaking rule, such as FIFO, would be needed to select among patients at the same acuity level.

Several *performance measures* (or *output metrics*) of queueing systems are often of interest:

- The *time in queue* is, as you'd guess, the time that an entity spends waiting in line (excluding service time). In a queueing network like Figure 2.1, we could speak of the time in queue at each station separately, or added up for each patient over all the individual times in queue from arrival to the system on the upper left, to exit on the far right.

- The *time in system* is the time in queue plus the time in service. Again, in a network, we could refer to time in system at each station separately, or overall from arrival to exit.

- The *number in queue* (or *queue length*) is the number of entities in queue (again, not counting any entities who might be in service), either at each station separately or overall in the whole system. Right now in Figure 2.1 there are two patients in queue at Sign In, one at Registration, four at Exam Rooms, and none at both Trauma Rooms and Treatment Rooms; there are seven patients in queue in the whole system.

- The *number in system* is the number of entities in queue *plus* in service, either at each station separately or overall in the whole system. Right now in Figure 2.1 there are four patients in system at Sign In, two at

Registration, seven at Exam Rooms, one at Trauma Rooms, and two at Treatment Rooms; there are 16 patients in the whole system.

- The *utilization* of a server (or group of parallel identical servers) is the time-average number of individual servers in the group who are busy, divided by the total number of servers in the group. For instance, at Exam Rooms there are three servers, and if we let $B_E(t)$ be the number of the individual exam rooms that are busy (occupied) at any time t, then the utilization is

$$\frac{\int_0^h B_E(t)dt}{3h}, \tag{2.1}$$

where h is the length (or *horizon*) of time for which we observe the system in operation.

With but very few exceptions, the results available from queueing theory are for *steady-state* (or *long-run*, or *infinite-horizon*) conditions, as time (real or simulated) goes to infinity, and often just for steady-state *averages* (or *means*). Here's common (though not universal) notation for such metrics, which we'll use:

- W_q = the steady-state average time in queue (excluding service times) of entities (at each station separately in a network, or overall across the whole network).

- W = the steady-state average time in system (including service times) of entities (again, at each station separately or overall).

- L_q = the steady-state average number of entities in queue (at each station separately or overall). Note that this is a *time* average, not the usual average of a discrete list of numbers, so might require a bit more explanation. Let $L_q(t)$ be the number of entities in queue at time instant t, so $L_q(t) \in \{0, 1, 2, \ldots\}$ for all values of continuous time t. Imagine a plot of $L_q(t)$ vs. t, which would be a piecewise-constant curve at levels 0, 1, 2, ... with jumps up and down at those time instants when entities enter and leave the queue(s). Over a finite time horizon $[0, h]$, the time-average number of entities in queue(s) (or the time-average *queue length* if we're talking about just a single queue) will be $\overline{L_q}(h) = \int_0^h L_q(t)dt/h$, which is a *weighted* average of the levels 0, 1, 2, ... of $L_q(t)$, with the weights being the proportion of time that $L_q(t)$ spends at each level. Then $L_q = \lim_{h \to \infty} \overline{L_q}(h)$.

- L = the steady-state average number of entities in system (at each station separately or overall). As with L_q, this is a *time* average, and is indeed defined similarly to L_q, with $L(t)$ defined as the number of entities in the *system* (in queue plus in service) at time t, and then dropping the subscript q throughout the discussion.

- ρ = the steady-state utilization of a server or group of parallel identical servers, typically for each station separately, as defined in equation (2.1) above for the Exam-Rooms example, but after letting $h \to \infty$.

Much of queueing theory over the decades has been devoted to finding the values of these five steady-state average metrics, and is certainly far beyond our scope here. We will mention, though, that due to its very special property (the memoryless property), the exponential distribution is critical in many derivations and proofs. For instance, in the simplest queueing models, interarrival times (times between the arrivals of two successive entities to the system) are often assumed to have an exponential distribution, as are individual service times. In more advanced models, those distributions can be generalized to variations of the exponential (like Erlang, an RV that is the sum of independent and identically distributed (IID) exponential RVs), or even to general distributions; with each step in the generalization, though, the results become more complex both to derive and to apply.

Finally, there's a standard notation to describe multiserver queueing stations, sometimes called *Kendall's notation*, which is compact and convenient, and which we'll use:

$$A/B/c/k.$$

An indication of the arrival process or interarrival-time distribution is given by A. The service-time RVs are described by B. The number of parallel identical servers is c (so $c = 3$ for the Exam Rooms in Figure 2.1, and $c = 1$ for Registration). The capacity (i.e., upper limit) on the system (including in queue plus in service) is denoted by k; if there's no capacity limit (i.e., $k = \infty$), then the $/\infty$ is usually omitted from the end of the notation. Kendall's notation is sometimes expanded to indicate the nature of the population of the input entities (the *calling population*), and the queue discipline, but we won't need any of that. Some particular choices for A and B are of note. Always, M (for Markovian, or maybe memoryless) means an exponential distribution for either the interarrival times or the service times. Thus the $M/M/1$ queue has exponential interarrival times, exponential service times (assumed independent of the interarrival times), and just a single server; $M/M/3$ would describe the Exam Rooms component in Figure 2.1 if we knew that the interarrival times to it (coming out of Registration) and the Exam-Room service times were exponential. For an Erlang RV composed of the sum of m IID exponential RVs, E_m is common notation, so an $M/E_3/2/10$ system would have exponential interarrival times, 3-Erlang service times, two parallel identical servers, and a limit of ten entities in the system at any one time (so a maximum of eight in the queue). The notation G is usually used for A or B when we want to allow a general interarrival-time or service-time distribution, respectively.

2.2 Little's Law and Other Relations

There are several important relationships among the steady-state average metrics W_q, W, L_q, and L, defined in Section 2.1, which make computing (or estimating) the rest of them fairly simple if you know (or can estimate) any one of them. In this section we're thinking primarily of just a single multiserver queueing station (like just the Exam Rooms in Figure 2.1 in isolation). Further notation we'll need is $\lambda =$ the *arrival rate* (which is the reciprocal of the expected value of the interarrival-time distribution), and $\mu =$ the *service rate* of an individual server, not a group of multiple parallel identical servers (so $\mu = 1/E(S)$ where S is an RV representing an entity's service time in an individual server).

The most important of these relationships is *Little's law*, the first version of which was derived in [36], but it's been generalized repeatedly, e.g. [54]; recently it celebrated its 50th anniversary ([37]). Stated simply, Little's law is just

$$L = \lambda W$$

with our notation from above. In our application we'll consider this just for a single multiserver queueing station, but be aware that it does apply more generally. The remarkable thing about Little's law is that it relates a *time* average (L on the left-hand side) to an *entity-based* observational average (W on the right-hand side).

Similar to Little's law, but now considering just the queue (and not the servers) at a queueing station, we have

$$L_q = \lambda W_q.$$

More physically intuitive than Little's law is the relationship

$$W = W_q + E(S),$$

where we assume that we know at least the expected value $E(S)$ of the service-time distribution, if not the whole distribution. This just says that your expected time in system is your expected time in queue plus your expected service time. This relationship, together with $L = \lambda W$ and $L_q = \lambda W_q$, allows you to get any of W_q, W, L_q, and L if you know just one of them. For instance, if you know (or can estimate) W_q, then just substituting yields $L = \lambda(W_q + E(S))$; likewise, if you know L, then after a little algebra you get $W_q = L/\lambda - E(S)$.

2.3   Specific Results for Some Multiserver Queueing Stations

In this section we'll just list formulas from the literature for a few multiserver queueing stations, i.e., something like the Exam Rooms in Figure 2.1, rather than an entire network of queues overall as shown across that entire figure. Remember, you can use Little's law and other relations from Section 2.2 to get the other typical steady-state average metrics. Throughout, $\rho = \lambda/(c\mu)$ is the utilization of the servers as a group (recall that c is the number of parallel identical servers at the queueing station); higher values of ρ generally portend more congestion. For all of these results (and throughout most of queueing theory), we need to assume that $\rho < 1$ ($\rho \leq 1$ isn't good enough) in order for any of these results to be valid, since they're all for steady state, so we must know that the system will not "explode" over the long run with the number of entities present growing without bound — the servers, as a group, need to be able to serve entities at least as fast as they're arriving.

In addition to just the four specific models considered here, there are certainly other such formulas and results for other queueing systems, available in the vast queueing-theory literature mentioned at the beginning of this chapter. However, they do tend to get quite complicated quite quickly, and in some cases are not really closed-form formulas, but rather equations involving the metrics you want, which you then need to "solve" using various numerical methods and approximations (as is actually the case with our fourth example below).

2.3.1 $M/M/1$

Perhaps the simplest of all queueing systems, we have exponential interarrival times, exponential service times, and just a single server.

The steady-state average number in system is $L = \rho/(1 - \rho)$. Remember that ρ is the mean arrival rate λ divided by the mean service rate μ if there's just a single server. Certainly, L can be expressed in different ways, like

$$L = \frac{\lambda}{1/E(S) - \lambda}.$$

If we know L, we can then use the relations in Section 2.2 to compute the other metrics W_q, W, and L_q.

2.3.2 $M/M/c$

Again, both interarrival times and services times follow exponential distributions, but now we have c parallel identical servers being fed by a single queue.

First, let $p(n)$ be the probability of having n entities in the system in steady state. The only one of these we really need is the steady-state probability that the system is empty, which turns out to be

$$p(0) = \frac{1}{\frac{(c\rho)^c}{c!(1-\rho)} + \sum_{n=0}^{c-1} \frac{(c\rho)^n}{n!}},$$

a little complicated but is just a formula that can be computed, maybe even in a spreadsheet, since c is finite, and often fairly small. (For any positive integer j, $j! = j \times (j-1) \times (j-2) \times \cdots \times 2 \times 1$, and is pronounced "$j$ factorial" — not by shouting "j." Mostly for convenience, 0! is just *defined* to be 1.) Then,

$$L_q = \frac{\rho(c\rho)^c p(0)}{c!(1 - \rho)^2},$$

from which you could get, for example, $L = L_q + \lambda/\mu$ along with the other metrics via the relationships in Section 2.2.

Though the formulas above for the $M/M/c$ are all closed-form (i.e., you can in principle plug numbers into them), they're a little complicated, especially the summation in the denominator in the expression for $p(0)$ unless c is very small. So we provide a small command-line program `mmc.exe`, which you can download (including the C source code) from the "students" section of this book's website (see the Preface for the URL and login/password), to do this for values that you specify for the arrival rate λ, the service rate μ, and the number of servers c. To invoke `mmc.exe` you need to run the Microsoft Windows Command Prompt window (usually via the Windows Start button, then Programs or All Programs,

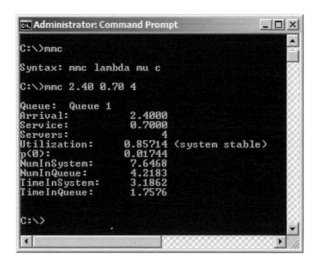

Figure 2.2: Using the `mmc.exe` command-line program to get steady-state queueing metrics for the $M/M/4$ queue with arrival rate $\lambda = 2.4$ per time unit and service rate $\mu = 0.7$ per time unit.

then in the Accessories folder). It's probably simplest to move the `mmc.exe` file to the "root" drive (usually `C:`) on your system, and then in the Command Prompt window, type `cd ..` repeatedly until the prompt reads just `C:\ >`. If you then type just `mmc` at the prompt you'll get a response telling you what the syntax is, as shown at the top of Figure 2.2. Then, type `mmc` followed by the values you want for λ, μ, and c, separated by any number of blank spaces. The program responds with the steady-state queueing metrics for this system, assuming that the values you specified result in a stable queue, i.e., values for which $\rho < 1$. The resulting metrics are all for steady-state, using whatever time units you used for your input parameters (of course, λ and μ must use the same time units, but that time unit could be anything and is not needed by `mmc.exe`).

2.3.3 $M/G/1$

This model once again has exponential interarrival times, but now the service-time distribution can be anything; however, we're back to just a single server. This might be more realistic than the $M/M/1$ since exponential service times, with a mode (most likely value) of zero, are usually far-fetched in reality.

Let σ be the standard deviation of the service-time RV S, so σ^2 is its variance; recall that $E(S) = 1/\mu$. Then

$$W_q = \frac{\lambda(\sigma^2 + 1/\mu^2)}{2(1 - \lambda/\mu)}.$$

From this (known as the *Pollaczek-Khinchine formula*), you can use Little's law and the other relations in Section 2.2 to get additional output metrics, like L = the steady-state average number in system.

Note that W_q here depends on the *variance* σ^2 of the service-time distribution, and not just on its mean $1/\mu$; larger σ^2 implies larger W_q. Even the occasional extremely long service time can tie up the single server, and thus the system, for a long time, with new entities arriving all the while, thus causing congestion to build up (as measured by W_q and the other metrics). And the higher the variance of the service time, the longer the right tail of the service-time distribution, and thus the better the chance of encountering some of those extremely long service times.

2.3.4   *G/M/1*

This final example is a kind of reverse of the preceding one, in that now inter-arrival times can follow any distribution (of course, they have to be positive to make sense), but service times are exponential. We just argued that exponential service times are far-fetched, but this assumption, especially its memoryless property, is needed to derive the results below. And, there's just a single server. This model is more complex to analyze since, unlike the prior three models, knowing the number of entities present in the system is not enough to predict the future probabilistically, because the non-exponential (and thus non-memoryless) interarrival times imply that we also need to know how long it's been since the most recent arrival. As a result, the "formula" isn't really a formula at all, in the sense of being a closed-form plug-and-chug situation. For more on the derivations, see, for example, [16] or [49].

Let $g(t)$ be the density function of the interarrival-time distribution; we'll assume a continuous distribution, which is reasonable, and again, taking on only positive values. Recall that μ is the service rate (exponential here), so the mean service time is $1/\mu$. Let z be a number between 0 and 1 that satisfies the *integral equation*

$$z = \int_0^\infty e^{-\mu t(1-z)} g(t)dt. \tag{2.2}$$

Then

$$W = \frac{1}{\mu(1-z)},$$

and from that, as usual, we can get the other metrics via the relations in Section 2.2.

So the question is: What's the value of z that satisfies equation (2.2)? Unfortunately, (2.2) generally needs to be solved for z by some sort of numerical method, such as Newton-Raphson iterations or another root-finding algorithm. And the task is complicated by the fact that the right-hand side of (2.2) involves an integral, which itself has under it the variable z we're seeking!

So in general, this is not exactly a "formula" for W, but in some particular cases of the interarrival-time distribution, we might be able to manage something. For instance, if interarrivals have a continuous uniform distribution on $[a, b]$ (with $0 \le a < b$), then their PDF is

$$g(t) = \begin{cases} 1/(b-a) & \text{if } a \le t \le b \\ 0 & \text{otherwise} \end{cases}.$$

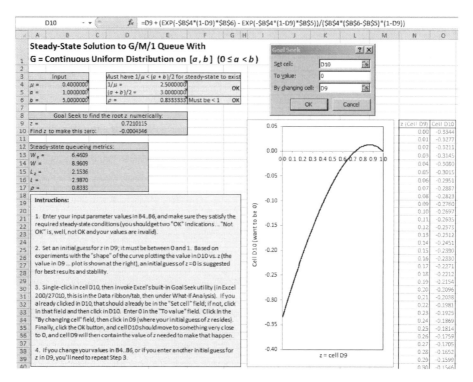

Figure 2.3: Spreadsheet `Uniform_M_1_queue_numerical_solution.xls` using Goal Seek to solve for steady-state Uniform/M/1 queueing metrics.

Substituting this into (2.2) results in

$$z = \int_a^b e^{-\mu t(1-z)} \frac{1}{b-a} dt,$$

which, after a little calculus and algebra, turns out to amount to

$$z = -\frac{1}{\mu(b-a)(1-z)} \left[e^{-\mu(1-z)b} - e^{-\mu(1-z)a} \right], \qquad (2.3)$$

and this can be "solved" for z by a numerical method.

Microsoft Excel's built-in Goal Seek capability, basically a numerical root-finder, might work. In Excel 2007/2010 Goal Seek is on the Data ribbon/tab, under the What-If Analysis button; in other versions of Excel it may be located elsewhere. In the file `Uniform_M_1_queue_numerical_solution.xls`, which you can download from the "students" section of this book's website (see the Preface for the URL and login/password), we've set this up for this Uniform/M/1 queue; see Figure 2.3. The formula in cell D10 (showing at the top of the figure) contains equation (2.3), except written as its left-hand side minus its right-hand side, so we use Goal Seek to find the value of z (in cell D9) that makes this difference in cell D10 (approximately) 0. You need to supply an initial "guess" at the root-finding value of z (which must be between 0 and 1) and enter your guess in cell D9; based on experimentation with the shape of the curve in cell D10 as a function of z in cell D9, it appears that initializing with $z = 0$ in cell D9 provides the most robust and stable results. The completed Goal Seek dialog is shown, and the results on the left are after it has run to completion. After approximating the required solution for z and confirming that cell D10 is (close to) 0, the spreadsheet also computes the standard steady-state queueing metrics. The text box at the bottom contains more detailed instructions for use of this spreadsheet model.

2.4 Queueing Networks

Queueing networks are composed of *nodes*, each being a multiserver queueing station (like Registration and Treatment Rooms in the urgent-care clinic of Figure 2.1), and connected by *arcs* representing possible paths of entity travel from one node to another (or from outside the system to a node, or from a node to exit the system). Entities can enter from outside the system into any node, though in our clinic they enter only to the Sign In node. Entities can also exit the system from any node, as they do from the Exam Rooms and Treatment Rooms nodes in our clinic. Internally, when an entity leaves a node it can go to any other node, with branching probabilities as in Figure 2.1; from Sign In they go to Registration with probability 0.9, and to Trauma Rooms with probability 0.1, etc. It's possible that all entities exiting a node go to the same place; such is the case out of the Registration and Trauma Rooms nodes.

We'll assume that all arrival processes from outside are independent of each other with exponential interarrival times (called a *Poisson process* since the number of entities arriving in a fixed time period follows the discrete Poisson distribution); in our clinic we have only one arrival source from outside, and all entities from it go to the Sign In node. We'll further assume that the service times everywhere are exponentially distributed, and independent of each other, and of the input arrival processes. And, we assume that all queue capacities are infinite. Finally, we'll assume that the utilization (a.k.a. *traffic intensity*) ρ at each node is strictly less than 1, so that the system does not explode anywhere in the long run. With all these assumptions, this is called a *Jackson network* (first developed in [19]), and a lot is known about it.

In our clinic, looking at just the Sign In node, this is an $M/M/2$ with arrival rate that we'll denote λ_{SignIn}, which would need to be given (so the interarrival times are IID exponential with mean $1/\lambda_{\text{SignIn}}$). Remarkably, in stable $M/M/c$ queueing stations, the output is also a Poisson process with the same rate as the input rate, λ_{SignIn} in this case. In other words, if you stand at the exit from the Sign-In station, you'll see the same probabilistic behavior that you see at the entrance to it. Now in this case that output stream is split by an independent toss of a (biased) coin, with each patient's going to Registration with probability 0.9, and to Trauma Rooms with probability 0.1. So we have two streams going out, and it turns out that each is also an independent Poisson process (this is called *decomposition* of a Poisson process), and at rate $0.9\lambda_{\text{SignIn}}$ into Registration and $0.1\lambda_{\text{SignIn}}$ into Trauma Rooms. Thus Registration can be analyzed as an $M/M/1$ queue with arrival rate $0.9\lambda_{\text{SignIn}}$, and Trauma Rooms can be analyzed as an $M/M/2$ queue with arrival rate $0.1\lambda_{\text{SignIn}}$. Furthermore, all three of the nodes we've discussed so far (Sign In, Registration, and Trauma Rooms) are probabilistically independent of each other.

Proceeding in this way through the network, we can analyze all the nodes as follows:

- Sign In: $M/M/2$ with arrival rate λ_{SignIn}.

- Registration: $M/M/1$ with arrival rate $0.9\lambda_{\text{SignIn}}$.

- Trauma Rooms: $M/M/2$ with arrival rate $0.1\lambda_{\text{SignIn}}$.

- Exam Rooms: $M/M/3$ with arrival rate $0.9\lambda_{\text{SignIn}}$.

- Treatment Rooms: $M/M/2$ with arrival rate

$$(0.9)(0.6)\lambda_{\text{SignIn}} + 0.1\lambda_{\text{SignIn}} = 0.64\lambda_{\text{SignIn}}.$$

The incoming process to Treatment Rooms is called the *superposition* of two independent Poisson processes, in which case we simply add their rates together to get the effective total incoming rate (and the superposed process is again a Poisson process).

So for each node separately, we could use the formulas for the $M/M/c$ case (or the command-line program `mmc.exe`) in Section 2.3 to find the steady-state average number and times in queue and system at each node separately.

You can do the same sort of thing to get the utilizations (traffic intensities) at each node separately. We know from the above list what the incoming rate is to each node, and if we know the service rate of each individual server at each node, we can compute the utilization there. For instance, let μ_{Exam} be the service rate at each of the three parallel independent servers at Exam Rooms. Then the "local" traffic intensity at Exam Rooms is $\rho_{\text{Exam}} = 0.9\lambda_{\text{SignIn}}/(3\mu_{\text{Exam}})$. Actually, these utilization calculations remain valid for any distribution of interarrival and service-times (not just exponential), in which case the arrival and service *rates* are defined as simply the reciprocals of the expected value of an interarrival and service-time RV, respectively.

As mentioned above, though our urgent-care-clinic example doesn't have it, there could be more Poisson input streams from outside the system arriving directly into any of the nodes; the overall effective input rate to a node is just the sum of the individual input rates coming into it. There could also be self-loops; for example out of Treatment Rooms, it could be that only 80% of patients exit the system, and the other 20% loop directly back to the input for Treatment Rooms (presumably for further treatment); in such a case, it may be necessary to solve a linear equation for, say, $\lambda_{\text{TreatmentRooms}}$. For the analysis of this section to remain valid, though, we do need to ensure that the "local" traffic intensity at each of the nodes stays strictly below 1.

2.5   Queueing Theory vs. Simulation

The nice thing about queueing-theoretic results is that they're exact, i.e., not subject to statistical variation (though in some cases where numerical analysis might be involved to find the solution, as in Section 2.3.4 for the $G/M/1$ queue, there could be roundoff error). As you'll soon see, simulation results are not exact, and *do* have statistical uncertainty associated with them, which we *do* need to acknowledge, measure, and address properly.

But queueing theory has other shortcomings of its own, mostly centered around the assumptions we need to make in order to derive formulas such as those in Section 2.3. In many real-world situations those assumptions will likely be simply incorrect, and it's hard to say what impact that might have on correctness of the results, and thus model validity. Assuming *exponential service times* seems particularly unrealistic in most situations, since the mode of the exponential distribution is *zero* — when you go to, say, airport security, do you think that it's most likely that your service time will be very close to zero, as opposed to having a most likely value of some number that's strictly greater than zero? And as we mentioned, such results are nearly always only for steady-state performance metrics, so don't tell you much if you're interested in *short-term* (or *finite-horizon* or *transient*) performance. And, queueing-theoretic results are not universally available for all interarrival-time and service-time distributions (recall the mathematical convenience of the memoryless exponential distribution), though approximation methods can be quite accurate.

Simulation, on the other hand, can most easily deal with short-term (or finite-horizon) time frames. In fact, steady-state is harder for simulation since we have to run for very long times, and also worry about biasing effects of initial conditions that are uncharacteristic of steady state. And when simulating, we can use whatever interarrival- and service-time distributions seem appropriate to fit the real system we're studying (see Sections 6.1 and 6.2), and in particular we have no need to assume that *anything* has an exponential distribution unless it happens to appear that way from the real-world data. Thus, we feel that simulation models stand a better chance of being more realistic, and thus more valid, models of reality, not to mention that their structure can go far beyond standard queueing models into great levels of complexity that render completely hopeless any kind of exact mathematical analysis. The only real downside to simulation is that you must remember that simulation results are statistical estimates, so must be analyzed with proper statistical techniques in order to be able to draw justified and precise conclusions, a point to which we'll return often in the rest of this book. In fact, the urgent-care clinic shown in Figure 2.1 and discussed throughout this chapter will be simulated (and properly statistically analyzed, of course), all in Simio, in a much more realistic framework in Chapter 9.

2.6 Problems

1. For an $M/M/1$ queue with mean interarrival time 1.25 minutes and mean service time 1 minute, find all five of W_q, W, L_q, L, and ρ. For each, interpret in words. Be sure to state all your units (always!), and the relevant time frame of operation.

2. Repeat Problem 1, except assume that the service times are not exponentially distributed, but rather (continuously) uniformly distributed be-

tween $a = 0.1$ and $b = 1.9$. Note that the expected value of this uniform distribution is $(a+b)/2 = 1$, the same as the expected service time in Problem 1. Compare all five of your numerical results to those from Problem 1 and explain intuitively with respect to this change in the service-time *distribution* (but with its *expected value* remaining at 1). *Hint*: In case you've forgotten, or your calculus has rusted completely shut, or you haven't already found it with a web search, the standard deviation of the continuous uniform distribution between a and b is $\sqrt{(b-a)^2/12}$ (that's right, you always divide by 12 regardless of what a and b are ... the calculus just works out that way).

3. Repeat Problem 1, except assume that the service times are triangularly distributed between $a = 0.1$ and $b = 1.9$, and with the mode at $m = 1.0$. Compare all five of your results to those from Problems 1 and 2. *Hint*: The expected value of a triangular distribution between a and b, and with mode m $(a < m < b)$, is $(a + m + b)/3$, and the standard deviation is $\sqrt{(a^2 + m^2 + b^2 - am - ab - bm)/18}$ ... do you think maybe it's time to dust off that calculus book (or, at least hone your web-search skills)?

4. In each of Problems 1, 2, and 3, suppose that we'd like to see what would happen if the arrival rate were to increase in small steps; maybe a single-server barbershop would like to increase business by some advertising or coupons. Create a spreadsheet or a computer program (if you haven't already done so to solve those problems), and re-evaluate all five of W_q, W, L_q, L, and ρ, except increasing the arrival rate by 5% over its original value, then by 15% over its original value, then by 20% over its original value, and so on up through increasing it by 100% over its original value (i.e., doubling the arrival rate). Make plots of each of the five metrics as functions of the percent increase in the arrival rate. Discuss.

5. Repeat Problem 1, except for an $M/M/3$ queue with mean interarrival time 1.25 minutes and mean service time 3 minutes at each of the three servers. *Hint*: You might want to consider creating a computer program or spreadsheet, or use `mmc.exe`.

6. In Problem 5, increase the arrival rate in the same steps as in Problem 4, and again evaluate all of W_q, W, L_q, L, and ρ at each step. If you run into any particular problems at some point as you increase the arrival rate, how many more servers (beyond the three that you already have) would you need to calm things down?

7. Show that the formula for L_q for the $M/M/c$ queue in Section 2.3 includes the $M/M/1$ as a special case.

8. Show that the formula for W_q for the $M/G/1$ queue in Section 2.3 includes the $M/M/1$ as a special case. *Hint*: The standard deviation of an exponential distribution with mean β is β ... again, quite discoverable by calculus or a web search (for the mathematically slothful).

9. Find all five of the steady-state queueing metrics for an $M/D/1$ queue, where D denotes a deterministic "distribution," i.e., the associated RV

(in this case representing service times) is a constant with no variation at all (also called a *degenerate* distribution). State parameter conditions for your results to be valid; use the same meaning for λ, μ, and ρ as we did in this chapter. Compare your results to those if D were replaced by a distribution with mean equal to the constant value from the original D distribution, except having at least some variation.

10. Consider a $D/D/1$ queue, where D represents a deterministic (or degenerate) distribution; see Problem 9. The interarrival times and service times do not have to be the same constant, of course. Find all five of the steady-state queueing metrics, and state parameter conditions for your results to be valid; use the same meaning for λ, μ, and ρ as we did in this chapter. *Hint*: This does not fit under any of the specific queueing models considered in Section 2.3 so don't bother looking for a formula there for plugging and chugging.

11. Show that the formula for L for the $G/M/1$ queue in Section 2.3 includes the $M/M/1$ as a special case. No numerical approximation is required, i.e., this can be done completely analytically. *Hint*: Remember that z must be between 0 and 1.

12. In the urgent-care clinic of Figure 2.1, suppose that the patients arrive from outside into the clinic (coming from the upper right corner of the figure and always into the Sign In station) with interarrival times that are exponentially distributed with mean 6 minutes. The number of individual servers at each station and the branching probabilities are all as shown in Figure 2.1. The service times at each node are exponentially distributed with means (all in minutes) of 3 for Sign In, 5 for Registration, 90 for Trauma Rooms, 16 for Exam Rooms, and 15 for Treatment Rooms. For each of the five stations, compute the "local" traffic intensity $\rho_{Station}$ there. Will this clinic "work," i.e., be able to handle the external patient load? Why or why not? If you could add a single server to the system, and add it to any of the five stations, where would you add it? Why? *Hint*: Unless you like using your calculator, a spreadsheet or computer program might be good, or perhaps use `mmc.exe`.

13. In Problem 12, for each of the five stations, compute each of W_q, W, L_q, L, and ρ, and interpret in words. Would you still make the same decision about where to add that extra single resource unit that you did in Problem 12? Why or why not? (Remember these are sick people, some of them seriously ill, not widgets being pushed through a factory.) *Hint*: Unless you really, *really*, **really** like using your calculator, a spreadsheet or computer program might be *very* good, or maybe you could use `mmc.exe`.

14. In Problems 12 and 13 (and without the extra resource unit), the inflow rate was ten per hour, the same as saying that the interarrival times have a mean of 6 minutes. How much higher could this inflow rate go and have the whole clinic still "work" (i.e., be able to handle the external patient load), even if only barely? Don't do any more calculations than necessary to answer this problem. If you'd like to get this really exact, you

might consider looking up a root-finding method and program it into your computer program (if you wrote one), or use the Goal Seek capability in Excel (if you created an Excel spreadsheet).

15. In Problems 12 and 13 (and without the extra resource unit), the budget's been cut and we need to eliminate one of the ten individual servers across the clinic; however, we have to keep at least one at each station. Where would this cut be least damaging to the system's operation? Could we cut two servers and expect the system still to "work?" More than two?

Chapter 3

Kinds of Simulation

Simulation is a very broad, diverse topic, and there are many different kinds of simulations, and approaches to carrying them out. In this chapter we describe some of these, and introduce terminology.

In Section 3.1 we classify the various kinds of simulations along several dimensions, and then in Section 3.2 we provide three examples of one kind of simulation, static stochastic (a.k.a. *Monte Carlo*) simulation. Section 3.3 considers two ways (manually and in a spreadsheet) of doing dynamic simulations without special-purpose dynamic-simulation software like Simio (neither is at all appealing, to say the least). Following that, Section 3.4 discusses the software options for doing dynamic simulations; Section 3.4.1 briefly describes how general-purpose programming languages like C++, Java, and Matlab might be used to carry out the kind of dynamic simulation logic done manually in Section 3.3.1, and Section 3.4.2 briefly describes special-purpose simulation software like Simio.

3.1 Classifying Simulations

"Simulation" is a wonderful word since it encompasses so many very different kinds of goals, activities, and methods. But this breadth can also sometimes render "simulation" a vague, almost-meaningless term. In this section we'll delineate the different kinds of simulation with regard to model structure, and provide examples. While there are other ways to categorize simulations, it's useful for us to consider *static* vs. *dynamic*; *continuous-change* vs. *discrete-change* (for dynamic models only); and *deterministic* vs. *stochastic*. Our focus in this book will be primarily on dynamic, discrete-change, stochastic simulation models.

3.1.1 Static vs. Dynamic Models

A *static* simulation model is one in which the passage of time plays no active or meaningful role in the model's operation and execution. Examples are using a random-number generator to simulate a gambling game or lottery, or to estimate the value of an integral or the inverse of a matrix, or to evaluate a financial profit-and-loss statement. While there may be a notion of time in the model, unless its passage affects the model's structure or operation, the model is static; three examples of this kind of situation, done via spreadsheets, are in Section 3.2.

In a *dynamic* simulation model, the passage of (simulated) time is an essential and explicit part of the model's structure and operation; it's impossible to build or execute the model without representing the passage of simulated time. Simulations of queueing-type systems are nearly always dynamic, since simulated time must be represented to make things like arrivals and service completions happen. An inventory simulation could be dynamic if there is logical linkage from one time to another in terms of, say, the inventory levels, re-stocking policy, or movement of replenishments. A supply-chain model including transportation or logistics would typically be dynamic since we'd need to represent the departure, movement, and arrival of shipments over time.

Central to any dynamic simulation is the simulation *clock*, just a variable representing the current value of simulated time, which is increased as the simulation progresses, and is globally available to the whole model. Of course, you need to take care to be consistent about the time units for the simulation clock. In modern simulation software, the clock is a *real-valued* variable (or, in computer-ese, a *floating-point* variable) so that the instants of time when things happen are represented exactly, or at least as exact as floating-point computer arithmetic can be (and that's *quite* exact even if not perfect). We bring this up only because some older simulation software, and unfortunately still some current special-purpose simulator templates for things like combat modeling, require that the modeler select a "small" *time step* (or time *slice* or time *increment*), say one minute for discussion purposes here. The simulation clock steps forward in these time steps only, then stops and evaluates things to see if anything has happened to require updating the model variables, so in our example minute by minute. There are two problems with this. First, a lot of computer time can be burned by just this checking, at each minute boundary, to see if something has happened; if not then nothing is done and this checking effort is wasted. Secondly, if, as is generally true, things can in reality happen at *any* point in continuous time, this way of operating forces them to occur instead on the one-minute boundaries, which introduces time-distortion error into the model. True, this distortion could be reduced by choosing the time step to be a second or millisecond or nanosecond, but this just worsens the run-time problem. Thus, time-stepped models, or integer-valued clocks, have serious disadvantages, with no particular advantages in most situations.

Static simulations can usually be executed in a spreadsheet, perhaps facilitated via a static-simulation add-in like @RISK or Crystal Ball (Section 3.2), or with a general-purpose programming language like C++, Java, or Matlab. Dynamic simulations, however, generally require more powerful software created specifically for them. Section 3.3.1 illustrates the logic of what for us is the most important kind of simulation, dynamic discrete-change simulation, in a manual simulation that uses a spreadsheet to track the data and do the arithmetic (but is by no means any sort of general simulation *model* per se). As you'll see in Section 3.3.2, it may be possible to build a general simulation model of a very, *very* simple queueing model in a spreadsheet, but not much more. For many years people used general-purpose programming languages to build dynamic simulation models (and still do on occasion, perhaps just for parts of a simulation model if very intricate nonstandard logic is involved), but this approach is just too tedious, cumbersome, time-consuming, and error-prone for the kinds of large, complex simulation models that people often want to build. These kinds of models really need to be done in dynamic-simulation software like Simio.

3.1.2 Continuous-Change vs. Discrete-Change Dynamic Models

In dynamic models, there are usually *state variables* that, collectively, describe the state of a simulated system at any point in (simulated) time. In a queueing system these state variables might include the lengths of the queues, the times of arrival of customers in those queues, and the status of servers (like idle, busy, or unavailable). In an inventory simulation the state variables might include the inventory levels and the amount of product on order.

If the state variables in a simulation model can change continuously over continuous time, then the model has *continuous-change* aspects. For example, the level of water in a tank (a state variable) can change continuously (i.e., in infinitesimally small increments/decrements), and at any point in continuous time, as water flows in the top and out the bottom. Often, such models are described by differential equations specifying the rates at which states change over time, perhaps in relation to their levels, or to the levels or rates of other states. Sometimes phenomena that are really discrete, like traffic flow of individual vehicles, are approximated by a continuous-change model, in this case viewing the traffic as a fluid. Another example of states that may, strictly-speaking, be discrete being modeled as continuous involves the famous *Lanchester combat differential equations*. In the Lanchester equations, $x(t)$ and $y(t)$ denote the sizes (or *levels*) of opposing forces (perhaps the number of soldiers) at time t in the fight. There are many versions and variations of the Lanchester equations, but one of the simpler ones is the so-called *modern* (or *aim-fire*) model where each side is shooting at the other side, and the attrition *rate* (so a derivative with respect to time) of each side at a particular point in time is proportional to the size (level) of the other side at that time:

$$\frac{dx(t)}{dt} = -ay(t)$$
$$\frac{dy(t)}{dt} = -bx(t)$$

(3.1)

where a and b are positive constants. The first equation in (3.1) says that the rate of change of the x force is proportional to the size of the y force, with a being a measure of effectiveness (lethality) of the y force — the larger the constant a, the more lethal is the y force, so the more rapidly the x force is attritted. Of course, the second equation in (3.1) says the same thing about the simultaneous attrition of the y force, with the constant b representing the lethality of the x force. As given here, these Lanchester equations can be solved analytically, so simulation isn't needed, but more elaborate versions of them cannot be analytically solved so would require simulation.

If the state variables can change only at instantaneous, separated, discrete points on the time axis, then the dynamic model is *discrete-change*. Most queueing simulation models are of this type, as state variables like queue lengths and server-status descriptors can change only *at* the times of occurrence of discrete events like customer arrivals, customer service completions, server breaks or breakdowns, etc. In discrete-change models, simulated time is realized only at the times of these discrete events — we don't even look at what's happening between successive events (because nothing is happening then). Thus, the simulation-clock variable jumps from the time of one event directly to the time of the next event, and at each of these points the simulation model must evaluate what's happening, and what changes are necessary to the state variables (and to other kinds of variables we'll describe in Section 3.3.1, like the event calendar and the statistical-tracking variables).

3.1.3 Deterministic vs. Stochastic Models

If all the input values driving a simulation model are fixed, non-random constants, then the model is *deterministic*. For example, a simple manufacturing line, represented by a queueing system, with fixed service times for each part, and fixed interarrival times between parts (and no breakdowns or other random events) would be deterministic. In a deterministic simulation, you'll obviously get the same results from multiple runs of the model, unless you decide to change some of the input constants.

Such a model may sound a bit far-fetched to you, and it does to us too. So most simulation models are *stochastic*, where at least some of the input values are not fixed constants, but rather random *draws* (or random *samples* or random *realizations*) from probability distributions, perhaps characterizing the variation in service and interarrival times in that manufacturing line. Specifying these input probability distributions and processes is part of model building, and will be discussed in Chapter 6, where methods for generating these random draws from probability distributions and processes during the simulation will also be covered. Since a stochastic model is driven, at root, by some sort of random-number generator, running the model just once gives you only a single sample of what *could* happen in terms of the output — much like running the real (not simulated) production line for one day and observing production on *that* day, or tossing a die once and observing that you got a 4 on *that* toss. Clearly, you'd never conclude from that single run of the production line that what happened on that particular day is what will happen every day, or that the 4 from that single toss of the die means that the other faces of the die are also 4. So you'd want to run the production line, or toss the die, many times to learn about the results' behavior (e.g., how likely it is that production will meet a target, or whether the die is fair or loaded). Likewise, you need to run a stochastic simulation model many times (or, depending on your goal, run it just once but for a very long time) to learn about the behavior of its results (e.g., what are the mean, standard deviation, and range of an important output performance measure?), a topic we'll discuss in Chapters 4 and 5.

This randomness or uncertainty in the output from stochastic simulation might seem like a nuisance, so you might be tempted to eliminate it by replacing the input probability distributions by their means, a not-unreasonable idea that has sometimes been called *mean-value analysis*. This would indeed render your simulation output non-random and certain, but it might also very well render it wrong. Uncertainty is often part of reality, including uncertainty on the output performance metrics. But beyond that, variability in model inputs can often have a major impact on even the *means* of the output metrics. For instance, consider a simple single-server queue having exponentially distributed interarrival times with mean 1 minute, and exponentially distributed service times with mean 0.9 minute. From queueing theory (see Chapter 2), we know that in steady state (i.e., in the long run), the expected customer waiting time in queue before starting service is 8.1 minutes. However, if we replaced the exponentially distributed interarrival and service times with their means (exactly 1 minute and exactly 0.9 minute, respectively), it's clear that no customer would ever have to wait in queue at all, so the mean waiting time in queue would be 0, which is very wrong compared to the correct answer of 8.1 minutes. Intuitively, variation in the interarrival and service times creates the possibility of many small interarrival times in a row, or many large service times in a row (or even just one really large service time), either of which leads to congestion and possibly long waits in queue. This is among the points made by Sam Savage in *The Flaw of Averages: Why We Underestimate Risk in the Face of Uncertainty* [50].

It often puzzles people starting to use simulation software that executing your stochastic model again actually gets you exactly the same numerical results. You'd think this shouldn't happen if a random-number generator is driving the model. But, as you'll see in Chapter 6, most random-number generators aren't really random at all, and will produce the same sequence of "random" numbers every time. What you need to do is *replicate* the model multiple times within the *same* execution, which causes you just to continue on in the fixed random-number sequence from one replication to the next, which will produce different (random) output from each replication and give you important information on the uncertainty in your results. Reproducibility of the basic random numbers driving the simulation is actually desirable for debugging, and for sharpening the precision of output by means of *variance-reduction techniques*, discussed in general simulation books like [3] and [30].

3.2 Static Stochastic Monte Carlo

Perhaps the earliest uses for stochastic computer simulation were in addressing models of systems in which there's no passage of time, so there's no time dimension at all. These kinds of models are sometimes called *Monte Carlo*, or *stochastic Monte Carlo*, an homage to the Mediterranean gambling paradise (or purgatory, depending on how you feel about gambling and casinos). In this section we'll discuss three simple models of this kind. The first (Model 3-1) is the sum of throwing two dice. The second (Model 3-2) is using this kind of simulation to estimate the value of an analytically-intractable integral. The third (Model 3-3) is a model of a classical single-period perishable inventory problem.

By the way, throughout this book, "Model x-y" will refer to the yth model in Chapter x. The corresponding files, which can be downloaded from the student area on textbook website as described in the Preface, will be named `Model_x_y.something`, where the x and y will include leading zeros pre-pended if necessary to make them two digits, so that they'll alphabetize correctly (we promise we'll never have more than 99 chapters or more than 99 models in any one chapter). And `something` will be the appropriate file-name extension, like `xls` in this chapter for Excel files, and `spfx` later for Simio project files. Thus, Model 3-1, which we're about to discuss, is accompanied by the file `Model_03_01.xls`.

We'll use Microsoft Excel® for the examples in this section due to its ubiquity and familiarity, but these kinds of simulations can also be easily done in programming languages like C++, Java, or Matlab.

3.2.1 Model 3-1: Throwing Two Dice

Throw two fair dice, and take the result as the sum of the faces that come up. Each die has probability $1/6$ of producing each of the integers $1, 2, \ldots, 6$ and so the sum will be one of the integers $2, 3, \ldots, 12$. Finding the probability distribution of the sum is easy, and is commonly covered in books covering elementary probability, so there's really no need to simulate this game, but we'll do so anyway to illustrate the idea of static Monte Carlo.

In `Model_03_01.xls` (see Figure 3.1) we provide a simulation of this in Excel, and simulate 50 throws of the pair of dice, resulting in 50 values of the sum. All the possible face values are in cells C4..C9, the probabilities of each of the faces are in cells A4..A9, and the corresponding probabilities that the die will be strictly less than that face value are in cells B4..B9 (as you'll see in a moment, it's convenient to have these strictly-less-than probabilities in B4..B9 for generating simulated throws of each die). The throw numbers are in column E, the results of the two individual dice from each throw are in columns F and G, and their sum for each throw is in column H. In cells J4..L4 are a few basic summary statistics on the 50 sums. If you have the spreadsheet `Model_03_01.xls` open yourself (which we hope you do), you'll notice that your results are different from those in Figure 3.1, since Excel's random-number generator (discussed in the next paragraph) is what's called *volatile*, so that it re-generates "fresh" random numbers whenever the spreadsheet is recalculated, which happens when it's opened.

The heart of this simulation is generating, or "simulating" the individual dice faces in cells F4..G53; we'll discuss this in two steps. The first step is to note that Excel comes with a built-in *random-number generator*, `RAND()` (that's right, the parentheses are required but there's nothing inside them), which is supposed to generate independent observations on a continuous uniform distribution between 0 and 1. We say "supposed to" since the underlying algorithm is a particular mathematical procedure that cannot, of course, generate "true" randomness, but only observations that "appear" (in the sense of formal statistical-inference tests) to be uniformly distributed between 0 and 1, and appear to be independent of each other. These are sometimes called *pseudo-random* numbers, a mouthful, so we'll just call them *random numbers*.

Static Monte Carlo Simulation of the Sum of two Dice

P(Face=i)	P(Face<i)	i	Toss	Die 1	Die 2	Sum	Mean Sum	Min Sum	Max Sum
0.1567	0.0000	1	1	5	6	11	7.3800	2	12
0.1567	0.1667	2	2	6	4	10			
0.1567	0.3333	3	3	3	3	6			
0.1567	0.5000	4	4	1	1	2			
0.1567	0.6667	5	5	2	5	7			
0.1567	0.8333	6	6	5	4	9			
1.0000	← check sum		7	6	2	8			
			8	3	3	6			
			9	3	4	7			
			10	6	6	12			
			11	3	2	5			
			12	5	2	7			
			13	3	4	7			
			14	5	6	11			
			15	4	6	10			
			16	5	4	9			
			17	5	2	7			
			18	2	2	4			
			19	3	6	9			
			20	5	5	10			
			21	5	1	6			
			22	1	6	7			
			23	5	4	9			
			24	5	3	8			
			25	5	1	6			
			26	5	2	7			
			27	5	2	7			
			28	4	1	5			
			29	3	1	4			
			30	5	3	8			
			31	5	4	9			
			32	2	6	8			
			33	1	5	6			
			34	3	2	5			
			35	4	6	10			
			36	5	1	6			
			37	4	4	8			
			38	2	5	7			
			39	3	3	6			
			40	4	5	9			
			41	3	4	7			
			42	3	2	5			
			43	2	5	7			
			44	5	4	9			
			45	4	4	8			
			46	6	1	7			
			47	6	2	8			
			48	4	3	7			
			49	1	4	5			
			50	6	2	8			

Figure 3.1: Static Monte Carlo simulation of the sum when throwing two fair dice, using the spreadsheet Model_03_01.xls.

We'll have more to say in Chapter 6 about how such random-number generators work, but suffice it to say here that they're complex algorithms that depend on solid mathematical research. Though the method underlying Excel's RAND() has been criticized, it will do for our demonstration purposes here, but we might feel more comfortable with a better random-number generator such as those supplied with Excel simulation add-ins like Palisade Corporation's @RISK® (www.palisade.com), or certainly with full-featured dynamic-simulation software like Simio.

The second step transforms a continuous observation, let's call it U, distributed continuously and uniformly on $[0, 1]$ from RAND(), into one of the integers $1, 2, \ldots, 6$, each with equal probability $1/6$ since the dice are assumed to be fair. Though the general ideas for generating observations on random variables (sometimes called *random variates*) is discussed in Chapter 6, it's easy to come up with an intuitive scheme for our purpose here. Divide the continuous unit interval $[0, 1]$ into six contiguous subintervals, each of width $1/6$: $[0, 1/6)$, $[1/6, 2/6), \ldots, [5/6, 1]$; since U is a continuous random variate it's not important whether the endpoints of these subintervals are closed or open, since U will fall exactly on a boundary only with probability 0, so we don't worry about it. Remembering that U is continuously uniformly distributed across $[0, 1]$, it will fall in each of these subintervals with probability equal to the subintervals' width, which is $1/6$ for each subinterval, and so we'll generate (or *return*) a value 1 if it falls in the first subinterval, or a value 2 if it falls in the second subinterval, and so on for all six subintervals. This is implemented in the formulas in cells F4..G53 (it's the same formula in all these cells),

$$= \texttt{VLOOKUP(RAND(), \$B\$4:\$C\$9, 2, TRUE)}.$$

This Excel function VLOOKUP, with its fourth argument set to TRUE, returns the value in the second column (that's what the 2 is for in the third argument) of the cell range B4..C9 (the second argument) in the row corresponding to the values between which RAND() (the first argument) falls in the first column of B4..C9. (The $s in $B$4:$C$9 are to ensure that reference will continue to be made to these fixed specific cells as we copy this formula down in columns F and G.) So we'll get C4 (which is 1) if RAND() is between B4 and B5, i.e. between 0 and 0.1667, which is our first subinterval of width (approximately) $1/6$, as desired. If RAND() is not in this first subinterval, VLOOKUP moves on to the second subinterval, between B5 and B6, so between 0.1667 and 0.3333, and returns C5 (which is 2) if RAND() is between these two values, which will occur with probability $0.3333 - 0.1667 = 1/6$ (approximately). This continues up to the last subinterval if necessary (i.e., if RAND() happened to be large, between 0.8333 and 1, a width of approximately $1/6$), in which case we get C9 (which is 6). If you're unfamiliar with the VLOOKUP function, it might help to, well, look it up online or in Excel's help.

The statistical summary results in J4..L4 are reasonable since we know that the range of the sum is 2 to 12, and the true expected value of the sum is 7. Since Excel's RAND() function is volatile, every instance of it re-generates a fresh random number any time the spreadsheet is recalculated, which can be forced by tapping the F9 key (there are also other ways to force recalculation). So try that a few times and note how the individual dice values change, as does the mean of their sum in J4 (and occasionally even the min sum and max sum in K4 and L4).

This idea can clearly be extended to any discrete probability distribution as long as the number of possible values is finite (we had six values here), and for unequal probabilities (maybe for dice that are "loaded," i.e., not "fair" with equal probabilities for all six faces). See Problem 1 for some extensions of this dice-throwing simulation.

3.2.2 Model 3-2: Monte-Carlo Integration

Say you want to evaluate the numerical value of the definite integral $\int_a^b h(x)dx$ (for $a < b$), which is the area under $h(x)$ between a and b, but suppose that, due to its form, it's hard (or impossible) to find the anti-derivative of $h(x)$ to do this integral exactly, no matter how well you did in calculus. To simplify the explanation, let's assume that $h(x) \geq 0$ for all x, but everything here still works if that's not true. The usual way to approach this is *numerically*, dividing the range $[a, b]$ into narrow vertical subintervals, then calculating the area in the strip above each subinterval under $h(x)$ by any of several approximations (rectangles, trapezoids, or fancier formulas), and then summing these areas. As the subintervals get more and more narrow, this numerical method becomes more and more accurate as an approximation of the true value of the integral. But it's possible to create a static Monte Carlo simulation to get a *statistical estimate* of this integral or area.

To do this, first, generate a random variate X that's distributed continuously and uniformly between a and b. Again, general methods of variate generation are discussed in Chapter 6, but here again there's easy intuition to lead to the correct formula to do this. Since $U = $ RAND() is distributed continuously and uniformly between 0 and 1, if we "stretch" the range of U by multiplying it by $b - a$, we'll not distort the uniformity of its distribution, and will likewise "stretch" its range to be uniformly distributed between 0 and $b - a$ (using the word "stretch" implies we're thinking that $b - a > 1$ but this all still works if $b - a < 1$, in which case we should say "compress" rather than "stretch"). Then if we add a, the range will be shifted right to $[a, b]$, still of width $b - a$, as desired, and still it's uniformly distributed on this range since this whole transformation is just linear (by saying "shifted right" we're thinking that $a > 0$, but this all still works if $a < 0$, in which case we should say "shifted left" when we add $a < 0$).

Next, take the value X you generated in the preceding paragraph such that it's continuously uniformly distributed on $[a, b]$, evaluate the function h at this value X, and finally multiply that by $b - a$ to get a random variate we'll call $Y = (b - a)h(X)$. If h were a flat (or constant) function, then no matter where X happened to fall on $[a, b]$, Y would be equal to the area under h between a and b (which is the integral $\int_a^b h(x)dx$ we're trying to estimate) since we'd be talking about just the area of a rectangle here. But if h were a flat function, you wouldn't be doing all this in the first place, so it's safe to assume that h is some arbitrary curve, and so Y will not be equal to $\int_a^b h(x)dx$ unless we were very lucky to have generated an X to make the rectangle of width $b - a$ and height $h(X)$ equal to the area under $h(x)$ over $[a, b]$. However, we know from the *first mean-value theorem for integration* of calculus that there exists (at least one) such "lucky" X somewhere between a and b so we could always hope. But better than hoping for extremely good luck would be to hedge our bets and generate (or *simulate*) a large number of independent realizations of X, compute $Y = (b - a)h(X)$, the area of a rectangle of width $b - a$ and height $h(X)$ for each one of them, and then average all of these rectangle-area values of Y. You can imagine that some rectangle-area values of Y will be bigger than the desired $\int_a^b h(x)dx$, while others will be smaller, and maybe on average they're about right. In fact, it can be shown by basic probability, and the definition of the *expected value* of a random variable, that the expected value of Y, denoted as $E(Y)$ is actually equal to $\int_a^b h(x)dx$; see Problem 3 for hints on how to prove this rigorously.

As a test example, let's take $h(x)$ to be the probability density function of a normal random variable with mean μ and standard deviation $\sigma > 0$, which is

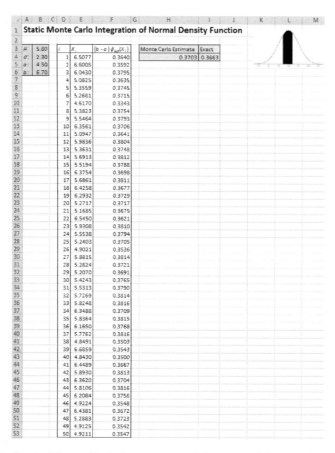

Figure 3.2: Static Monte Carlo integration of the normal density function using Model_03_02.xls.

often denoted as $\phi_{\mu,\sigma}(x)$, and is given by

$$\phi_{\mu,\sigma}(x) = \frac{1}{\sigma\sqrt{2\pi}} e^{-\frac{(x-\mu)^2}{2\sigma^2}}$$

for any real value of x (positive, negative, or zero). There's no simple closed-form anti-derivative of $\phi_{\mu,\sigma}(x)$ (which is why old-fashioned printed statistics books used to have tables of normal probabilities in the back or on the inside covers), so this will serve as a good test case for Monte-Carlo integration, and we can get a very accurate benchmark from those old normal tables (or from more modern methods, like Excel's built-in NORMDIST function). Purely arbitrarily, let's take $\mu = 5.8$, $\sigma = 2.3$, $a = 4.5$, and $b = 6.7$. In Model_03_02.xls (see Figure 3.2) we provide a simulation of this in Excel, and simulate 50 draws of X (denoted X_i for $i = 1, 2, \ldots, 50$) distributed continuously and uniformly between $a = 4.5$ and $b = 6.7$ (column E), resulting in 50 values of $(6.7 - 4.5)\phi_{5.8,2.3}(X_i)$ in column F; we used Excel's NORMDIST function to evaluate the density $\phi_{5.8,2.3}(X_i)$ in column F. We leave it to you to examine the formulas in columns E and F to make sure you understand them. The average of the 50 values in column F is in cell H4, and in cell I4 is the exact benchmark value using `NORMDIST` again. Tap the F9 key a few times to observe the randomness in the simulated results.

Monte Carlo integration is seldom actually used for one-dimensional integrals like this, since the non-stochastic numerical area-estimating methods mentioned earlier are usually more efficient and quite accurate; Monte Carlo integration is, however, used to estimate integrals in higher dimensions, and over irregularly shaped domains of integration. See Problems 3 and 4 for analysis and extension of this example of Monte Carlo integration.

3.2.3 Model 3-3: Single-Period Inventory Profits

Your local sports team has just won the championship, so as the owner of several sporting-goods stores in the area you need to decide how many hats with the team's logo you should order. Based on past experience, your best estimate of demand D within the next month (the peak period) is that it will be between $a = 1000$ and $b = 5000$ hats. Since that's about all you think you can sell, you assume a uniform distribution; because hats are discrete items, this means that selling each of $a = 1000, 1001, 1002, \ldots, 5000 = b$ hats is an event with probability $1/(b + 1 - a) = 1/(5000 + 1 - 1000) = 1/4001 \approx 0.000249938$. Your supplier will sell you hats at a wholesale unit price of $w = \$11.00$, and during the peak period you'll sell them at a retail price of $r = \$16.95$. If you sell out during the peak period you have no opportunity to re-order so you just miss out on any additional sales (and profit) you might have gotten if you'd ordered more. On the other hand, if you still have hats after the peak period, you mark them down to a clearance price of $c = \$6.95$ and sell them at a loss; assume that you can eventually sell all remaining hats at this clearance price without further mark-downs. So, how many hats h should you order to maximize your profit?

Each hat you sell during the peak period turns a profit of $r - w = \$16.95 - \$11.00 = \$5.95$, so you'd like to order enough to satisfy all peak-period demand. But each hat you *don't* sell during the peak period loses you $w - c = \$11.00 - \$6.95 = \$4.05$, so you don't want to over-order either and incur this downside risk. Clearly, you'd like to wait until demand occurs and order just that many hats, but customers won't wait and that's not how production works, so you must order before demand is realized. This is a classical problem in the analysis of perishable inventories, known generally as the *newsvendor problem* (thinking of daily *hardcopy* newspapers [if such still exist when you read this book] instead of hats, which are almost worthless at the end of that day), and has other applications like fresh food, gardening stores, and even blood banks. Much research has investigated analytical solutions in certain special cases, but we'll develop a simple spreadsheet simulation model that will be more generally valid, and could be easily modified to different problem settings.

We first need to develop a general formula for profit, expressed as a function of h = the number of hats ordered from the supplier. Now D is the peak-period demand, which is a discrete random variable distributed uniformly on the integers $\{a = 1000, 1001, 1002, \ldots, b = 5000\}$. The number of hats sold at the peak-period retail price of $r = \$16.95$ will be $\min(D, h)$, and the number of hats sold at the post-peak-period clearance price of $c = \$6.95$ will be $\max(h - D, 0)$. To verify these formulas, try out specific values of h and D, both for when $D \geq h$ so you sell out, and for when $D < h$ so you have a surplus left over after the peak period that you have to dump at the clearance price. Profit $Z(h)$ is (total revenue) – (total cost):

$$Z(h) \;=\; \underbrace{r\min(D, h)}_{\text{Peak-period revenue}} \;+\; \underbrace{c\max(h - D, 0)}_{\text{Clearance revenue}} \;-\; \underbrace{wh}_{\text{Total cost}} \;. \qquad (3.2)$$

$$\underbrace{}_{\text{Total revenue}}$$

Note that the profit $Z(h)$ is itself a random variable since it involves the random variable D as part of its definition; it also depends on the number of hats ordered, h, so that's part of its notation.

Before moving to the spreadsheet, there's one more thing we need to figure out — how to generate (or "simulate") observations on the demand random variable D. We'll again start with Excel's built-in (pseudo)random-number generator RAND() to provide independent observations on a continuous uniform distribution between 0 and 1, as introduced earlier in Section 3.2.1. So for arbitrary integers a and b with $a < b$ (in our example, $a = 1000$ and $b = 5000$), how do we transform a random number U continuously and uniformly distributed between 0 and 1, to an observation on a discrete uniform random variable D on the integers $\{a, a + 1, \ldots, b\}$? Chapter 6 discusses this, but for now we'll just rely on some intuition, as we did for the two earlier spreadsheet simulations in Sections 3.2.1 and 3.2.2. First, $a + (b + 1 - a)U$ will be continuously and uniformly distributed between a and $b + 1$; plug in the extremes $U = 0$ and $U = 1$ to confirm the range, and since this is just a linear transformation, the uniformity of the distribution is preserved. Then round this down to the next lower integer, an operation that's sometimes called the *integer-floor function* and denoted $\lfloor \cdot \rfloor$, to get

$$D = \lfloor a + (b + 1 - a)U \rfloor. \qquad (3.3)$$

This will be an integer between a and b inclusively on both ends (we'd get $D = b + 1$ only if $U = 1$, a probability-zero event that we can safely ignore). And the probability that D is equal to any one of these integers, say i, is the probability that $i \leq a + (b + 1 - a)U < i + 1$, which, after a bit of algebra, is the same thing as

$$\frac{i - a}{b + 1 - a} \leq U < \frac{i + 1 - a}{b + 1 - a}. \qquad (3.4)$$

Note that the endpoints are both on the interval $[0, 1]$. Since U is continuously uniformly distributed between 0 and 1, the probability of the event in (3.4) is just the width of this subinterval,

$$\frac{i + 1 - a}{b + 1 - a} - \frac{i - a}{b + 1 - a} = \frac{1}{b + 1 - a} = \frac{1}{5000 + 1 - 1000} = \frac{1}{4001} \approx 0.000249938,$$

exactly as desired! Excel has an integer-floor function called INT, so the Excel formula for generating an observation on peak-period demand D in (3.3) is

$$= \text{INT}(a \;+\; (b \;+\; 1 \;-\; a)*\text{RAND}()) \qquad (3.5)$$

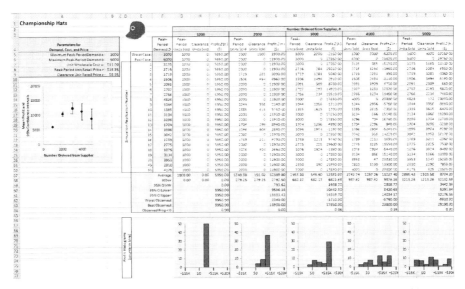

Figure 3.3: Single-period inventory spreadsheet simulation using only built-in Excel features (first sheet of `Model_03_03.xls`).

where a and b will be replaced by absolute references to the cells containing their numerical values (absolute since we'll be copying this formula down a column).

The first sheet (tab at the bottom) of the file `Model_03_03.xls` contains our simulation using only the tools that come built-in with Excel. Figure 3.3 shows most of the spreadsheet (to save space, rows 27-54 are hidden and the numerical frequency-distribution tables at the bottom have been cropped out), but it's best if you could just open the Excel file itself and follow along in it as you read, paying particular attention to the formulas entered in the cells. The blue-shaded cells in A4..B9 contain the model input parameters. These should be set apart like this and then used in the formulas rather than "hard-wiring" these constants into the Excel model, to facilitate possible "what-if" experiments later to measure the impact of changing their values; that way, we need only change their values in the blue cells, rather than throughout the spreadsheet wherever they're used. We decided, essentially arbitrarily, to make 50 IID replications of the model (each row is a replication), and the replication number is in column E. Starting in row 7 and going down, the light-orange-shaded cells in Column F contain the 50 peak-period demands D, as generated via (3.5), which in each of these cells is the formula

$$\texttt{=INT(\$B\$5 + (\$B\$6 + 1 - \$B\$5)*RAND())}$$

since a is in B5 and b is in B6. We use the \$ symbols to force these to be absolute references since we typed this formula into F7 and then copied it down. Above the realized demands, we entered the worst possible ($a = 1000$) and best possible ($b = 5000$) cases for peak-period demand D in order to see the worst and best possible values of profit for each trial value of $h =$ number of hats ordered, as well as the 50 realized values of profit for each of the 50 realized values of peak-period demand D and for each trial value of h. If you have this spreadsheet open you'll doubtless see numerical results that differ from those in Figure 3.3 because Excel's RAND() function is *volatile*, meaning that it re-generates every single random number in the spreadsheet upon each spreadsheet recalculation. One way to force a recalculation is to tap the F9 key on your keyboard, and you'll see all the numbers (and the plots) change. Even though we have 50 replications, there is still considerable variation from one set of 50 replications to the next, with each tap of the F9 key. This only serves to emphasize the importance of proper statistical design and analysis of simulation experiments, and to do something to try to ascertain an adequate sample size.

We took the approach of simply trying several values for h (1000, 2000, 3000, 4000, and 5000) spanning the range of possible peak-period demands D under our assumptions. For each of these values of h, we computed the profit on each of the 50 replications of peak-period demand if h had been each of these values, and then summarized the statistics numerically and graphically.

If we order just $h = 1000$ hats (columns G-I), then we sell 1000 hats at the peak-period price for sure since we know that $D \geq 1000$ for sure, leading to a profit of exactly $(r - w)D = (\$16.95 - \$11.00) \times 1000 = \$5950$ every time — there's no possibility of having anything left over to dump at the clearance price (column H contains only 0s), and there's no risk of profiting any less than this. There's also no chance of profiting any *more* than this $5950 since, of course, if $D > 1000$, which it will be with very high probability $(1 - 1/(5000 + 1 - 1000) \approx 0.999750062)$, you're leaving money on the table with such a small, conservative, totally risk-averse choice of h.

Moving further to the right in the spreadsheet, we see that for higher order values h, the mean profits (row 57) rise to a maximum point at somewhere around $h = 3000$ or 4000, then fall off for $h = 5000$ due to the relatively large numbers of hats left unsold after the peak period; this is depicted in the plot in column A where the heavy dots are the sample means and the vertical I-beams depict 95% confidence intervals for the population means. As h rises there is also more variation in profit (look at the standard deviations in row 58), since there is more variation in both peak-period units sold and clearance units sold (note that their sum is always h, which it of course must be in this model). This makes the 95% confidence intervals on expected profit (rows 59-61 and the plot in column A) larger, and also makes the histograms of profit (rows 66-77) more spread out (it's important that they all be scaled identically), which is both good and bad — the good is that there's a chance, albeit small, to turn very large profits; but the bad is that there's an increasing chance of actually losing money (row 64, and the areas to the left of the red zero marker in the histograms). So, how do you feel about risk, both upside and downside?

The second sheet of Model_03_03.xls has the same simulation model but using the simulation add-in @RISK from Palisade Corporation; see Figure 3.4 with the @RISK ribbon showing at the top, but again it's better if you just open the file to this sheet. To execute (or even properly view) this sheet you must already have @RISK loaded. We won't attempt here to give a detailed account of @RISK, as that's well documented elsewhere, e.g. [1]. But here are

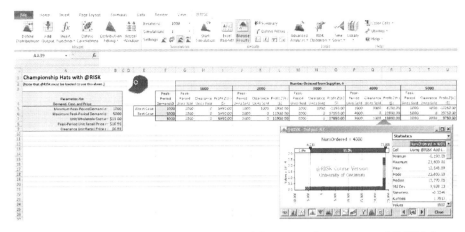

Figure 3.4: Single-period inventory spreadsheet simulation using @RISK (second sheet of `Model_03_03.xls`).

the major points and differences from simulating using only the built-in Excel capabilities:

1. We still have to create the basic model, the most important parts of which are the input (blue-shaded) areas, and the formulas in row 7 for the outputs (profits in this model).

2. However, we no longer need to copy this single-row simulation down to many more rows, one for each replication.

3. @RISK provides the capability for a large number of input distributions, including the discrete uniform we want for peak-period demands. This is entered in cell F7 via the @RISK formula `=RiskIntUniform(1000,5000)`, which can be entered by first single-clicking in this cell, and then clicking the Define Distributions button on the left end of the @RISK ribbon to find a gallery of distributions from which to select and parameterize your inputs (we have only one random input in this model).

4. The number of replications (or "Iterations" as they're called by @RISK) is simply entered (or selected from the pull-down) in that field near the center of the @RISK ribbon. We selected 1000, which is the same thing as having 1000 explicit rows in our original spreadsheet simulation using only the built-in Excel tools.

5. We need to identify the cells that represent the outputs we want to watch; these are the profit cells I7, L7, O7, R7, and U7. To do so, just click on these cells, one by one, and then click the Add Output button near the left end of the @RISK ribbon and optionally change the Name for output labeling. We've shaded these cells light green.

6. To run the simulation, click the Start Simulation button just to the right of the Iterations field in the @RISK ribbon.

7. We don't need to do anything to collect and summarize the output (as we did before with the means, confidence intervals, and histograms) since @RISK keeps track of and summarizes all this behind the scenes.

8. There are many different ways to view the output from @RISK. Perhaps one of the most useful is an annotated histogram of the results. To get this (as shown in Figure 3.4), single-click on the output cell desired (such as R7, profit for the case of $h = 4000$), then click on the Browse Results button in the @RISK ribbon. You can drag the vertical splitter bars left and right and get the percent of observations between them (like 15.6% negative profits, or losses, in this case), and in the left and right tails beyond the splitter bars. Also shown to the right of the histogram are some basic statistics like mean, standard deviation, minimum, and maximum.

It's clear that add-ins like @RISK greatly ease the task of carrying out static simulations in spreadsheets. Another advantage is that they provide and use far better random-number generators than the RAND() function that comes with Excel. Competing static-simulation add-ins for Excel include Oracle's Crystal Ball® (www.oracle.com/crystalball/), and Frontline Systems' Risk Solver® (www.solver.com/risk-solver-pro).

3.3   Dynamic Simulation Without Special Software

In this section we'll consider how dynamic simulations might be carried out without the benefit of special-purpose dynamic-simulation software like Simio. Section 3.3.1 carries out a manual (i.e., no computer at all) simulation of a simple single-server queue, using an Excel spreadsheet to help with the record-keeping and arithmetic, but that is by no means any kind of general manual-simulation "model." Then Section 3.3.2 develops a spreadsheet model of *part* of a single-server queue by exploiting a very special relationship that is not available for more complex dynamic queueing simulations.

3.3.1   Model 3-4: Manual Dynamic Simulation

Though you won't see it using high-level simulation-modeling software like Simio, all dynamic discrete-change simulations work pretty much the same way under the hood. In this section we'll describe how this goes, since even with high-level software like Simio, it's still important to understand these concepts even if you'll (thankfully) never actually have to do a simulation this way. The logic for this is also what you'd need to code a dynamic discrete-change simulation using a general-purpose programming language like C++, Java, or Matlab.

Our system here is just a simple single-server queue. *Entities* (which could represent customers in a service system, parts in a manufacturing system, or patients in a health-care system) show up at arrival times that, for now, will be assumed to be simply given from some magical[1] outside source (we'll assume the first arrival to be at time 0), and require service times that are also just given;

---

[1] Actually, see Section 4.2.1.

all times are in minutes and the queue discipline is first-in, first-out (FIFO). The simulation will just end abruptly at time 8 minutes, even if entities are in queue or in service at that time. For output, we'll compute the utilization of the server (proportion or percent of time busy), as well as the average, minimum, and maximum of each of (a) the total entity time in system, (b) the entity time in queue (just waiting in line, exclusive of service time), (c) the number of entities in the system, and (d) the number of entities in the queue. The averages in (a) and (b) are just simple averages of lists of numbers, but the averages in (c) and (d) are *time*-averages, which are the total areas under the curves tracking these variables over continuous time, divided by the final simulation-clock time, 8. For example, if $L_q(t)$ is the number of entities waiting in the queue at time t, then the time-average number of entities in queue is $\int_0^8 L_q(t)dt/8$, which is just the weighted average of the possible queue-length values $(0, 1, 2, \ldots)$, with the weights being the proportions of time the curve is at level $0, 1, 2, \ldots$. Likewise, if $B(t)$ is the server-busy function (0 if idle at time t, 1 if busy at time t), and $L(t)$ is the number of entities in system (in queue plus in service) at time t, then the server-utilization proportion is $\int_0^8 B(t)dt/8$, and the time-average number of entities in system is $\int_0^8 L(t)dt/8$.

We'll keep track of things, and do the arithmetic, in an Excel spreadsheet called `Model_03_04.xls`, with the files that accompany this book and can be downloaded as described in the Preface, shown in Figure 3.5. The light-blue-shaded cells B5..B14 and D5..D14 contain the given arrival and service times. It turns out to be more convenient to work with – rather than the given arrival times – the *interarrival* times, which are calculated via the Excel "=" formulas in C6..C14; type `CTRL-`` (while holding down the Control key, type the back-quote character) to toggle between viewing the values and formulas in all cells. Note that an interarrival time in a row is the time between the arrival of that row's entity and the arrival of the prior entity in the preceding row (explaining why there's no interarrival time for entity 1, which arrives at time 0).

In a dynamic discrete-change simulation, you first identify the *events* that occur at instants of simulated time, and that may change the system state. In our simple single-server queue, the events will be the *arrival* of an entity, the end of an entity's service and immediate *departure* from the system, and the *end of the simulation* (which might seem like an "artificial" event but it's one way to stop a simulation that's to run for a fixed simulated-time duration — we'll "run" ours for 8 minutes). In more complex simulations, additional events might be the breakdown of a machine, the end of that machine's repair time, a patient's condition improves from one acuity level to a better one, the time for an employee break rolls around, a vehicle departs from an origin, that vehicle arrives at its destination, etc. Identifying the events in a complex model can be difficult, and there may be alternative ways to do so. For instance, in the simple single-server queue you *could* introduce another event, when an entity begins service. However, this would be unnecessary since this "begin-service" event would occur only when an entity ends service and there are other entities in queue, or when a lucky entity arrives to find the server idle with no other entities already waiting in line so goes directly into service; so, this "event" is already covered by the arrival and departure events.

Manual Single-Server-Queue Simulation (spreadsheet-assisted for arithmetic)

Given Arrival, Service Times (Minutes)

Entity Number	Arrival Time	Interarrival Time	Service Time
1	0.000000		0.486165
2	1.965834	1.965834	0.665121
3	2.275418	0.309584	0.354917
4	2.909957	0.634539	0.563653
5	4.033363	1.123406	0.051575
6	5.136229	1.102866	2.652734
7	6.539427	1.403198	0.379540
8	6.744513	0.205086	0.330272
9	7.845616	1.101103	2.985174
10	8.330407	0.484791	0.290218

The last three service times are not used.

Pink-shaded cells below are not completely general, i.e., the formulas in them depend on these particular given arrival and service times, so this spreadsheet will not in general update correctly with different arrival and service times.

			System State Variables			Arrival Times of Customers			Event Calendar			Entity-Based Observations		Areas of Individual Rectangles Under Curves		
Event Time	Event Type	Entity Number	Server Status	in System	in Queue	in Service	1st in Queue	2nd in Queue	Next Arrival	Next Departure	End Simulation	Time in System	Time in Queue	Server Busy	Number in System	Number in Queue
0.000000	Initialize		0	0	0				0.000000		8.000000			0.000000	0.000000	0.000000
0.000000	Arrival	1	1	1	0	0.000000			1.965834	0.486165	8.000000		0.000000	0.000000	0.000000	0.000000
0.486165	Departure	1	0	0	0				1.965834		8.000000	0.486165		0.486165	0.486165	0.000000
1.965834	Arrival	2	1	1	0	1.965834			2.275418	2.630955	8.000000		0.000000	0.000000	0.000000	0.000000
2.275418	Arrival	3	1	2	1	1.965834	2.275418		2.909957	2.630955	8.000000			0.309584	0.309584	0.000000
2.630955	Departure	2	1	1	0	2.275418			2.909957	2.985872	8.000000	0.665121	0.355537	0.355537	0.711074	0.355537
2.909957	Arrival	4	1	2	1	2.275418	2.909957		4.033363	2.985872	8.000000			0.279002	0.279002	0.000000
2.985872	Departure	3	1	1	0	2.909957			4.033363	3.543525	8.000000	0.710454	0.075915	0.075915	0.151830	0.075915
3.549525	Departure	4	0	0	0				4.033363		8.000000	0.639568		0.563653	0.563653	0.000000
4.033363	Arrival	5	1	1	0	4.033363			5.136229	4.084938	8.000000		0.000000	0.000000	0.000000	0.000000
4.084938	Departure	5	0	0	0				5.136229		8.000000	0.051575		0.051575	0.051575	0.000000
5.136229	Arrival	6	1	1	0	5.136229			6.539427	7.788963	8.000000		0.000000	0.000000	0.000000	0.000000
6.539427	Arrival	7	1	2	1	5.136229	6.539427		6.744513	7.788963	8.000000			1.403198	1.403198	0.000000
6.744513	Arrival	8	1	3	2	5.136229	6.539427	6.744513	7.845616	7.788963	8.000000			0.205086	0.410172	0.205086
7.788963	Departure	6	1	2	1	6.539427	6.744513		7.845616	8.168503	8.000000	2.652734	1.249536	1.044450	3.133350	2.088900
7.845616	Arrival	9	1	3	2	6.539427	6.744513	7.845616	8.330407	8.168503	8.000000			0.056653	0.113306	0.056653
8.000000	End Sim		1	3	2	6.539427	6.744513	7.845616	8.330407	8.168503	8.000000			0.154384	0.463152	0.308768
		Average:	Need time-weighted averages, not simple averages (O36 .. Q36).									0.8676	0.2401	0.6232	1.0095	0.3864
		Minimum:	0	0	0									0.0516	0.0000	The averages in O36..Q36 are time averages, not simple averages.
		Maximum:	1	3	2									2.6527	1.2495	

Figure 3.5: Manual discrete-change simulation of a single-server queue, using the spreadsheet Model_03_04.xls to hold data and do calculations. The pink-shaded cells contain formulas that are specific to this particular input set of arrival and service times, so this spreadsheet isn't a general model per se that will work for any set of arrival and service times.

As mentioned in Section 3.1.2, the simulation clock in a dynamic discrete-change model is a variable that jumps from one event time to the next in order to move the simulation forward. The *event calendar* (or *event list*) is a data structure of some sort that contains information on upcoming events. While the form and specific type of data structure vary, the event calendar will include information on when in simulated time a future event is to occur, what type of event it is, and often which entity is involved. The simulation proceeds by identifying the next (i.e., soonest) event in the event calendar, advancing the simulation-clock variable to that time, and executing the appropriate event logic. This effects all the necessary changes in the system state variables and the statistical-tracking variables needed to produce the desired output, and updates the event calendar as required.

For each type of event, you need to identify the logic to be executed, which sometimes depends on the state of the system at the moment. Here's basically what we need to do when each of our event types happens in this particular model:

- **Arrival.** First, schedule the next arrival for now (the simulation-clock value in Column A) plus the next interarrival time in the given list, by updating the next-arrival spot in the event calendar (column J in Model_03_04.xls and Figure 3.5). Compute the continuous-time output areas of the rectangles under $B(t)$, $L(t)$, and $L_q(t)$ between the time of the last event and now, based on their values that have been valid since that last event and before any possible changes in their values that will be made in this event; record the areas of these rectangles in columns O-Q. What we do next depends on the server status:

- If the server is idle, put the arriving entity immediately into service (put the time of arrival of the entity in service in the appropriate row in column G, to enable later computation of the time in system), and schedule this entity's service completion for now plus the next service time in the given list. Also, record into the appropriate row in column N a time in queue of 0 for this lucky entity.

- If the server is busy, put the arriving entity's arrival time at the end of a list representing the queue of entities awaiting service, in the appropriate row of columns H-I (it turns out that the queue length in this particular simulation is never more than two, so we need only two columns to hold the arrival times of entities in queue, but clearly we could need more in other cases if the queue becomes longer). We'll later need these arrival times to compute times in queue and in system. Since this is the beginning of this entity's time in queue, and we don't yet know when it will leave the queue and start service, we can't yet record a time in queue into Column N for this entity.

- **Departure** (upon a service completion). As in an arrival event, compute and record the continuous-time output areas since the last event in the appropriate row of columns O-Q. Also, compute and record into the appropriate row in column M the time in system of the departing entity (current simulation clock, minus the time of arrival of this entity, which now resides in column G). What we do next depends on whether there are other entities waiting in queue:

  - If other entities are currently waiting in the queue, take the first entity out of queue and place it into service (copy the arrival time from column H into the next row down in column G). Schedule the departure of this entity (column K) for now plus the next service time. Also compute the time in queue of this entity (current simulation clock in column A, minus the arrival time just moved into column G) into the appropriate row in column N.

  - If there are no other entities waiting in queue, simply make the server-status variable (column D below row 18) 0 for idle, and blank out the event-calendar entry for the next departure (column K).

- **End simulation** (at the fixed simulation-clock time 8 minutes). Compute the final continuous-time output areas in columns O-Q from the time of the last event to the end of the simulation, and compute the final output summary performance measures in rows 36-38.

It's important to understand that, though processing an event will certainly take *computer* time, there is no *simulation* time that passes during an event execution — the simulation-clock time stands still. And as a practical matter in this manual simulation, you probably should check off in some way the inter-arrival and service times as you use them, so you can be sure to move through the lists and not skip or re-use any of them.

Rows 19-35 of the spreadsheet Model_03_04.xls and in Figure 3.5 trace the simulation from initialization at time 0 (row 19), down to termination at time 8 minutes (row 35), with a row for each event execution. Column A has the simulation-clock time t of the event, column B indicates the event type, and column C gives the entity number involved. Throughout, the cell entries in a row represent the appropriate values *after* the event in that row has been executed, and blank cells are places where an entry isn't applicable. Columns D-F give respectively the server status $B(t)$ (0 for idle and 1 for busy), the number of entities in the system, and the number of entities in the queue (which in this model is always the number in system minus the server status ... think about it for a minute). Column G contains the time of arrival of the entity currently in service, and columns H-I have the times of arrival in the first two positions in the queue (we happen never to need a queue length of more than two in this example).

The event calendar is in columns J-L, and the next event is determined by taking the minimum of these three values, which is what the =MIN function in column A of the following row does. An alternative data structure for the event calendar is to have a list (or more sophisticated data structure like a tree or heap) of records or rows of the form [entity number, event time, event type], and when scheduling a new event insert it so that you keep this list of records ranked in increasing order on event time, and thus the next event is always at the top of this list. This is closer to the way simulation software actually works, and has the advantage that we could have multiple events of the same type scheduled, as well as being much faster (especially when using the heap data structure), which can have a major impact on run speed of large simulations with many different kinds of events and long event lists. However, the simple structure in columns J-L will do for this example.

When a time in system is observed (upon a departure), or a time in queue is observed (upon going into service), the value is computed (the current simulation clock in column A, minus the time of arrival) and recorded into column M or N. Finally, columns O-Q record the areas of the rectangles between the most recent event time and the event time of this row, in order to compute the time averages of these curves.

The simulation starts with initialization in row 19 at time 0. This should be mostly self-explanatory, except note that the end-simulation event time in L19 is set for time 8 minutes here (akin to lighting an 8-minute fuse), and the next-departure time is left blank (luckily, Excel interprets this blank as a large number so that the =MIN in A20 works).

The formula =MIN(J19:L19) in A20 finds the smallest (soonest) value in the event calendar in the preceding row, which is 0, signifying the arrival of entity 1 at time 0. We change the server status to 1 for busy in D20, and the number in system is set to 1 in E20; the number in queue in F20 remains at 0. We enter 0 in G20 to record the time of arrival of this entity, which is going directly into service, and since the queue is empty there are no entries in H20..I20. The event calendar in J20..L20 is updated for the next arrival as now (0) plus the next interarrival time (1.965834), and for the next departure as now (0) plus the next service time (0.486165); the end-simulation event time is unchanged, so it is just copied down from directly above. Since this is an arrival event, we're not observing the end of a time in system so no entry is made in M20; however, since this is an arrival of a lucky entity directly into service, we record a time in queue of 0 into N20. Cells O20..Q20 contain the areas of rectangles composing the areas under the $B(t)$, $L(t)$, and $L_q(t)$ curves since the last event, which are all zero in this case since it's still time 0.

Row 21 executes the next event, the departure of entity 1 at time 0.486165. Since this is a departure, we have a time in system to compute and enter in M21. The next rectangle areas are computed in O21..Q21; for instance, P21 has the formula =E20*($A21-$A20), which multiplies the prior number-in-system value 1 in E20 by the duration $A21-$A20 of the time interval between the prior event (A21) and the current time (A20); the $s are to keep the column A fixed so that we can copy such formulas across columns O-Q (having originally been entered in column O). Note that the time of the next arrival in J21 is unchanged from J20, since that arrival is still scheduled to occur in the future; since there is nobody else in queue at this time to move into service, the time of the next departure in K21 is blanked out.

The next event is the arrival of entity 2 at time 1.965834, in row 22. Since this is an arrival of an entity to an empty system, the action is similar to that of the arrival of entity 1 at time 0, so we'll skip the details and encourage you to peruse the spreadsheet, and especially examine the cells that have formulas rather than constants.

After that, the next event is another arrival, of entity 3 at time 2.275418, in row 23. Since this entity is arriving to a non-empty system, the number in queue, for the first time, becomes nonzero (1, actually), and we record in H23 the time of arrival (now, so just copy from A23) of this entity, to be used later to compute the time in queue and time in system of this entity. We're observing the completion of neither a time in queue nor a time in system, so no entries are made in M23 or N23. The new rectangle areas under the curves in O23..Q23 are computed, as always, and based on their levels in the preceding row.

The next event is the departure of entity 2 at time 2.630955, in row 24. Here, entity 3 moves from the front of the queue into service, so that entity's arrival time is copied from H23 into G24 to represent this, and entity 3's time in queue is computed in N24 as =A24-G24 = $2.630955 - 2.275418 = 0.355537$, in N24.

The simulation proceeds in this way (we've covered all the different kinds of things that can happen, so leave it to you to follow things the rest of the way through), until eventually the end-simulation event at time 8 becomes the next event (neither the scheduled arrival of entity 10 at time 8.330407, nor the departure of entity 7 at time 8.168503, ever happens). At that time, the only thing to do is compute the final rectangle areas in O35..Q35. Figure 3.6 shows plots of the $B(t)$, $L(t)$, and $L_q(t)$ curves over the simulation.

In rows 36-38 we compute the output performance measures, using the built-in Excel functions =AVERAGE, =MIN, =MAX, and =SUM (have a look at the formulas in those cells). Note that the time averages in O36..Q36 are computed by

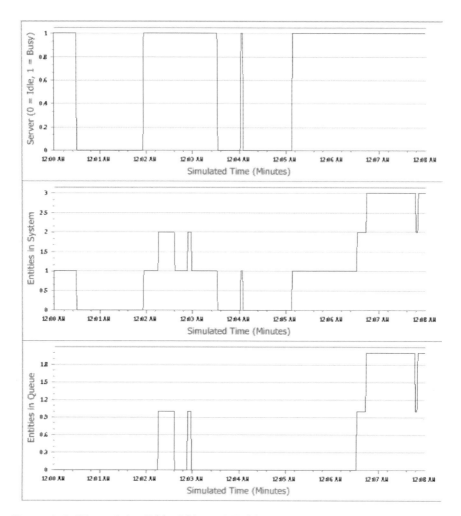

Figure 3.6: Plots of the $B(t)$, $L(t)$, and $L_q(t)$ curves for the manual simulation.

summing the individual rectangle areas and then dividing by the final simulation clock time, 8, which is in A35. As examples of the final output performance measures, the server was busy 62.32% of the time (O36), the average time in system and in queue were respectively 0.8676 minute and 0.2401 minute (M36 and N36), and the time-average length of the queue is 0.3864 entities (Q36). In Chapter 4 we'll do this same model in Simio (the actual source of the "given" input interarrival and service times) and we'll confirm that we get all these same output performance measures.

If you're alert (or even awake after the numbing narrative above), you might wonder why we didn't compute standard deviations, perhaps using Excel's =STDEV function, for the observations of times in system and times in queue in columns M and N, to go along with their means, minima, and maxima as we did in Model_03_03.xls for the monthly profits (doing so for the entries in columns O-Q wouldn't make sense since they need to be weighted by time-durations). After all, we've mentioned output variation, and you can just see from looking at columns M and N that there's variation in these observed times in system and times in queue. Actually, there's nothing stopping us (or you) from adding a row to compute the standard deviations, but this would be an example of doing something just because you *can*, without thinking about whether it's *right* (like making those gratuitously fancy three-dimensional bar charts of simple two-dimensional data, which serves only to obfuscate). So what's the problem with computing standard deviations here? Well, while you've probably forgotten it from your basic statistics class, the sample variance (of which the standard deviation is just the square root) is an *unbiased* estimator of the population variance *only if* the individual observations are statistically *independent* of each other — there comes a time in the proof of that fact where you must factor the expected value of the product of two observations into the product of their individual expectations, and that step requires independence. Our times in system (and in queue) are, however, *correlated* with each other, likely *positively* correlated, since they represent entities in sequence that simply physically depend on each other (so this kind of within-sequence correlation is called *autocorrelation* or sometimes *serial correlation*). If this is a single-teller bank and you're unlucky enough to be right behind (or even several customers behind) some guy with a bucket of loose coins to deposit, that guy is going to have a long time in system, *and so are you* since you have to wait for him! This is precisely a story of positive correlation. So a sample standard deviation, while it could be easily computed, would tend to underestimate the actual population standard deviation in comparison to what you'd get from independent samples, i.e., your sample standard deviation would be biased *low* (maybe severely) and thus understate the true variability in your results. The problem is that your observations are kind of tied to each other, not tied *exactly* together with a rigid steel beam, but maybe with rubber bands, so they're not free to vary as much from each other as they would be if you cut the rubber bands and made them independent; the stronger the correlation (so the thicker the rubber bands), the more severe the downward bias in the sample-variance estimator. This could lull you into believing that you have more precise results than you really do — a dangerous place to be, especially since you wouldn't know you're there. And for this same reason (autocorrelation), we've also refrained from making histograms of the times in system and in queue (as we did with the independent and identically distributed monthly profits in `Model_03_03.xls`), since the observed times' being positively autocorrelated create a risk that the histogram would be uncharacteristically shifted left or right unless you knew that your run was long enough to have "sampled" appropriately from all parts of the distribution.

One more practical comment about the way we did things in our spreadsheet: In columns M-Q we computed *and saved*, in each event row, the *individual* values of observed times in system and in queue, and the *individual* areas of rectangles under the $B(t)$, $L(t)$, and $L_q(t)$ curves between each pair of successive events. This worked just fine in this little example, since it allowed us to use the built-in Excel functions in rows 36-38 to get the summary output performance measures, which was easy and convenient. Computationally, though, this could be quite inefficient in a long simulation (hundreds of thousands of event rows instead of our 17), or one that's replicated many times, since we'd have to store and later recall all of these *individual* values to get the output summary measures. What actually happens instead in simulation software (or if you used a general-purpose programming language rather than a spreadsheet), is that statistical *accumulator* variables are set up, into which we just add each value as it's observed, and that way we don't have to store and later recall each individual value. We also would put in logic to keep a running value of the current maximum and minimum of each output, and just compare each new observation to the current extreme values so far to see if this new observation is a new extreme. You can see how this would be far more efficient computationally, and in terms of storage use.

Finally, the spreadsheet `Model_03_04.xls` in Figure 3.5 isn't completely general, i.e., it will *not* necessarily remain valid if we change the blue-shaded input arrival times and service times to different values. The reason for this lack of complete generality is that the formulas in many cells (the pink-shaded ones) are valid only for that type of event, and in that situation (e.g., a departure event's logic depends on whether the queue is empty). Spreadsheets are actually poorly suited for carrying out dynamic discrete-change simulations, which is why they are seldom used in real settings; our use of a spreadsheet here is just to hold the data and do the arithmetic, not for building any kind of robust or flexible general model per se. In Section 3.4.1 we'll briefly discuss using general-purpose programming languages for dynamic discrete-change simulation, but really, software like Simio is needed for models of any complexity, which is what's covered in most of the rest of this book.

3.3.2 Model 3-5: Single-Server Queueing Delays

Spreadsheets are everywhere, and everybody knows at least a little about them, so you might wonder about building complete, general, and flexible *dynamic* simulation *models* inside spreadsheets. We used spreadsheets for the static stochastic Monte Carlo simulations in Section 3.2, but they could have been done more efficiently in a programming language. In Section 3.3.1 we *used* a spreadsheet to hold our numbers and do our arithmetic in a manual dynamic simulation, but because the formulas in that spreadsheet depended on the particular input data (interarrival and service times), it was not at all like a *real model* of any completeness, generality, or flexibility. While it's difficult, if not practically impossible, to use a spreadsheet to do anything but the very simplest of dynamic simulations in a general way, spreadsheets have been used extensively for static simulations like those in Section 3.2 (see Section 3.1.1 for the difference between static and dynamic simulation).

While the single-period inventory problem in Section 3.2.3 discussed the passage of time (the first month was the peak period, and the time thereafter was the clearance-price period), time did not play an explicit role in the model. You could just as easily think of the peak period as a single instant, followed immediately by a single clearance-price instant. So that really was a *static* model in terms of the role played by the passage of time (it played no role). In this section we'll simulate a *very* simple *dynamic* single-server queueing model in a spreadsheet model where the passage of time plays a central role in the operation of the model. However, being able to do this within the confines of a spreadsheet depends heavily on the fact that there happens to be a simple recurrence relation for this one special case that enables us to simulate the output process we want: the waiting times in queue (excluding their service time) of entities. Unlike the manual simulation in Section 3.3.1, we won't be able to compute directly any of the other outputs here.

For a single-server queue starting empty and idle at time 0, and with the first entity arrival at time 0, let A_i be the interarrival time between entity $i-1$ and entity i (for $i = 2, 3, \ldots$). Further, let S_i be the service-time requirement of entity i (for $i = 1, 2, \ldots$). Note that the A_i's and S_i's will generally be random variables from some interarrival-time and service-time probability distributions; unlike in Chapter 2, though, these distributions could really be anything since now we're simulating rather than trying to derive analytical solutions. Our output process of interest is the *waiting time* (or *delay*) in queue $W_{q,i}$ of the ith entity to arrive; this waiting time does not include the time spent in service, but just the (wasted) time waiting in line. Clearly, the waiting time in queue of the first entity is 0, as that lucky entity arrives (at time 0) to find the system empty of other entities, and the server idle, so we know that $W_{q,1} = 0$ for sure. We don't know, however, what the subsequent waiting times in queue $W_{q,2}, W_{q,3}, \ldots$ will be, since they depend on what the A_i's and S_i's will happen to be. In this case, though, once the interarrival and service times are observed from their distributions, there is a relationship between subsequent $W_{q,i}$'s, known as *Lindley's recurrence* ([35]):

$$W_{q,i} = \max(W_{q,i-1} + S_{i-1} - A_i, 0), \text{ for } i = 2, 3, \ldots, \tag{3.6}$$

and this is anchored by $W_{q,1} = 0$. So we could "program" this into a spreadsheet, with a row for each entity, once we have a way of generating observations from the interarrival and service-time distributions.

Let's assume that both of these distributions are exponential, with mean interarrival time 1.25 minutes, and mean service time 1 minute; we'll also assume that all observations within and across these distributions are independent of each other. So we next need a way to generate observations from an exponential distribution with given mean, say, $\beta > 0$. It turns out (as will be discussed in Section 6.4) that a general way to do this (or at least *try* to do this) for continuous distributions is to set $U = F(X)$, where U is a random number distributed continuously and uniformly between 0 and 1, and F is the cumulative distribution function of the random variable X desired. Then, try to solve this for X to get a formula for generating an observation on X, given a generated random number U. In the case of the exponential distribution with mean β, the cumulative distribution function is

$$F(x) = \begin{cases} 1 - e^{-x/\beta} & \text{if } x \geq 0 \\ 0 & \text{otherwise} \end{cases}. \tag{3.7}$$

Setting $U = F(X)$ in (3.7) and solving for X yields, after a few lines of algebra,

$$X = -\beta \ln(1 - U) \tag{3.8}$$

where ln is the natural (base e) logarithm function.

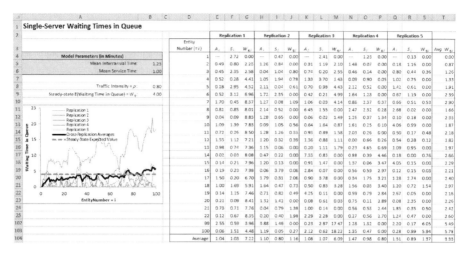

Figure 3.7: Waiting times in queue in single-server queue with Excel (`Model_03_05.xls`).

So, armed with Lindley's recurrence in (3.6) and a formula for generating observations from an exponential distribution in (3.8), plus of course our handy (if crude) Excel random-number generator `RAND()`, we're ready to simulate in Excel the waiting-time process (for a little while, anyway) from a single-server queue with exponential interarrival and service times. Figure 3.7 shows the file `Model_03_05.xls`, with rows 26-101 hidden to save space (open that file to see all the rows, and most importantly, the formulas in the cells). The blue-shaded cells A4..B6 contain the model's input parameters, and the traffic intensity ρ and steady-state expected waiting time in queue W_q are computed from them in A8..B9 (see Chapter 2 for these formulas for this $M/M/1$ queue). The entity numbers $i = 1, 2, ..., 100$ are in column D (we arbitrarily decided to "run" this for 100 entities' waiting times in queue), so each of these rows represents an entity finishing its waiting time in queue. Column E generates the interarrival times A_i via the Excel formula `=-$B$5*LN(1 - RAND())` where cell B5 contains the mean interarrival time; column F similarly generates the service times (`LN` is the built-in Excel function for the natural logarithm). Column G then implements Lindley's recurrence from (3.6); for example, cell G20 (for entity $i = 17$) contains the formula `=MAX(G19 + F19 - E20, 0)` for $W_{q,17}$ from $W_{q,16}$ in cell G19, S_{16} in cell F19, and A_{17} in cell E20. Note that we do *not* have the \$s on these cell references since we *want* their rows to change as we copy these formulas down; this indicates the dependence (autocorrelation) between the waiting times in queue, a behavior that will be important to developing proper statistical-analysis methods for simulation output (we can't treat successive observations in simulation output as being independent, unlike traditional classical statistical analysis, as discussed near the end of Section 3.3.1).

Columns E-G are then simply copied across to the right four times, to get a total of five independent *replications* of this 100-entity waiting-time-in-queue spreadsheet simulation; these blocks of three columns each are not changing in any way, except of course that they are all using independent draws from the volatile random-number generator `RAND()`. Column T averages, for each $i = 1, 2, \ldots, 100$, the five values of each $W_{q,i}$ across the five replications, so this column contains a *cross-replication* average of the waiting times of the first, second, ..., 100th arriving entities in the five replications. Row 104 contains column averages for the A_i's and S_i's; these merely confirm (estimate) the mean interarrival and service times in cells B5 and B6, but for the $W_{q,i}$'s they are *within-replication* averages of the waiting times for each of the five replications separately. (We refrain from computing standard deviations of the waiting times within the replications, and we likewise don't show histograms of these within-replication waiting times, due to the biasing effect of the autocorrelation, as discussed in Section 3.3.1.) The plot on the left shows the waiting times for each replication (light wiggly lines) as a function of the entity number, the cross-replication waiting-time-average process (heavy wiggly line, but not as wiggly as the light lines), and the steady-state expected waiting time in queue (non-wiggly horizontal dashed line).

Tap the F9 key a few times to exercise the volatility of `RAND()`, and watch all the interarrival, service, and waiting times change, and watch the wiggly lines in the plot jump around (the vertical axis on the plot auto-scales in response to the maximum waiting time with each tap of F9 and re-generation of all the random numbers). Notice:

- All the simulation-based curves are "anchored" on the left at 0, merely confirming that the first lucky entity to arrive has a zero wait in queue every time. But generally, as time goes by (as measured by the entity number on the horizontal axis, at least ordinally), the waiting times tend to creep up, though certainly not smoothly or monotonically due to the random noise in the interarrival and service times.

- The heavy wiggly line doesn't wiggle as much as the light wiggly lines do, since it's the average of those light lines, and the large and small values for each entity number tend to dampen each other out.

- All the wiggly lines, both light and heavy, have a "meandering" character to them. That is, they're certainly random, but as you follow each line to the right, the new values are "tied," at least loosely, to those that came not too long before them. This confirms the (positive) autocorrelation of the waiting times within a replication, as you can see algebraically from Lindley's recurrence in (3.6), and as discussed near the end of Section 3.3.1.

- While some taps of F9 might give you a reasonable convergence of the waiting times to the steady-state expected value (horizontal dashed line), most don't, indicating that (*a*) there's a lot of noise in this system, and (*b*) it could take a lot longer than 100 entities to "approach" long-run steady-state conditions.

Go ahead and play around with the mean interarrival and mean service times in B5 and B6, but make sure that the mean interarrival time is always more than the mean service time, or else the formula in B9 for the steady-state expected waiting time in queue W_q will be invalid (though the simulation will still be valid). See how things change, especially as the mean interarrival and mean service times get close to each other (so the traffic intensity ρ in B8 creeps upward toward 1, and W_q just keeps getting bigger and bigger).

We want to emphasize that being able to do this *particular* dynamic simulation in a spreadsheet depends *completely* on the availability of Lindley's recurrence in (3.6); and note as well that this is just for the waiting-time-in-queue output process, and not for anything else we might like to get out of even this simple system (like queue length or server utilization, as we did in the manual simulation in Section 3.3.1). In general, for complex dynamic systems we'll have nothing like Lindley's recurrence available, so simulating realistic dynamic systems in a spreadsheet is completely impractical, and almost always outright impossible. That's why you need special dynamic-simulation software like Simio.

3.4   Software Options for Dynamic Simulation

In Section 3.3 we struggled through two different approaches (manual and in a spreadsheet) for carrying out dynamic simulations, neither of which is at all acceptable for anything serious. In this section we generally describe two more realistic ways to approach execution of dynamic simulations.

3.4.1   General-Purpose Programming Languages

The manual-simulation logic described and painstakingly carried out in Section 3.3.1 can be much more efficiently and generally expressed in a general-purpose procedural programming language like C++, Java, or Matlab, rather than in a spreadsheet. Historically, this is how all simulations were done before the advent of special-purpose simulation software. It's still done today in some situations, where the model is at least relatively simple or there is need for great computer speed. However, it's still quite painstaking (though not as bad as manual simulation with a spreadsheet as in Section 3.3.1), slow to develop, and error-prone; it does have the advantage, though, of affording almost complete flexibility to model even the quirkiest of behaviors that can't be envisioned by developers of high-level special-purpose simulation software.

When using a general-purpose programming language for a dynamic discrete-change simulation, you still need to identify the events, and define the logic for each of them. Then, you write a subprogram (or function) to carry out the action for each type of event (actually, even the indenting of the multi-level bullet points in Section 3.3.1 for the general logic of the Arrival and Departure events suggests how these subprograms might be structured). The event subprograms are tied together with some sort of main program or function that determines the next event and directs execution to the corresponding event subprogram, and takes care of everything else we did in the manual simulation of Section 3.3.1, like event-calendar updating. At the end, you need to program the appropriate statistical summarization and create the output. You also need to be concerned with random-number generators (for stochastic simulations), how to generate observations from the desired input probability distributions, and keep track of the model state variables in some sort of data structure. As you're seeing, this will be quite a lot of work for models of much complexity, so it can take a lot of analyst time, and is prone to errors, some of which can be very hard to detect.

Nonetheless, people still do this sometimes, perhaps as quirky modules of larger, more standard simulations. For more on this kind of programming see, for example, [3], [30], or [53].

3.4.2 Special-Purpose Simulation Software

In this chapter you've seen various approaches to carrying out different kinds of simulations. This has included "manual" simulation, aided by a spreadsheet to keep track of things and do the arithmetic, indication of how general-purpose programming languages might be used to program a simulation along the lines of the logic in the manual simulation, and finally building general simulation models within a spreadsheet, perhaps aided by simulation-specific add-ins.

You've also seen that all of these approaches to doing simulation have substantial drawbacks, in different ways. Manual simulation is obviously just completely impractical for any actual application. While people have used general-purpose programming languages for simulation (and to some extent still do), which has the appeals of almost complete modeling flexibility as well as usually good execution speed, programming nontrivial dynamic simulations from scratch is hardly practical for wide use, due to the analyst time required, and the high chance of making programming errors. Spreadsheets will likely continue to be widely used for static simulations, especially with an appropriate simulation add-in, but they simply cannot accommodate the logic, data-structure, or generality needs of large-scale dynamic simulations.

Actually, people realized long ago (and long before spreadsheets) that there was a real need for relatively easy-to-use, high-level software for dynamic simulation, and specifically for discrete-change stochastic dynamic simulations. Going back into the 1960s, special-purpose simulation languages like GPSS ([51], [52]), SIMSCRIPT ([27]), and SIMULA ([5]) were developed, some of which were early examples of the *object-oriented* paradigm used by Simio, though it wasn't called that at the time. Extensions to general-purpose programming languages like FORTRAN were developed, notably GASP ([46]). These led to more modern, self-contained simulation languages like SLAM ([47]) and SIMAN ([44]). All of these simulation languages, though, were really just that, *languages*, with text-line entry and picky syntax, so were not especially user-friendly or intuitive either to use or interpret. Graphical user interfaces, often with integrated animation, were then developed, which opened the door to far more people; these software packages include Arena ([25]), ExtendSim (`www.extendsim.com`), Promodel (`www.pmcorp.com`), Flexsim (`www.flexsim.com`), Simul8 (`www.simul8.com`), and others. For more on simulation software, see [3], [30], and especially [40] for comprehensive reviews; for the latest (ninth) in a sequence of biennial simulation-software surveys, see [57].

Simio is a comparatively recent product (this book uses Simio Version 5.91) that takes advantage of the latest in software-development and modeling capabilities. Although relatively new, Simio has been developed by the same people who pioneered the development of SIMAN and Arena, so there is quite a lot of experience behind it. It centers around *intelligent objects* and provides a new object-based paradigm that radically changes the way objects are built and used. Simio objects are created using simple graphical process flows that require no programming. This makes it easier to build your own modeling objects and application-focused libraries. Most of this book will show you how to use simulation, and specifically Simio, to solve practical problems effectively.

3.5   Problems

1. Extend the simulation of throwing two dice in Section 3.2.1 in each of the following ways (one at a time, not cumulatively):

    a) Instead of 50 throws, extend the spreadsheet to make 500 throws, and compare your results. Tap the F9 key to get a "fresh" set of random numbers and thus a fresh set of results.

    b) Load the dice by changing the probabilities of the faces to be something other than uniform at 1/6 for each face. Be careful to ensure that your probabilities sum to 1.

    c) Use @RISK (see Section 3.2.3), or another Excel add-in for static Monte Carlo spreadsheet simulation, to make 10,000 throws of the pair of fair dice, and compare your results to the true probabilities of getting the sum equal to each of $2, 3, \ldots, 12$ as well as to the true expected value of 7.

2. In the simulation of throwing two dice in Section 3.2.1, derive (from elementary probability theory) the probability that the sum of the two dice will be each of $2, 3, \ldots, 12$. Use these probabilities to find the (true) expected value and standard deviation of the sum of the two dice, and compare with the (sample) mean in cell J4; add to the spreadsheet calculation of the (sample) standard deviation of the sum of the two dice and compare with your exact analytical result for the standard deviation. Keep the number of throws at 50.

3. Prove rigorously, using probability theory and the definition of the expected value of a random variable, that in Section 3.2.2, $E(Y) = \int_a^b h(x)dx$. Start by writing $E(Y) = E[(b-a)h(X)] = (b-a)E[h(X)]$, then use the definition of the expected value of a function of a random variable, and finally remember that X is continuously uniformly distributed on $[a, b]$ so has density function $f(x) = 1/(b-a)$ for $a \leq x \leq b$.

4. Use @RISK, or another Excel add-in for static spreadsheet simulation, to extend the example in Section 3.2.2 to 10,000 values of X_i, rather than just the 50 values in `Model_03_02.xls` in Figure 3.2. Compare your results to those in `Model_03_02.xls` as well as to the (almost) exact numerical integral.

5. In the Monte Carlo integration of Section 3.2.2, add to the spreadsheet calculation of the standard deviation of the 50 individual values, and use

that, together with the mean already in cell H4, to compute a 95% confidence interval on the exact integral in cell I4; does your confidence interval contain, or "cover" the exact integral? How often (tap F9 repeatedly and keep track manually)? Repeat all this, but with a 90% confidence interval, and then with a 99% confidence interval.

6. In the manual simulation of Section 3.3.1 and the Excel file Model_03_04.xls, speed up the service times by decreasing them all by 20%. What's the effect on the same set of output performance measures? Discuss.

7. Repeat Problem 6, except now slow down the service times by increasing them all by 20%. Discuss.

8. In the manual simulation of Section 3.3.1 and the Excel file Model_03_04.xls, change the queue discipline from FIFO to *shortest-job-first* (SJF), i.e., when the server becomes idle and there's a queue, the entity with the shortest service time will enter service next. This is also called *shortest-processing-time* (SPT); see Section 2.1. To do this, you'll need to "assign" the (possibly-future) service time to each entity as it arrives, and then use that to change your queue discipline. What's the effect on the same set of output performance measures? You could do this by either (*a*) inserting new entities into the queue so that it stays ranked in increasing order on service time (so the entity with the shortest service time will always be at the front of the queue), or (*b*) leaving the queue order as it is, with new arrivals always joining the end of the queue (if they can't go right into service) regardless of their service time, and then searching through it each time the server becomes idle and is ready to choose the correct next entity from somewhere in the queue — the results should be identical, but which would be faster computationally (think of a long queue, not necessarily just this little example)?

9. In the inventory spreadsheet simulation of Section 3.2.3 and the Excel file Model_03_03.xls, we decided, essentially arbitrarily, to make 50 replications (50 rows) and our estimates of mean profit just had whatever confidence-interval half-widths (row 59) they happened to have at the end, from that sample size of 50. Change Model_03_03.xls to make 200 replications (four times as many), and as you'd expect, those half-widths get smaller — by what factor? Measure this as best you can from your output (which will be challenging due to the volatility of Excel's random-number generator). Look up in any standard statistics book the formula for confidence-interval half widths (on the mean, using the usual normal- and *t*-distribution approaches), and interpret your empirical results in light of that formula. About how many replications would you need to reduce these half widths (in comparison with the original 50-replication spreadsheet) by a factor of ten (i.e., your mean estimates would have one more digit of statistical precision)?

10. In the inventory spreadsheet simulation of Section 3.2.3 and the Excel file Model_03_03.xls, the *same* numerical realization of peak-period demand in column F for a given replication (row) was used for computing profit

for all five of the trial values of the order amount h. Instead, you could have a separate, independent column of demands for each value of h — change Model_03_03.xls to do it this way instead, which is just as valid. Since our *real* interest is in the *differences* in profit for different values of h, is this new method "better?" Remember, there's "noise" (statistical variability) associated with your average-profit estimates, and the less noise there is, the more precise will be your estimates, of both the profit values themselves for a given value of h, and especially of the differences between profits across different values of h.

11. In the inventory spreadsheet simulation of Section 3.2.3 and the Excel file Model_03_03.xls, suppose you had the choice between either hiring a really big person to do some negotiating with your supplier to get the unit wholesale cost down to $9.00 from $11.00, or hiring a slick advertising agency for some "image" hype and thus be able to charge $18.95 unit retail rather than $16.95 during the peak period (assume that the clearance retail price stays at $6.95 either way). Which would be better? Are you sure? Base your comparison on what seems to be the best (most profitable) value of h from each of the two alternatives (which may or may not be the same values of h). Try to do at least some sort of statistical analysis of your results.

12. In Problem 11, instead of asking the advertising agency to hype toward higher price, would it be better to ask them to hype toward increasing the high end of peak-period demand from 5000 to 6000? Assume that your unit wholesale cost is the original $11.00, and that your unit peak-period and clearance prices are also at their original levels of $16.95 and $6.95, respectively. Base your comparison on what seems to be the best (most profitable) value of h from each of the two alternatives (which may or may not be the same values of h).

13. In the inventory spreadsheet simulation of Section 3.2.3 and the Excel file Model_03_03.xls, instead of *selling* hats (at a loss) left over after the peak period, you could *donate* them to a charity and be able to deduct your wholesale costs from your taxable corporate income; assume you're in the 34% marginal-tax bracket. Aside from the certain good feelings of helping out the charity, would this help out your bottom line? Base your comparison on what seems to be the best (most profitable) value of h from each of the two alternatives (which may or may not be the same values of h).

14. In the queueing spreadsheet simulation of Section 3.3.2 and the Excel file Model_03_05.xls, "run" the model out to 1000 customers' waiting times in queue, rather than just 100. Compare your results with the steady-state expected waiting time in queue W_q, as opposed to those from the shorter 100-customer run. Explain.

15. In the queueing spreadsheet simulation of Section 3.3.2 and the Excel file Model_03_05.xls (with a "run" length of 100, as originally done), push the arrival rate up by decreasing the mean interarrival time from

1.25 minutes to 1.1 minutes and compare your results with the original model. Then, push the arrival rate up more by decreasing the mean interarrival time more, down to 1.05, then 1.01, and see how this affects your results, both the steady-state expected waiting time in queue W_q, and your simulation results.

16. Continue right on over the brink with Problem 15, and go ahead and reduce the mean interarrival time on down to 1.00, then 0.99, then 0.95, then 0.88, and finally 0.80 (be brave!). What's happening? In light of the "stability" discussions in Chapter 2, are your results even "meaningful" (whatever that means in this context)? You might want to consider this "meaningfulness" question separately for the steady-state expected waiting time in queue W_q, and for your 100-customer simulation results.

17. Walther has a roadside produce stand where he sells oats, peas, beans, and barley. He buys these products at per-pound wholesale prices of, respectively, $1.05, $3.17, $1.99, and $0.95; he sells them at per-pound retail prices of, respectively, $1.29, $3.76, $2.23, and $1.65. Each day the amount demanded (in pounds) could be as little as zero for each product, and as much as 10, 8, 14, and 11 for oats, peas, beans, and barley, respectively; he sells only whole-pound amounts, no partial pounds. Assume a discrete uniform distribution for daily demand for each product over its range; assume as well that Walther always has enough inventory to satisfy all demand. The summer selling season is 90 days, and demand each day is independent of demand on other days. Create a spreadsheet simulation that will, for each day as well as for the whole season, simulate Walther's total cost, total revenue, and total profit.

18. A new car has a sticker price (the full retail asking price) of $22,000. However, there are several customer-chosen options (automatic transmission, upgraded speakers, fog lights, auto-dimming rear-view mirror, moonroof) that can add between $0 and $2000 to the selling price of the car depending on which options the customer chooses; assume that the total cost of all options is distributed uniformly and continuously on that range. Also, many buyers bargain down the price of the car; assume that this bargain-down amount is distributed uniformly and continuously between $0 and $3000, and is independent of the options. The selling price of the car is $22,000 plus the total of the options cost, minus the bargain-down amount.

    a) Create an Excel spreadsheet simulation that will compute the actual selling price for the next 100 cars sold, using only the capabilities that come built-in with Excel, i.e., don't use @RISK or any other third-party add-ins. Find the sample mean, standard deviation, minimum, and maximum of your 100 selling prices. Form a 95% confidence interval for the expected selling price of infinitely many future sales (not just the next 100), and do this within Excel (*Hint*: check out the built-in Excel function TINV and be careful to understand its input parameters). Also, create a histogram of your 100 selling prices,

Table 3.1: New-car options for Problem 19.

Option	Price	% Customers choosing option
Automatic transmission	$1000	60
Upgraded speakers	$120	10
Fog lights	$200	30
Auto-dimming rear-view mirror	$180	20
Moonroof	$500	40

    either by mimicking what was done in Model 3-4, or by using the Data Analysis ToolPak that comes with Excel (you may need to install it via File > Options > Add-Ins, then in the Manage window at the bottom select Excel Add-ins, then the Go button just to the right, then check the box for Analysis ToolPak, then OK) and its Histogram capability (you may want to explore Excel help for this). Do everything within Excel, i.e., no side hand/calculator calculations or printed statistical tables, etc.

b) In a second sheet (tab) of your same Excel file, repeat part a) above, except now use @RISK or another Excel add-in for static spreadsheet simulation, including its built-in variate-generation functions rather than your own formulas to generate the uniform random variates needed; also, don't compute the confidence interval on expected selling price. Instead of 100 cars, do 10,000 cars to get both a better estimate of the expected selling price, and a smoother histogram (use the histograms that come with @RISK). Again, do everything within Excel (other than using @RISK).

19. To put a finer point on Problem 18, the prices of the five available individual options and the percent of customers who choose each option are in Table 3.1 (e.g., there's a probability of 0.6 that a given customer will choose the automatic transmission, etc.). Assume that customers choose each option independently of whether they choose any other option. Repeat part a) of Problem 18, but now with this pricing structure, where each customer either will or won't choose each option; the bargain-down behavior here is the same as in Problem 18, and is independent of customer choice of options. Do this within Excel, i.e., do not use @RISK. *Hint 1*: The probability that a continuous random number on [0, 1] is less that 0.6 is 0.6. *Hint 2*: The built-in Excel function =IF(condition, x, y) evaluates to x if the condition is true, and to y if the condition is false.

Chapter 4

First Simio Models

The primary goal of this chapter is to introduce the simulation model-building process using Simio. Hand-in-hand with simulation-model building goes the statistical analysis of simulation output results, so as we build our models we'll also exercise and analyze them to see how to make valid inferences about the system being modeled. The chapter first will build a complete Simio model and introduce the concepts of model verification, experimentation, and statistical analysis of simulation output data. Although the basic model-building and analysis processes themselves aren't specific to Simio, we'll focus on Simio as an implementation vehicle.

The initial model used in this chapter is very simple, and except for run length is basically the same as Model 3-4 done manually in Section 3.3.1 and Model 3-5 in a spreadsheet model in Section 3.3.2. This model's familiarity and simplicity will allow us to focus on the process and the fundamental Simio concepts, rather than on the model. We'll then make some easy modifications to the initial model to demonstrate additional Simio concepts. Then, in subsequent chapters we'll successively extend the model to incorporate additional Simio features and simulation-modeling techniques to support more comprehensive systems. This is a simple single-server queueing system with arrival rate $\lambda = 48$ entities/hour and service rate $\mu = 60$ entities/hour (Figure 4.1). This system could represent a machine in a manufacturing system, a teller at a bank, a cashier at a fast-food restaurant, or a triage nurse at an emergency room, among many other settings. For our purposes, it really does't matter what is being modeled — at least for the time being. Initially, assume that

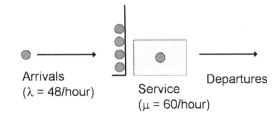

Arrivals
(λ = 48/hour)

Service
(μ = 60/hour)

Departures

Figure 4.1: Example single-server queueing system.

the arrival process is Poisson (i.e., the interarrival times are exponentially distributed and independent of each other), the service times are exponential and independent (of each other and of the interarrival times), the queue has infinite capacity, and the queue discipline will be first-in first-out (FIFO). Our interest is in the typical queueing-related metrics such as the number of entities in the queue (both average and maximum), the time an entity spends in the queue (again, average and maximum), utilization of the server, etc. If our interest is in long-run or steady-state behavior, this system is easily analyzed using standard queueing-analysis methods (as described in Chapter 2), but our interest here is in modeling this system using Simio.

This chapter actually describes two alternative methods to model the queuing system using Simio. The first method uses the *Facility Window* and Simio objects from the *Standard Library* (Section 4.2). The second method uses Simio *Processes* (Section 4.3) to construct the model at a lower level, which is sometimes needed to model things properly or in more detail. These two methods are not completely separate — the Standard Library objects are actually built using Processes. The pre-built Standard-Library objects generally provide a higher-level, more natural interface for model building, and combine animation with the basic functionality of the objects. Custom-constructed Processes provide a lower-level interface to Simio and are typically used for models requiring special functionality or faster execution. In Simio, you also have access to the Processes that comprise the Standard Library objects, but that's a topic for a future chapter.

The chapter starts with a tour around the Simio window and user interface in Section 4.1. As mentioned above, Section 4.2 guides you through how to build a model of the system in the Facility Window using the Standard Library objects. We then experiment a little with this model, as well as introduce the important concepts of statistically independent replications, warmup, steady-state vs. terminating simulations, and verify that our model is correct. Section 4.3 re-builds the first model with Simio Processes rather than objects. Section 4.4 adds context to the initial model and modifies the interarrival and service-time distributions. Sections 4.5 and 4.6 show how to use innovative approaches enabled by Simio for effective statistical analysis of simulation output data. Section 4.7 describes the basic Simio animation features and adds animation to the models. As your models start to get more interesting you will start finding unexpected behavior. So we will end this chapter with Section 4.8 describing the basic procedure to find and fix model problems. Though the systems being modeled in this chapter are quite simple, after going through this material you should be well on your way to understanding not only how to build models in Simio, but also how to use them.

4.1   The Basic Simio User Interface

Before we start building Simio models, we'll take a quick tour in this section through Simio's user interface to introduce what's available and how to navigate to various modeling components.

When you first load Simio you'll see either a new Simio model — the default behavior — or the most recent model that you had previously opened if you

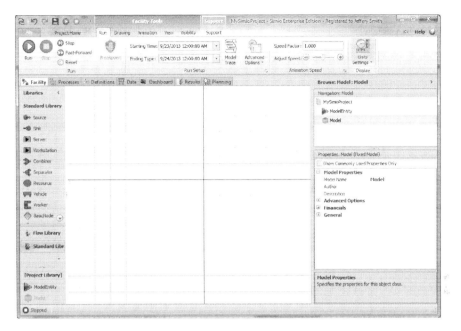

Figure 4.2: Facility Window in the new model.

have the "Load most recent project at startup" checkbox checked on the File page. Figure 4.2 shows the default initial view of a new Simio model. Although you may have a natural inclination to start model building immediately, we encourage you to take time to explore the interface and the Simio-related resources provided through the Support ribbon (described below). These resources can save you an enormous amount of time.

Ribbons

Ribbons are the innovative interface components introduced with Microsoft® Office 2007 to replace the older style of menus and toolbars. Ribbons help you quickly complete tasks through a combination of intuitive organization and automatic adjustment of contents. Commands are organized into logical groups, which are collected together under *tabs*. Each tab relates to a type of activity, such as running a model or drawing symbols. Tabs are automatically displayed or brought to the front based on the context of what you're doing. For example, when you're working with a symbol, the Symbols tab becomes prominent. Note that which specific ribbons are displayed depends on "where you are" in the project (i.e., what items are selected in the various components of the interface).

Support Ribbon

The Simio Support ribbon (see Figure 4.3) includes many of the resources available to learn and get the most out of Simio, as well as how to contact the Simio people with ideas, questions, or problems. Additional information is available

Figure 4.3: Simio Support ribbon.

through the link to Simio Technical Support (http://www.simio.com/resources/
technical-support/) where you will find a description of the technical-support
policies and links to the Simio User Forum and other Simio-related groups.
Simio version and license information is also available on the Support ribbon.
This information is important whenever you contact Support.

Simio includes comprehensive help available at the touch of the F1 key or
the "?" icon in the upper right of the Simio window. If you prefer a printable
version, you'll find a link to the *Simio Reference Guide* (a .pdf file). The help
and reference guides provide an indexed searchable resource describing basic
and advanced Simio features. For additional training opportunities you'll also
find links to training videos and other on-line resources. The Support ribbon
also has direct links to open example projects and SimBits (covered below), and
to access Simio-related books, release and compatibility nodes, and the Simio
user forum.

Project Model Tabs

In addition to the ribbon tabs near the top of the window, if you have a Simio
project open, you'll see a second set of tabs just below the ribbon. These are the
project model tabs used to select between multiple windows that are associated
with the active model or experiment. The windows that are available depend on
the object class of the selected model, but generally include Facility, Processes,
Definitions, Data, Dashboard, and Results. If you are using an Enterprise Simio
license, you will also see the Planning tab. Each of these will be discussed in
detail later, but initially you'll spend most of your time in the Facility Window
where the majority of model development, testing, and interactive runs are
done.

Object Libraries

Simio *object libraries* are collections of object definitions, typically related to
a common modeling domain or theme. Here we give a brief introduction to
Libraries — Section 5.1.1 provides additional details about objects, libraries,
models and the relationships between them. Libraries are shown on the left
side of the Facility Window. In the standard Simio installation, the Standard
Library and the Flow Library are attached by default and the Project Library is
an integral part of the project. The Standard and Flow libraries can be opened
by clicking on their respective names at the bottom of the libraries window
(only one can be open at a time). The Project Library remains open and can
be expanded/condensed by clicking and dragging on the '....' separator. Other

Table 4.1: Simio Standard Library objects.

Object	Description
Source	Generates entity objects of a specified type and arrival pattern.
Sink	Destroys entities that have completed processing in the model.
Server	Represents a capacitated process such as a machine or service operation.
Workstation	Models a complex workstation with setup, processing, and teardown phases, as well as secondary resource and material requirements.
Combiner	Combines multiple entities together with a parent entity (e.g., a pallet).
Separator	Splits a batched group of entities or makes copies of a single entity.
Resource	A generic object that can be seized and released by other objects.
Vehicle	A transporter that can follow a fixed route or perform on-demand pickups/dropoffs.
Worker	Models activities associated with people. Can be used as a moveable object or a transporter and can follow a shift schedule.
BasicNode	Models a simple intersection between multiple links.
TransferNode	Models a complex intersection for changing destination and travel mode.
Connector	A simple zero-time travel link between two nodes.
Path	A link over which entities may independently move at their own speeds.
TimePath	A link that has a specified travel time for all entities.
Conveyor	A link that models both accumulating and non-accumulating conveyor devices.

libraries can be added using the Load Library button on the Project Home ribbon.

The Standard Object Library on the left side of the Facility Window is a general-purpose set of objects that comes standard with Simio. Each of these objects represents a physical object, device, or item that you might find if you looked around a facility being modeled. In many cases you'll build most of your model by dragging objects from the Standard Library and dropping them into your Facility Window. Table 4.1 lists the objects in the Simio Standard Library.

The Project Library includes the objects defined in the current project. As such, any new object definitions created in a project will appear in the Project Library for that project. Objects in the Project Library are defined/updated via the Navigation Window (described below) and they are used (placed in the Facility Window) via the Project Library. In order to simplify modeling, the

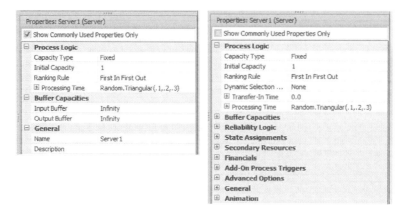

Figure 4.4: 'Show Commonly Used Properties Only' option.

Project Library is pre-populated with a ModelEntity object. The Flow Library includes a set of objects for modeling flow processing systems. Refer to the Simio Help for more information on the use of this library. Other domain-specific libraries are available on the Simio User Forum and can be accessed using the Shared Items button on the Support ribbon. The methods for building your own objects and libraries will be discussed in Chapter 10.

Properties Window

The Properties Window on the lower right side displays the properties (characteristics) of any object or item currently selected. For example, if a Server has been placed in the Facility Window, when it's selected you'll be able to display and change its properties in the Properties Window. The gray bars indicate categories or groupings of similar properties. By default the most commonly changed categories are expanded so you can see all the properties. The less commonly changed categories are collapsed by default, but you can expand them by clicking on the "+" sign to the left. If you change a property value it will be displayed in bold and its category will be expanded to make it easy to discern changes from default values. To return a property to its default value, right click on the property name and select Reset.

In version 5.91, Simio introduced an option to display only the 'commonly used properties' for the selected object. Figure 4.4 shows the respective properties windows for the Server object with the option selected and not selected. Note that this option does not change the behavior of the object instances in any way. Instead, it simply hides some of the properties in order to simplify the interface and highlight the commonly used properties. While modeling, if you are looking for an object property and can't find it, make sure that this box is unchecked so that you can view the complete set of object properties.

Navigation Window

A Simio *project* consists of one or more *models* or *objects*, as well as other components like *symbols* and *experiments*. You can navigate between the components using the Navigation Window on the upper right. Each time you select a new component you'll see the tabs and ribbons change accordingly. For example, if you select ModelEntity in the navigation window, you'll see a slightly different set of project model tabs available, and you might select the Definitions tab to add a state to that entity object. Then select the original Model object in the navigation window to continue editing the main model. If you get confused about what object you're working with, look to the title bar at the top of the navigation window or the highlighted bar within the navigation window. Note that objects appearing in a library *and* in the Navigation Window are often confusing to new users — just remember that objects are *placed* from the library and are *defined/edited* from the Navigation Window. Additional details are provided in Section 5.1.1.

SimBits

One feature you'll surely want to exploit is the SimBits collection. *SimBits* are small, well-documented models that illustrate a modeling concept or explain how to solve a common problem. The full documentation for each can be found in an accompanying automatically loaded .pdf file, as well as in the on-line help. Although they can be loaded directly from the Open menu item (replacing the currently open model), perhaps the best way to find a helpful SimBit is to look for the SimBit button on the Support ribbon. On the target page for this button you will find a categorized list of all of the SimBits with a filtering mechanism that lets you quickly find and load SimBits of interest (in this case, loading into a second copy of Simio, preserving your current workspace). SimBits are a helpful way to learn about new modeling techniques, objects, and constructs.

Moving/Configuring Windows and Tabs

The above discussions refer to the default window positions, but some window positions are easily changed. Many design-time and experimentation windows and tabs (for example the Process window or individual data table tabs) can be changed from their default positions by either right-clicking or dragging. While dragging, you'll see two sets of arrows called *layout targets* appear: a set near the center of the window and a set near the outside of the window. For example Figure 4.5 illustrates the layout targets just after you start dragging the tab for a table. Dropping the table tab onto any of the arrows will cause the table to be displayed in a new window at that location.

You can arrange the windows into vertical and horizontal tab groups by right clicking any tab and selecting the appropriate option. You can also drag some windows (Search, Watch, Trace, Errors, and object Dashboards) outside of the Simio application, even to another monitor, to take full advantage of your screen real estate. If you ever regret your custom arrangement of the windows or you lose a window (that is, it should be displayed but you can't find it), use the *Reset button* on the Project Home ribbon to restore the default window configuration.

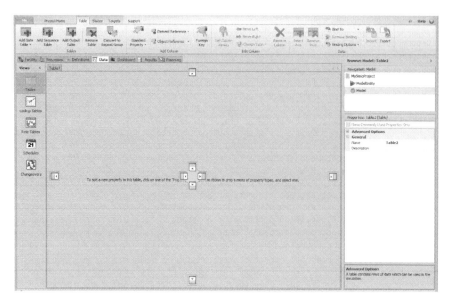

Figure 4.5: Dragging a tabbed window to a new display location.

4.2 Model 4-1: First Project Using the Standard Library Objects

In this section we'll build the basic model described above in Simio, and also do some experimentation and analysis with it, as follows: Section 4.2.1 takes you through how to build the model in Simio, in what's called the Facility Window using the Standard Library, run it (once), and look through the results. Next, in Section 4.2.2 we'll use it to do some initial informal experimentation with the system to compare it to what standard queueing theory would predict. Section 4.2.3 introduces the notions of statistically replicating and analyzing the simulation output results, and how Simio helps you do that. In Section 4.2.4 we'll talk about what might be roughly described as long-run vs. short-run simulations, and how you might need to "warm up" your model if you're interested in how things behave in the long run. Section 4.2.5 revisits some of the same questions raised in Section 4.2.2, specifically trying to verify that our model is correct, but now we are armed with better tools like warmup and statistical analysis of simulation output data. All of our discussion here is for a situation when we have only one scenario (system configuration) of interest; we'll discuss the more common goal of comparing alternative scenarios in Sections 5.5 and 9.1.1, and will introduce some additional statistical tools in those sections for such goals.

4.2.1 Building the Model

Using Standard Library objects is the most common method for building Simio models. These pre-built objects will be sufficient for many common types of models. Figure 4.6 shows the completed model of our queueing system using

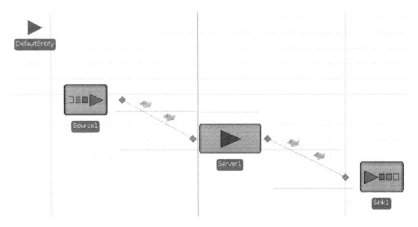

Figure 4.6: Completed Simio model (Facility Window) of the single-server queueing system — Model 4-1.

Simio's Facility Window (note that the Facility tab is highlighted in the Project Model Tabs area). We'll describe how to construct this model step by step in the following paragraphs.

The queueing model includes entities, an entity-arrival process, a service process, and a departure process. In the Simio Facility Window, these processes can be modeled using the *Source*, *Server*, and *Sink* objects. To get started with the model, start the Simio application and, if necessary, create a new model by clicking on the "New" item in the File page (accessible from the File ribbon). Once the default new model is open, make sure that the Facility Window is open by clicking on the Facility tab, and that the Standard Library is visible by clicking on the Standard Library section heading in the Libraries bar on the left; Figure 4.2 illustrates this. First, add a ModelEntity object by clicking on the ModelEntity object in the ProjectLibrary panel, then drag and drop it onto the Facility Window (actually, we're dragging and dropping an *instance* of it since the object definition stays in the ProjectLibrary panel). Next, click on the *Source* object in the Standard Library, then drag and drop it into the Facility Window. Similarly, click, drag, and drop an instance of each of the *Server* and *Sink* objects onto the Facility Window. The next step is to connect the Source, Server, and Sink objects in our model. For this example, we'll use the standard *Connector* object, which transfers entities between *nodes* in zero simulation time. To use this object, click on the Connector object in the Standard Library. After selecting the Connector, the cursor changes to a set of cross hairs. With the new cursor, click on the *Output Node* of the Source object (on its right side) and then click on the *Input Node* of the Server object (on its left side). This tells Simio that entities flow (instantly, i.e., in zero simulated time) out of the Source object and into the Server object. Follow the same process to add a connector between the Output Node of the Server object to the Input Node of the Sink object. Figure 4.7 shows the model with the connector in place between the Sink and Server objects.

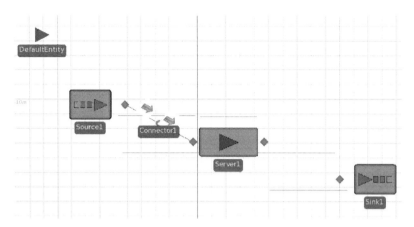

Figure 4.7: Model 4-1 with the Source and Server objects linked by a Connector object.

By the way, now would be a good time to save your model ("save early, save often," is a good motto for every simulationist). We chose the name `Model_04_01.spfx` (`spfx` is the default file-name extension for Simio project files), following the naming convention for our example files given in Section 3.2; all our completed example files are available on the book's website, as described in the Preface.

Before we continue constructing our model, we need to mention that the Standard Library objects include several default *queues*. These queues are represented by the horizontal green lines in Figure 4.7. Simio uses queues where entities[1] potentially *wait* — i.e., remain in the same logical place in the model for some period of simulated time. Model 4-1 includes the following queues:

- Source1 OutputBuffer.Contents — Used to store entities waiting to move out of the Source object.

- Server1 InputBuffer.Contents — Used to store entities waiting to enter the Server object.

- Server1 Processing.Contents — Used to store entities currently being processed by the Server object.

- Server1 OutputBuffer.Contents — Used to store entities waiting to exit the Server object.

- Sink1 InputBuffer.Contents — Used to store entities waiting to enter the Sink object.

In our simple single-server queueing system in Figure 4.1, we show only a single queue and this queue corresponds to the InputBuffer.Contents queue for the Server1 object. The Processing.InProcess queue for the Server1 object

---

[1]Technically, *tokens* rather than entities wait in Simio queues, but we'll discuss this issue in more detail in Chapter 5.

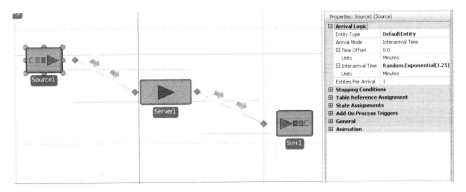

Figure 4.8: Setting the interarrival-time distribution for the Source object.

stores the entity that's being processed at any point in simulated time. The other queues in the Simio model are not used in our simple model (actually, the entities simply move through these queues instantly, in zero simulated time).

Now that the basic structure of the model is complete, we'll add the model parameters to the objects. For our simple model, we need to specify probability distributions governing the interarrival times and service times for the arriving entities. The Source object creates arriving entities according to a specified arrival process. We'd like a Poisson arrival process at rate $\lambda = 48$ entities per hour, so we'll specify that the entity interarrival times are exponentially distributed with a mean of 1.25 minutes (a *time* between entities of $= 60/48$ corresponds to a *rate* of 48/hour). In the formal Simio object model, the interarrival time is a *property* of the Source object. Object properties are set and edited in the Properties Window — select the Source object (click on the object) and the Properties Window will be displayed on the right panel (see Figure 4.8). The Source object's interarrival-time distribution is set by assigning the Interarrival Time property to *Random.Exponential*(1.25) and the Units property to Minutes; click on the + just to the left of "Interarrival Time" to expose the Units property and use the pull-down on the right to select Minutes. This tells Simio that each time an entity is created, it needs to sample a random value from an exponential distribution with mean 1.25, and to create the *next* entity that far into the future for an arrival rate of $\lambda = 60 \times (1/1.25) = 48$ entities/hour, as desired. The random-variate functions available via the keyword "Random" are discussed further in Section 4.4. The Time Offset property (usually set to 0) determines when the initial entity is created. The other properties associated with the Arrival Logic can be left at their defaults for now. With these parameters, entities are created recursively for the duration of the simulation run.

The default object name (Source1, for the first source object), can be changed by either double-clicking on the name tag below the object with the object selected, or through the Name property in the General properties section. Or, like most items in Simio, you can rename by using the F2 key. Note that the General section also includes a Description property for the object, which can be quite useful for model documentation. You should get into the habit of including a meaningful description for each model object because whatever you enter there will be displayed in a tool tip popup note when you hover the mouse over that object.

In order to complete the queueing logic for our model, we need to set up the service process for the Server object. The Processing Time property of the Server module is used to specify the processing times for entities. This property should be set to *Random.Exponential*(1) with the Units property being Minutes. The final step for our initial model is to tell Simio to run the model for 10 hours. To do this, click on the Run ribbon/tab, then in the Ending Type pull-down, select the Run Length option and enter 10 Hours. Before running our initial model, we'll set the running speed for the model.

The Speed Factor is used to control the speed of the interactive execution of the model explicitly. Changing the Speed Factor to 50 (just type it into the Speed Factor field in the Run ribbon) will speed up the run to a speed that's more visually appealing for this particular model. The "optimal" Speed Factor for an interactive run will depend on the model and object parameters and the individual preferences, as well as the speed of your computer, so you should definitely experiment with the Speed Factor for each model[2].

At this point, we can actually run the model by clicking on the Run icon in the upper left of the ribbon. The model is now running in *interactive mode*. As the model runs, the simulated time is displayed in the footer section of the application, along with the percentage complete. Using the default *Speed Factor*, simulation time will advance fairly slowly, but this can be changed as the model runs. When the simulation time reaches 10 (the run length that we set), the model run will automatically pause.

In Interactive Mode, the model results can be viewed at any time by stopping or pausing the model and clicking on the *Results* tab on the tab bar. Run the model until it reaches 10 hours and view the current results. Simio provides two different ways to view the basic model results: Pivot Grid, and Reports (the options are on the left panel — click on the corresponding icon to switch between the views). Figure 4.9 shows the Pivot Grid for Model 4-1 paused at time 10 hours[3]. The Pivot Grid format is extremely flexible and provides a very quick method to find specific results. If you're not used to this type of report, it can look a bit overwhelming at first, but you'll quickly learn to appreciate it as you begin to work with it. The Pivot Grid results can also be easily exported to a CSV (comma-separated values) text file, which can be imported into Excel and other applications. Each row in the default Pivot Grid includes an output value based on:

---

[2]Technically, the Speed Factor is the amount of simulation time, in tenths of a second, between each animation frame

[3]As mentioned in the Preface, Simio uses an agile development process with frequent minor updates, and occasional major updates. It's thus possible that the output values you get when you run our examples interactively may not always *exactly* match the output numbers we're showing here, which we got as we wrote the book. This could, as noted at the end of Section 1.4, be due to small variations across releases in low-level behavior, such as the order in which simultaneous events are processed. Regardless of the reason for these differences, their existence just emphasizes the need to do proper statistical design and analysis of simulation experiments, and not just run it once to get "the answer," a point that we'll make repeatedly throughout this book

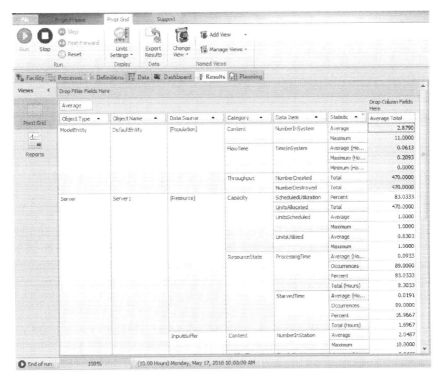

Figure 4.9: Pivot Grid report for the interactive run of Model 4-1.

- Object Type

- Object Name

- Data Source

- Category

- Data Item

- Statistic

So, in Figure 4.9, the Average (Statistic) value for the TimeInSystem (Data Item) of the DefaultEntity (Object Name) of the ModelEntity type (Object Type) is 0.0613 hours (0.0613 hours × 60 minutes/hour = 3.6780 minutes. Note that units for the Pivot Grid times, lengths, and rates can be set using the Time Units, Length Units, and Rate Units items in the Pivot Grid ribbon; if you switch to Minutes the Average TimeInSystem is 3.6753, so our hand-calculated value of 3.6780 minutes has a little round-off error in it. Further, the TimeInSystem data item belongs to the FlowTime Category and since the value is based on entities (dynamic objects), the Data Source is the set of Dynamic Objects.

If you're looking at this on your computer (as you should be!), scrolling through the Pivot Grid reveals a lot of output performance measures even

from a small model like this. For instance, just three rows below the Average
TimeInSystem of 0.0613 hours, we see under the Throughput Category that a
total of 470 entities were created (i.e., entered the model through the Source
object), and in the next row that 470 entities were destroyed (i.e., exited the
model through the Sink object). Though not always true, in this particular
run of this model all of the 470 entities that arrived also exited during the 10
hours, so that at the end of the simulation there were no entities present. You
can confirm this by looking at the animation when it's paused at the 10-hour
end time. (Change the run time to, say, 9 hours, and then 11 hours, to see that
things don't always end up this way, both by looking at the final animation as
well as the NumberCreated and NumberDestroyed in the Throughput Category
of the Pivot Grid)[4]. So in our 10-hour run, the output value of 0.0613 hours for
average time in system is just the simple average of these 470 entities' individual
times in system.

The Pivot Grid supports three basic types of data manipulation:

1. Grouping: Dragging column headings to different relative locations will
   change the grouping of the data.

2. Sorting: Clicking on an individual column heading will cause the data to
   be sorted based on that column.

3. Filtering: Hovering the mouse over the upper right corner of a column
   heading will expose a funnel-shaped icon. Clicking on this icon will bring
   up a dialog that supports data filtering. If a filter is applied to any column,
   the funnel icon is displayed (no mouse hover required). Filtering the data
   allows you quickly to view the specific data in which you're interested
   regardless of the amount of data included in the output.

Pivot Grids also allow the user to store multiple *views* of the filtered, sorted,
and grouped Pivot Grids. Views can be quite useful if you are monitoring a
specific set of performance metrics. The Simio documentation section on Pivot
Grids includes much more detail about how to use these specific capabilities.
The Pivot Grid format is extremely useful for finding information when the
output includes many rows.

The *Reports* format gives the interactive run results in a formatted, detailed
report format, suitable for printing, exporting to other file formats, or emailing
(the formatting, printing, and exporting options are available from the Print
Preview tab on the ribbon). Figure 4.10 shows the Reports format with the
Print Preview tab open on the ribbon. Scrolling down to the "TimeInSystem
- Average (Hours)" heading on the left will show a Value of 0.06126, the same
(up to roundoff) as we saw for this output performance measure in the Pivot
Grid in Figure 4.9.

---

[4]While you're playing around with the simulation run length, try changing it to 8 *minutes*
and compare some of the Pivot Grid results with what we got from the manual simulation in
Section 3.3.1 given in Figure 3.5. Now we can confess that those "magical" interarrival and
service times for that manual simulation were generated in this Simio run, and we recorded
them via the Model Trace capability.

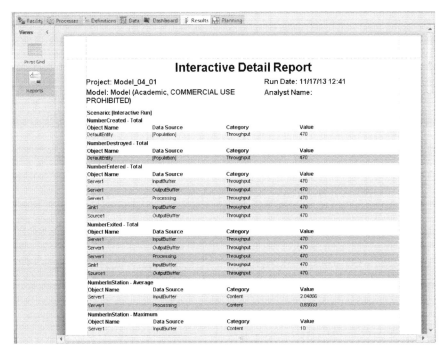

Figure 4.10: Standard report view for Model 4-1.

Table 4.2: Comparison of the queueing analysis and initial model results for the first model.

Metric	Queueing	Model
Utilization (ρ)	0.800	0.830
Number in system (L)	4.000	2.879
Number in queue (L_q)	3.200	2.049
Time in system (W)	0.083	0.061
Time in queue (W_q)	0.067	0.044

4.2.2   Initial Experimentation and Analysis

Now that we have our first Simio model completed, we'll do some initial, informal experimenting and analysis with it to understand the queueing system it models. As we mentioned earlier, the long-run, steady-state performance of our system can be determined analytically using queueing analysis (see Chapter 2 for details). Note that for any but the simplest models, this type of exact analysis will not be possible (this is why we use simulation, in fact). Table 4.2 gives the steady-state queueing results and the simulation results taken from the Pivot Table in Figure 4.9.

You'll immediately notice that the numbers in the Queueing column are not equal to the numbers in the Model column, as we might expect. Before discussing the possible reasons for the differences, we first need to discuss one more important and sometimes-concerning issue. If you return to the Facility

Window (click on the Facility tab just below the ribbon), reset the model (click on the Reset icon in the Run ribbon), re-run the model, allow it to run until it pauses at time 10 hours, and view the Pivot Grid, you'll notice that the results are identical to those from the previous run (displayed in Figure 4.9). If you repeat the process again and again, you'll always get the same output values. To most people new to simulation, and as mentioned in Section 3.1.3, this seems a bit odd given that we're supposed to be using random values for the entity interarrival and service times in the model. This illustrates the following critical points about computer simulation:

1. The "random" numbers used are not truly random in the sense of being unpredictable, as mentioned in Section 3.1.3 and discussed in Section 6.3 — instead they are *pseudo-random*, which, in our context, means that the precise sequence of generated numbers is deterministic (among other things); and

2. Through the random-variate-generation process discussed in Section 6.4, some simulation software can control the pseudo-random number generation and we can exploit this control to our advantage.

The concept that the "supposedly random numbers" are actually predictable can initially cause great angst for new simulationists (that's what you're now becoming). However, for simulation, this predictability is a good thing. Not only does it make grading simulation homework easier (important to the authors), but (more seriously) it's also useful during model debugging. For example, when you make a change in the model that *should* have a predictable effect on the simulation output, it's very convenient to be able to use the same "random" inputs for the same purposes in the simulation, so that any changes (or lack thereof) in output can be directly attributable to the model changes, rather than to different random numbers. As you get further into modeling, you'll find yourself spending significant time debugging your models so this behavior will prove useful to you (see Section 4.8 for detailed coverage of the debugging process and Simio's debugging tools). In addition, this predictability can be used to reduce the required simulation run time through a variety of techniques called *variance reduction*, which are discussed in general simulation texts (such as [3] or [30]). Simio's default behavior is to use the same sequence of random variates (draws or observations on model-driving inputs like interarrival and service times) each time a model is run. As a result, running, resetting, and re-running a model will yield identical results unless the model is explicitly coded to behave otherwise.

Now we can return to the question of why our initial simulation results are not equal to our queueing results in Table 4.2. There are three possible explanations for this mismatch:

1. Our Simio model is wrong, i.e., we have an error somewhere in the model itself;

2. Our expectation is wrong, i.e., our assumption that the simulation results *should* match the queueing results is wrong; or

3. Sampling error, i.e., the simulation model results match the expectation in a probabilistic sense, but we either haven't run the model long enough, or for enough replications (separate independent runs starting from the same state but using separate random numbers), or are interpreting the results incorrectly.

In fact, if the results are not equal when comparing simulation results to our expectation, it's *always* one or more of these possibilities, regardless of the model. In our case, we'll see that our expectation is wrong, and that we have not run the model long enough. Remember, the queueing-theory results are for long-run steady-state, i.e., after the system/model has run for an essentially infinite amount of time. But we ran for only 10 hours, which for this model is evidently not sufficiently close to infinity. Nor have we made enough replications (items 2 and 3 from above). Developing expectations, comparing the expectations to the simulation-model results, and iterating until these converge is a very important component of model verification and validation (we'll return to this topic in Section 4.2.5).

4.2.3 Replications and Statistical Analysis of Output

As just suggested, a *replication* is a run of the model with a fixed set of starting and ending conditions using a specific and separate, non-overlapping sequence of input random numbers and random variates (the exponentially distributed interarrival and service times in our case). For the time being, assume that the starting and ending conditions are dictated by the starting and ending simulation time (although as we'll see later, there are other possible kinds of starting and ending conditions). So, starting our model empty and idle, and running it for 10 hours, constitutes a replication. Resetting and re-running the model constitutes running *the same* replication again, using *the same* input random numbers and thus random variates, so obviously yielding *the same* results (as demonstrated above). In order to run a *different* replication, we need a different, separate, non-overlapping set of input random numbers and random variates. Fortunately, Simio handles this process for us transparently, but we can't run multiple replications in Interactive mode. Instead, we have to create and run a Simio *Experiment*.

Simio Experiments allow us to run our model for a user-specified number of replications, where Simio guarantees that the generated random variates are such that the replications are statistically independent from one another, since the underlying random numbers do not overlap from one replication to the next. This guarantee of independence is critical for the required statistical analysis we'll do. To set up an experiment, go to the Project Home ribbon and click on the New Experiment icon. Simio will create a new experiment and switch to the Experiment Design view as shown in Figure 4.11 after we changed both Replications Required near the top, and Default Replications on the right, to 5 from their default values of 10. To run the experiment, select the row corresponding to Scenario1 (the default name) and click the Run icon (the one with two white right arrows in it in the Experiment window, not the one with one right arrow in it in the Model window). After Simio runs the five replications, select the Pivot Grid report (shown in Figure 4.12). Compared

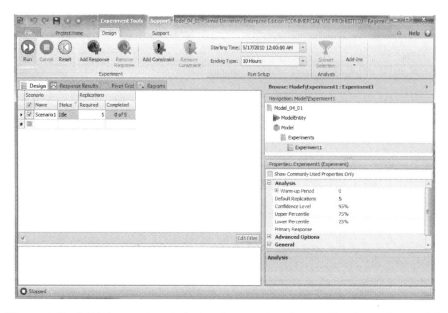

Figure 4.11: Initial experiment design for running five replications of a model.

Object Type	Object Name	Data Source	Category	Data Item	Statistic	Average	Minimum	Maximum	Half Width
ModelEntity	DefaultEntity	[DynamicObjects]	Content	NumberInSystem	Average	3.6592	2.2958	6.7441	2.2029
					Maximum	15.8000	11.0000	28.0000	8.6636
			FlowTime	TimeInSystem	Average (Ho...	0.0762	0.0489	0.1306	0.0395
					Maximum (H...	0.2888	0.2018	0.5096	0.1601
					Minimum (Ho...	0.0003	0.0000	0.0009	0.0005
			Throughput	NumberCreated	Total	471.8000	448.0000	507.0000	26.7909
				NumberDestroyed	Total	464.4000	446.0000	479.0000	15.5208
Server	Server1	[Resource]	Capacity	ScheduledUtilization	Percent	78.7339	76.3323	83.0333	3.2689
				UnitsAllocated	Total	465.2000	447.0000	480.0000	15.3912
				UnitsScheduled	Average	1.0000	1.0000	1.0000	0.0000
					Maximum	1.0000	1.0000	1.0000	0.0000
				UnitsUtilized	Average	0.7873	0.7633	0.8303	0.0327
					Maximum	1.0000	1.0000	1.0000	0.0000
			ResourceState	ProcessingTime	Average (Ho...	0.0794	0.0578	0.0933	0.0166
					Occurrences	101.4000	89.0000	132.0000	21.7982
					Percent	78.7339	76.3323	83.0333	3.2689
					Total (Hours)	7.8734	7.6332	8.3033	0.3269
				StarvedTime	Average (Ho...	0.0214	0.0181	0.0243	0.0034
					Occurrences	100.6000	89.0000	131.0000	21.5851
					Percent	21.2661	16.9667	23.6677	3.2689
					Total (Hours)	2.1266	1.6967	2.3668	0.3269
		InputBuffer	Content	NumberInStation	Average	2.8718	1.5325	5.9590	2.2020
					Maximum	14.8000	10.0000	27.0000	8.6636
			HoldingTime	TimeInStation	Average (Ho...	0.0593	0.0327	0.1148	0.0401
					Maximum (H...	0.2668	0.1500	0.5058	0.1764
					Minimum (Ho...	0.0000	0.0000	0.0000	0.0000

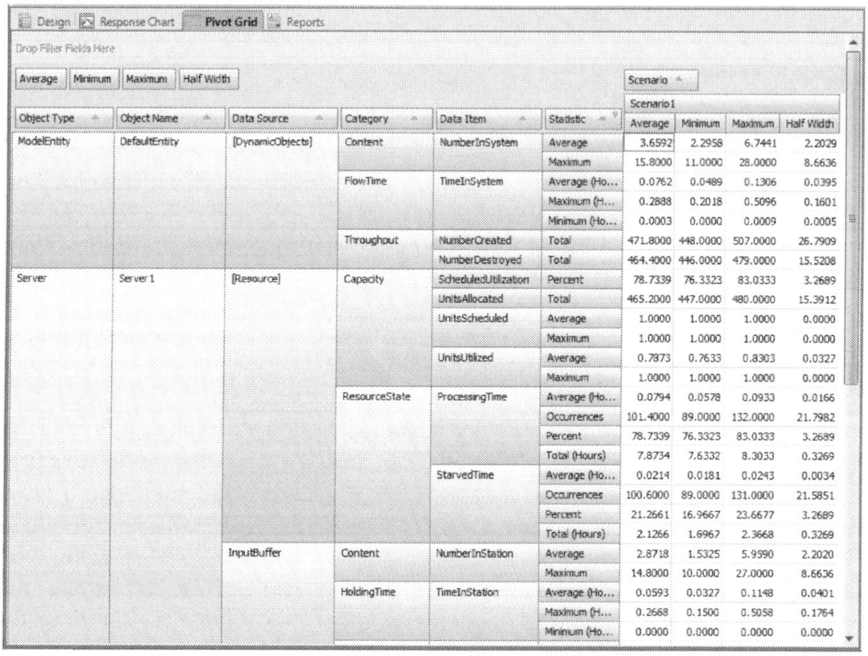

Figure 4.12: Experiment Pivot Grid for the five replications of the model.

to the Pivot Grid we saw while running in Interactive Mode (Figure 4.9), we see the additional results columns for Minimum, Maximum, and Half Width (of 95% confidence intervals on the expected value, with the Confidence Level being editable in the Experiment Design Properties), reflecting the fact that we now have five independent observations of each output statistic.

To understand what these *cross-replication* output statistics are, focus on the entity TimeInSystem values in rows 3-5. For example:

- The 0.0762 for the Average of Average (Hours) TimeInSystem (yes, we meant to say "Average" twice there) is the average of five numbers, each of which is a *within-replication* average time in system (and the first of those five numbers is 0.0613 from the single-replication Pivot Grid in Figure 4.9. The 95% confidence interval 0.0762 ± 0.0395, or $[0.0367, 0.1157]$ (in hours), contains, with 95% confidence , the *expected* within-replication average time in system, which you can think of as the result of making an infinite number of replications (not just five) of this model, each of duration 10 hours, and averaging all those within-replication average times in system. Another interpretation of what this confidence interval covers is the expected value of the probability distribution of the simulation output random variable representing the within-replication average time in system. (More discussion of output confidence intervals appears below in the discussion of Table 4.4.)

- Still in the Average (Hours) row, 0.1306 is the maximum of these five average within-replication times in system, instead of their average. In other words, across the five replications, the largest of the five Average TimeInSystem values was 0.1306, so it is the maximum average.

- Average maximum, anyone? In the next row down for Maximum (Hours), the 0.2888 on the left is the average of five numbers, each of which is the maximum individual-entity time in system within that replication. And the 95% confidence interval 0.2888 ± 0.1601 is trying to cover the expected maximum time in system, i.e., the maximum time in system averaged over an infinite number of replications rather than just five.

- Maybe more meaningful as a really bad worst-case time in system, though, would be the 0.5096 hour, being the maximum of the five within-replication maximum times in system

Table 4.3 gives the queueing metrics for each of the five replications of the model, as well a the sample mean (Avg) and sample standard deviation (StDev) across the five replications for each metric. To access these individual-replication output values, click on the Export Details icon in the Pivot Grid ribbon; click Export Summaries to get cross-replication results like means and standard deviations, as shown in the Pivot Grid itself. The exported data file is in CSV format, which can be read by a variety of applications, such as Excel. The first thing to notice in this table is that the values can vary significantly between replications (L and L_q, in particular). This variation is specifically why we cannot draw inferences from the results of a single replication.

Table 4.3: Five replications of data for the first model.

Metric being estimated	Replication					Avg	StDev
	1	2	3	4	5		
Utilization (ρ)	0.830	0.763	0.789	0.769	0.785	0.787	0.026
Number in system (L)	2.879	2.296	3.477	2.900	6.744	3.659	1.774
Number in queue (L_q)	2.049	1.532	2.688	2.131	5.959	2.872	1.774
Time in system (W)	0.061	0.049	0.075	0.065	0.131	0.076	0.032
Time in queue (W_q)	0.044	0.033	0.058	0.048	0.115	0.059	0.032

Table 4.4: Comparing 5 replications with 50 replications.

Metric being estimated	5 Replications		50 Replications	
	Avg	h	Avg	h
Utilization (ρ)	0.787	0.033	0.789	0.014
Number in system (L)	3.659	2.203	3.794	0.433
Number in queue (L_q)	2.872	2.202	3.004	0.422
Time in system (W)	0.076	0.040	0.078	0.008
Time in queue (W_q)	0.059	0.040	0.062	0.008

Since our model inputs (entity interarrival and service times) are random, the simulation-output performance metrics (simulation-based estimates of ρ, L, L_q, W, and W_q, which we could respectively denote as $\widehat{\rho}$, $\widehat{L}$, $\widehat{L_q}$, $\widehat{W}$, and $\widehat{W_q}$) are random variables . The queueing analysis gives us the exact steady-state values of ρ, L, L_q, W, and W_q. Based on how we run replications (the same model, but with separate independent input random variates), each replication generates one observation on each of $\widehat{\rho}$, $\widehat{L}$, $\widehat{L_q}$, $\widehat{W}$, and $\widehat{W_q}$. In statistical terms, running n replications yields n independent, identically distributed (IID) observations of each random variable. This allows us to estimate the mean values of the random variables using the sample averages across replications. So, the values in the Avg column from Table 4.3 are estimates of the corresponding random-variable expected values. What we don't know from this table is how *good* our estimates are. We do know, however, that as we increase the number of replications, our estimates get better, since the sample mean is a *consistent* estimator (its own variance decreases with n), and from the *strong law of large numbers* (as $n \rightarrow \infty$, the sample mean across replications $\rightarrow$ the expected value of the respective random variable, with probability 1).

Table 4.4 compares the results from running five replications with those from running 50 replications. Since we ran more replications of our model, we expect the estimates to be better, but averages still don't give us any specific information about the *quality* (or *precision*) of these estimates. What we need is an *interval estimate* that will give us insight about the sampling error (the averages are merely *point estimates*). The h columns give such an interval estimate. These columns give the half-widths of 95% confidence intervals on the means constructed from the usual normal-distribution approach (using the sample standard deviation and student's t distribution with $n-1$ degrees of freedom, as given in any beginning statistics text).

Consider the 95% confidence intervals for L based on five and 50 replications:

$$5 \text{ replications} : 3.659 \pm 2.203 \text{ or } [1.456, 5.862]$$
$$50 \text{ replications} : 3.794 \pm 0.433 \text{ or } [3.361, 4.227]$$

Based on five replications, we're 95% *confident* that the true mean (expected value or population mean) of $\widehat{L}$ is between 1.456 and 5.862, while based on 50 replications, we're 95% *confident* that the true mean is between 3.361 and 4.227. (Strictly speaking, the interpretation is that 95% of confidence intervals formed in this way, from replicating, will cover the unknown true mean.) So the confidence interval on the mean of an output statistic provides us a measure of the sampling error and, hence, the quality (precision) of our estimate of the true mean of the random variable. By increasing the number of replications (samples), we can make the half-width increasingly small. For example, running 250 replications results in a CI of $[3.788, 4.165]$ — clearly we're more comfortable with our estimate of the mean based on 250 replications than we are based five replications. In cases where we make independent replications, the confidence-interval half-widths therefore give us guidance as to how many replications we should run if we're interested in getting a precise estimate of the true mean; due to the $\sqrt{n}$ in the denominator of the formula for the confidence-interval half width, we need to make about four times as many replications to cut the confidence interval half-width in half, compared to its current size from an initial number of replications, and about 100 times as many replications to make the interval $1/10$ its current size. Unfortunately, there is no specific rule about "how close is close enough" — i.e., what values of h are acceptably small for a given simulation model and decision situation. This is a judgment call that must be made by the analyst or client in the context of the project. There is a clear trade-off between computer run time and reducing sampling error. As we mentioned above, we can make h increasingly small by running enough replications, but the cost is computer run time. When deciding if more replications are warranted, two issues are important:

1. What's the cost if I make an incorrect decision due to sampling error?

2. Do I have time to run more replications?

So, the first answer as to why our simulation results shown in Table 4.2 don't match the queueing results is that we were using the results from a *singe replication* of our model. This is akin to rolling a die, observing a 4 (or any other single value) and declaring that value to be the *expected value* over a large number of rolls. Clearly this would be a poor estimate, regardless of the individual roll. Unfortunately, using results from a single replication is quite common for new simulationists, despite the significant risk. Our general approach going forward will be to run multiple replications and to use the sample averages as estimates of the means of the output statistics, and to use the 95% confidence-interval half-widths to help determine the appropriate number of replications if we're interested in estimating the true mean. So, instead of simply using the averages (point estimates), we'll also use the confidence intervals (interval estimates) when analyzing results. The standard Simio Pivot Grid report for experiments (see Figure 4.12) automatically supports this approach by providing the sample average and 95% confidence-interval half-widths for all output statistics.

The second reason for the mismatch between our expectations and the model results is a bit more subtle and involves the need for a *warm-up period* for our model. We will discuss that in the next section.

4.2.4   Steady-State vs. Terminating Simulations

Generally, when we start running a simulation model that includes queueing or queueing-network components, the model starts in a state called "empty and idle," meaning that there are no entities in the system and all servers are idle. Consider our simple single-server queueing model. The first entity that arrives will *never* have to wait for the server. Similarly, the second arriving entity will likely spend less time in the queue (on average) than the 100th arriving entity (since the only possible entity in front of the second entity will be the first entity). Depending on the characteristics of the system being modeled (the expected server utilization, in our case), the distribution and expected value of queue times for the third, fourth, fifth, etc. entities can be significantly different from the distribution and expected value of queue times at *steady state*, i.e., after a long time that is sufficient for the effects of the empty-and-idle initial conditions to have effectively worn off. The time between the start of the run and the point at which the model is determined to have reached (another one of those judgment calls) steady state is called the *initial transient period*, which we'll now discuss.

The basic queueing analysis that we used to get the results in Table 4.2 (see Chapter 2) provides exact expected-value results for systems *at steady state*. As discussed above, most simulation models involving queueing networks go through an initial-transient period before effectively reaching steady state. Recording model statistics during the initial-transient period and then using these observations in the replication summary statistics tabulation can lead to *startup bias*, i.e., $E(\widehat{L})$ may not be equal to L. As an example, we ran four experiments where we set the run length for our model to be 2, 5, 10, 20, and 30 hours and ran 500 replications each. The resulting estimates of L (along with the 95% confidence intervals, of course) were:

$$2 \text{ hours} : 3.232 \pm 0.168 \text{ or } [3.064, 3.400]$$
$$5 \text{ hours} : 3.622 \pm 0.170 \text{ or } [3.622, 3.962]$$
$$10 \text{ hours} : 3.864 \pm 0.130 \text{ or } [3.734, 3.994]$$
$$20 \text{ hours} : 3.888 \pm 0.096 \text{ or } [3.792, 3.984]$$
$$30 \text{ hours} : 3.926 \pm 0.080 \text{ or } [3.846, 4.006]$$

For the 2, 5, 10, and 20 hour runs, it seems fairly certain that the estimates are still biased downwards with respect to steady state (the steady-state value is $L = 4.000$). At 30 hours, the mean is still a little low, but the confidence interval covers 4.000, so we're not sure. Running more replications would likely reduce the width of the confidence interval and 4.000 *may* be outside it so that we'd conclude that the bias is still significant with a 30-hour run, but we're still not sure. It's also possible that running additional replications wouldn't provide the evidence that the startup bias is significant — such is the nature of statistical sampling. Luckily, unlike many scientific and sociological experiments, we're in total control of the replications and run length and can experiment until we're satisfied (or until we run out of either computer time or human patience). Before continuing we must point out that you can't "replicate away" startup bias. The transient period is a characteristic of the system and isn't an artifact of randomness and the resulting sampling error.

Instead of running the model long enough to wash out the startup bias through sheer arithmetic within each run, we can use a *warm-up period*. Here, the model run period is divided so that statistics are not collected during the initial (warm-up) period, though the model is running as usual during this period. After the warm-up period, statistics are collected as usual. The idea is that the model will be in a state close to steady-state when we start recording statistics if we've chosen the warm-up period appropriately, something that may not be especially easy in practice. So, for our simple model, the expected number of entities in the queue when the first entity arrives after the warm-up period would be 3.2 ($L_q = 3.2$ at steady state). As an example, we ran three additional experiments where we set the run lengths and warm-up periods to be $(20, 10)$, $(30, 10)$, and $(30, 20)$, respectively (in Simio, this is done by setting the Warm-up Period property for the Experiment to the length of the desired warm-up period). The results when estimating $L = 4.000$ are:

$$\text{(Run length, warmup)} = (20, 10) : 4.033 \pm 0.155 \text{ or } [3.978, 4.188]$$
$$\text{(Run length, warmup)} = (30, 10) : 4.052 \pm 0.103 \text{ or } [3.949, 4.155]$$
$$\text{(Run length, warmup)} = (30, 20) : 3.992 \pm 0.120 \text{ or } [3.872, 4.112]$$

It seems that the warm-up period has helped reduce or eliminate the startup bias in all cases and we have not increased the overall run time beyond 30 hours. So, we have improved our estimates without increasing the computational requirements by using the warm-up period. At this point, the natural question is "How long should the warm-up period be?" In general, it's not at all easy to determine even approximately when a model reaches steady state. One heuristic but direct approach is to insert dynamic animated Status Plots in the Simio model's Facility Window (in the Model's Facility Window, select the Animation ribbon under Facility Tools — see Chapter 8 for animation details) and just make a judgment about when they appear to stop trending systematically; however, these can be quite "noisy" (i.e., variable) since they depict only one replication at a time during the animation. We'll simply observe the following about specifying warm-up periods:

- If the warm-up period is too short, the results will still have startup bias (this is potentially bad); and

- If the warm-up period is too long, our sampling error will be higher than necessary (as we increase the warm-up period length, we decrease the amount of data that we actually record).

As a result, the "safest" approach is to make the warm-up period long and increase the overall run length and number of replications in order to achieve

Table 4.5: Comparison of the queueing analysis and our final experiment.

Metric being estimated	Queueing	Simulation
Utilization (ρ)	0.800	0.800 ± 0.004
Number in system (L)	4.000	3.992 ± 0.120
Number in queue (L_q)	3.200	3.192 ± 0.117
Time in system (W)	0.083	0.083 ± 0.002
Time in queue (W_q)	0.067	0.066 ± 0.002

acceptable levels of sampling error (measured by the half-widths of the confidence intervals). Using this method we may expend a bit more computer time than is absolutely necessary, but computer time is cheap these days (and bias is insidiously dangerous since in practice you can't measure it)!

Of course, the discussion of warm-up in the previous paragraphs assumes that you actually *want* steady-state values; but maybe you don't. It's certainly possible (and common) that you're instead interested in the "short-run" system behavior during the transient period; for example, same-day ticket sales for a sporting event open up (with an empty and idle system) and stop at certain pre-determined times, so there is no steady state at all of any relevance. In these cases, often called *terminating simulations*, we simply ignore the warm-up period in the experimentation (i.e., default it to 0), and what the simulation produces will be an unbiased view of the system's behavior during the time period of interest, and relative to the initial conditions in the model.

The choice of whether the steady-state goal or the terminating goal is appropriate is usually a matter of what your study's intent is, rather than a matter of what the model structure may be. We will say, though, that terminating simulations are much easier to set up, run, and analyze, since the starting and stopping rules for each replication are just part of the model itself, and not up to analysts' judgment; the only real issue is how many replications you need to make in order to achieve acceptable statistical precision in your results.

4.2.5 Model Verification

Now that we've addressed the replications issue, and possible warm-up period if we want to estimate steady-state behavior, we'll revisit our original comparison of the queueing analysis results to our updated simulation model results (500 replications of the model with a 30-hour run length and 20-hour warm-up period). Table 4.5 gives both sets of results. As compared to the results shown in Table 4.2, we're much more confident that our model is "right." In other words, we have fairly strong evidence that our model is *verified* (i.e., that it behaves as we expect it to). Note that it's not possible *provably* to verify a model. Instead, we can only collect evidence until we either find errors or are convinced that the model is correct.

Recapping the process that we went through:

1. We developed a set of *expectations* about our model results (the queueing analysis).

2. We developed and ran the model and compared the model results to our expectations (Table 4.2).

3. Since the results didn't match, we considered the three possible explanations:

   a) Our Simio model is wrong (i.e., we have an error somewhere in the model itself) — we skipped over this one.

   b) Our expectation is wrong (i.e., our assumption that the simulation results *should* match the queueing results is wrong) — we found that we needed to warm up the model to get it close to steady state in order effectively to eliminate the startup bias (i.e., our expectation that our analysis including the transient period should match the steady-state results was wrong). Adding a warm-up period corrected for this.

   c) Sampling error (i.e., the simulation-model results match the expectation in a probabilistic sense, but we either haven't run the model long enough or are interpreting the results incorrectly) — we found that we needed to *replicate* the model and increase the run length to account appropriately for the randomness in the model outputs.

4. We finally settled on a model that we feel is correct.

It's a good idea to try to follow this basic verification process for all simulation projects. Although we'll generally not be able to compute the exact results that we're looking for (otherwise, why would we need simulation?), we can always develop *some* expectations, even if they're based on an abstract version of the system being modeled. We can then use these expectations and the process outlined above to converge to a model (and set of expectations) about which we're highly confident.

Now that we've covered the basics of model verification and experimentation in Simio, we'll switch gears and discuss some additional Simio modeling concepts for the remainder of this chapter. However, we'll definitely revisit these basic issues throughout the book.

4.3 Model 4-2: First Model Using Processes

Although modeling strictly with high-level Simio objects (such as those from the Standard Library) is fast, intuitive, and (almost) easy for most people, there are often situations where you'll want to use the lower-level Simio Processes. You may want to construct your model or augment existing Simio objects, either to do more detailed or specialized modeling not accommodated with objects, or improve execution speed if that's a problem. Using Simio Processes requires a fairly detailed understanding of Simio and discrete-event simulation methodology, in general. This section will only demonstrate a simple, but fundamental Simio Process model of our example single-server queueing system. In the following chapters, we'll go into much more detail about Simio Processes where called for by the modeling situation.

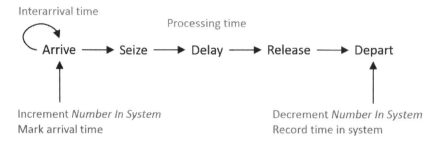

Figure 4.13: Basic process for the seize-delay-release model.

In order to model systems that include finite-capacity resources for which entities compete for service (such as the server in our simple queueing system), Simio uses a *Seize-Delay-Release* model. This is a standard discrete-event-simulation approach and many other simulation tools use the same or a similar model. Complete understanding of this basic model is essential in order to use Simio Processes effectively. The model works as follows:

- Define a resource with capacity c. This means that the resource has up to c arbitrary units of capacity that can be simultaneously allocated to one or more entities at any point in simulated time.

- When an entity requires service from the resource, the entity *seizes* some number s of units of capacity from the resource.

- At that point, if the resource has s units of capacity not currently allocated to other entities, s units of capacity are immediately allocated to the entity and the entity begins a *delay* representing the service time, during which the s units remain allocated to the entity. Otherwise, the entity is automatically placed in a queue where it waits until the required capacity is available.

- When an entity's service-time delay is complete, the entity *releases* the s units of capacity of the resource and continues to the next step in its process. If there are entities waiting in the resource queue and the resource's available capacity (including the units freed by the just-departed entity) is sufficient for one of the waiting entities, the first such entity is removed from the queue, the required units of capacity are immediately allocated to that entity, and that entity begins its delay.

From the modeling perspective, each entity simply goes through the *Seize-Delay-Release* logic and the simulation tool manages the entity's queueing and allocation of resource capacity to the entities. In addition, most simulation software, including Simio, automatically records queue, resource, and entity-related statistics as the model runs. Figure 4.13 shows the basic seize-delay-release process. In this figure, the "Interarrival time" is the time between successive entities and the "Processing time" is the time that an entity is delayed for processing. The *Number In System* tracks the number of entities in the

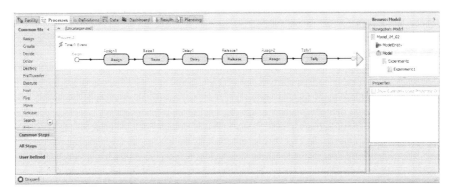

Figure 4.14: Process view of Model 4-2.

system at any point in simulated time and the *marking* and recording of the arrival time and time in system tracks the times that all entities spend in the system.

For our single-server queueing model, we simply set $c = 1$ and $s = 1$ (for all entities). So our single-server model is just an implementation of the basic seize-delay-release logic illustrated in Figure 4.13. Creating this model using processes is a little bit more involved than it was using Standard Library Objects, but it's instructive to go through the model development and see the mechanisms for collecting user-defined statistics. The steps to implement this model in Simio (see Figure 4.14) are as follows:

1. Open Simio and create a new model

2. Create a Resource object in the Facility Window by dragging a Resource object from the Standard Library onto the Facility Window. In the Process Logic section of the object's properties, verify that the Initial Capacity Type is Fixed and that the Capacity is 1 (these are the defaults). Note the object Name in the General section (the default is Resource1).

3. Make sure that the Model is highlighted in the Navigation Window and switch to the Definitions Window by clicking on the Definitions tab and choose the States section by clicking on the corresponding panel icon on the left. This prepares us to add a state to the model.

4. Create a new discrete (integer) state by clicking on the Integer icon in the States ribbon. Change the default Name property of IntegerState1 for the state to "WIP". Discrete States are used to record numerical values. In this case, we're creating a place to store the current number of entities in the model by creating an Integer Discrete State for the model (the *Number In System* from Figure 4.13).

5. Switch to the Elements section by clicking on the panel icon and create a *Timer* element by clicking on the Timer icon in the Elements ribbon (see Figure 4.15). The Timer element will be used to trigger entity arrivals (the loop-back arc in Figure 4.13). In order to have Poisson arrivals

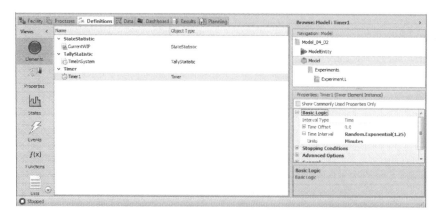

Figure 4.15: Timer element for the Model 4-2.

at the rate of $\lambda = 48$ entities/hour, or equivalently exponential interarrivals with mean $1/0.8 = 1.25$ minutes, set the Time Interval property to "Random.Exponential(1.25)" and make sure that the Units are set to Minutes.

6. Create a State Statistic by clicking on the State Statistic icon in the Statistics section of the Elements ribbon. Set the State Variable Name property to WIP (the previously defined model Discrete State so it appears on the pull-down there) and set the Name property to "CurrentWIP". We're telling Simio to track the value of the state over time and record a time-dependent statistic on this value.

7. Create a Tally Statistic by clicking on the Tally Statistic icon in the Statistics section of the Elements ribbon. Set the Name property to "TimeIn-System" and set the Unit Type property to "Time". Tally Statistics are used to record observational (i.e., discrete-time) statistics.

8. Switch to the Process Window by clicking on the Processes tab and create a new Process by clicking on the Create Process icon in the Process ribbon.

9. Set the Triggering Event property to be the newly created timer event (see Figure 4.16). This tells Simio to execute the process whenever the timer goes off.

10. Add an Assign step by dragging the Assign step from the Common Steps panel to the process, placing the step just to the right of the Begin indicator in the process. Set the State Variable Name property to WIP and the New Value property to $WIP + 1$, indicating that when the event occurs, we want to increment the value of the state variable to reflect the fact that an entity has arrived to the system (the "Increment" in Figure 4.13).

11. Next, add the Seize step to the process just to the right of the Assign step. To indicate that Resource1 should be seized by the arriving entity, click the "..." button on the right and select the Seizes property in the

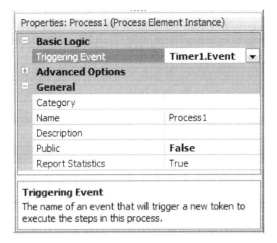

Figure 4.16: Setting the triggering event for the process.

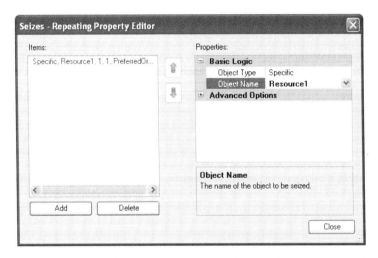

Figure 4.17: Setting the Seize properties to indicate that Resource1 should be seized.

Basic Logic section, click Add, and then indicate that the specific object Resource1 should be seized (see Figure 4.17).

12. Add the Delay step immediately after the Seize step and set the Delay Time property to $Random.Exponential(1)$ minutes to indicate that the entity delays should be exponentially distributed with mean 1 minute (equivalent to the original service rate of 60/hour).

13. Add the Release step immediately after the Delay step and set the Releases property to Resource1.

14. Add another Assign step next to the Release step and set the State Variable Name property to WIP and the New Value property to $WIP - 1$,

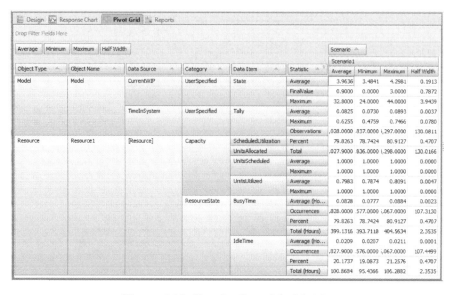

Figure 4.18: Results from Model 4-2.

indicating that when the entity releases the resource, we want to decrement the value of the state variable to reflect the fact that an entity has left the system.

15. Add a Tally step and set the TallyStatisticName property to TimeInSystem (the Tally Statistic was created earlier so is available on the pull-down there), and set the Value property to $TimeNow - Token.TimeCreated$ to indicate that the recorded value should be the current simulation time minus the time that the current token was created. This time interval represents the time that the current entity spent in the system. The Tally step implements the "Record" function shown in Figure 4.13. Note that we used the token state Token.TimeCreated instead of *marking* the arrival time as shown in Figure 4.13.

16. Finally, switch back to the Facility Window and set the run parameters (e.g., set the Ending Type to a fixed run length of 1000 hours).

Note that we'll discuss the details of *States*, *Properties*, *Tokens*, and other components of the Simio Framework in Chapter 5.

To test the model, create an Experiment by clicking on the New Experiment icon in the Project Home ribbon. Figure 4.18 shows the Pivot Grid results for a run of 10 replications of the model. Notice that the report includes the *User-Specified* category including the *CurrentWIP* and *TimeInSystem* statistics. Unlike the ModelEntity statistics *NumberInSystem* and *TimeInSystem* that Simio collected automatically in the Standard Library object model from Section 4.2, we explicitly told Simio to collect these statistics in the process model. Understanding user-specified statistics is important, as it's very likely that you'll want more than the default statistics as your models become larger and more complex. The *CurrentWIP* statistic is an example of a *time-dependent* statistic. Here, we defined a Simio state (step 4), used the process logic to update the value of the state when necessary (step 10 to increment and step 14 to decrement), and told Simio to keep track of the value as it evolves over simulated time and to report the summary statistics (of $\widehat{L}$, in this case — step 4). The *TimeInSystem* statistic is an example of an *observational* or *tally* statistic. In this case, each arriving entity contributes a single observation (the time that entity spends in the system) and Simio tracks and reports the summary statistics for these values ($\widehat{W}$, in this case). Step 7 sets up this statistic and step 15 records each observation.

Another thing to note about our processes model is that it runs significantly faster than the corresponding Standard Library objects model (to see this, simply increase the run length for both models and run them one after another). The speed difference is due to the overhead associated with the additional functionality provided by the Standard Library objects (such as automatic collection of statistics, animation, collision detection on paths, resource failures, etc.).

As mentioned above, most Simio models that you build will use the Standard Library objects and it's unlikely that you'll build complete models using only Simio processes. However, *processes* are fundamental to Simio and it is important to understand how they work. We'll revisit this topic in more detail in Section 5.1.4, but for now we'll return to our initial model using the Standard Library objects.

4.4 Model 4-3: Automated Teller Machine (ATM)

In developing our initial Simio models, we focused on an arbitrary queueing system with *entities* and *servers* — very boring. Our focus for this section and for Section 4.7 is to add some context to the models so that they'll more closely represent the types of "real" systems that simulation is used to analyze. We'll continue to enhance the models over the remaining chapters in this Part of the book as we continue to describe the general concepts of simulation modeling and explore the features of Simio. In Models 4-1 and 4-2 we used observations from the exponential distribution for entity inter-arrival and service times. We did this so that we could exploit the mathematical "niceness" of the resulting $M/M/1$ queueing model in order to demonstrate the basics of randomness in simulation. However, in many modeling situations, entity inter-arrivals and service times don't follow nice exponential distributions. Simio and most other simulation packages can sample from a wide variety of distributions to support general modeling. Models 4-3 and 4-4 will demonstrate the use of a triangular distribution for the service times, and the models in Chapter 5 will demonstrate the use of many of the other standard distributions. Section 6.1 discusses how to specify such input probability distributions in practice so that your simulation model will validly represent the reality you're modeling.

Model 4-3 models the automated teller machine (ATM) shown in Figure 4.19. Customers enter through the door marked Entrance, walk to the ATM, use the ATM, and walk to the door marked Exit and leave. For this model, we'll assume that the room containing the ATM is large enough to handle any number

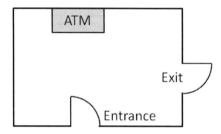

Figure 4.19: ATM example.

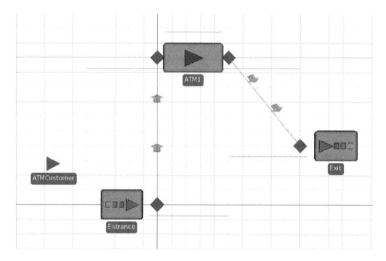

Figure 4.20: Model 4-3: ATM example.

of customers waiting to use the ATM (this will make our model a bit easier, but is certainly not required and we'll revisit the use of limited-capacity queues in future chapters). With this assumption, we basically have a single-server queueing model similar to the one shown in Figure 4.1. As such, we'll start with Model 4-1 and modify the model to get our ATM model (be sure to use the Save Project As option to save Model 4-3 initially so that you don't over-write your file for Model 4-1). The completed ATM model (Model 4-3) is shown in Figure 4.20. The required modifications are as follows:

1. Update the object names to reflect the new model context (ATMCustomer for entities, Entrance for the Source object, ATM1 for the Server object, and Exit for the Sink object);

2. Rearrange the model so that it "looks" like the figure;

3. Change the Connector and entity objects so that the model includes the customer walk time; and

4. Change the ATM processing-time distribution so that the ATM transaction times follow a triangular distribution with parameters (0.25, 1.00,

1.75) minutes (that is, between 0.25 and 1.75 minutes, with a mode of
1.00 minute).

Updating the object names doesn't affect the model's running characteristics
or performance, but naming the objects can greatly improve model readability
(especially for large or complicated models). As such, you should get into the
habit of naming objects and adding meaningful descriptions using the Description property. Renaming objects can be done by either selecting the object,
hitting the F2 key, and typing the new name; or by editing the Name property
for the object. Rearranging the model to make it look like the system being
modeled is very easy — Simio maintains the connections between objects as you
drag the object around the model. Note that in addition to moving objects,
you can also move the individual object input and output nodes.

In our initial queueing model (Model 4-1) we assumed that entities simply
"appeared" at the server upon arrival. The Simio Connector object supported
this type of entity transfer. This is clearly not the case in our ATM model
where customers walk from the entrance to the ATM and from the ATM to
the exit (most models of "real" systems involves some type of similar entity
movement). Fortunately, Simio provides several objects from the Standard
Library to facilitate modeling entity movements:

- Connector — Transfers entities between objects in zero simulation time
  (i.e., instantly, at infinite speed);

- Path — Transfers entities between objects using the distance between
  objects and entity speed to determine the movement time;

- TimePath — Transfers entities between objects using a user-specified
  movement-time expression; and

- Conveyor — Models physical conveyors.

We'll use each of these methods over the next few chapters, but we'll use Paths
for the ATM model (note that the *Simio Reference Guide*, available via the F1
key or the "?" icon in the upper right of the Simio window, provides detailed
explanations of all of these objects). Since we're modifying Model 4-1, the
objects are already connected using Connectors. The easiest way to change
a Connector to a Path is to right-click on the Connector and choose the Path
option from the Convert to Type sub-menu. This is all that's required to change
the connection type. Alternatively, we could delete the Connector object and
add the Path object manually by clicking on the Path in the Standard Library
and then selecting the starting and ending nodes for the Path.

The entity-movement time along a Path object is determined by the path
length and the entity speed. Simio models are drawn to scale by default so
when we added a path between two nodes, the length of the path was set
as the distance between the two nodes[5]. The Length property in the Physical

---

[5]Whenever you *input* lengths or other properties with units, the + will expand a field
where you can specify the input units. The Unit Settings button on the Run ribbon allows you
to change the units displayed on *output*, such as in the facility-window labels, the pivot-grid
numbers, and trace output.

Characteristics/Size group of the General section gives the current length of the Path object. The length of the path can also be estimated using the drawing grid. The logical path length can also be manually set if it's not convenient to draw the path to scale. To set the logical length manually, set the Drawn to Scale property to False and set the Logical Length property to the desired length. The entity speed is set through the Initial Desired Speed property in the Travel Logic section of the entity properties. In Model 4-3, the path length from the entrance to the ATM is 10 meters, the path length from the ATM to the exit is 7 meters, and the entity speed is 1 meter/second. With these values, an entity requires 10 seconds of simulated time to move from the entrance to the ATM, and 7 seconds to move from the ATM to the exit. The path lengths and entity speed can be easily modified as dictated by the system being modeled.

The final modification for our ATM model involves changing the processing-time distribution for the server object. The characteristics of the exponential distribution probably make it ill-suited for modeling the transaction or processing time at an ATM. Specifically, the exponential distribution is characterized by lots of relatively small values and a few extremely large values, since the mode of its density function is zero. Given that all customers must insert their ATM card, correctly enter their personal identification number (PIN), and select their transaction, and that the number of ATM transaction types is generally limited, a bounded distribution is likely a better choice. We'll use a triangular distribution with parameters 0.25, 1.00, and 1.75 minutes. Determining the appropriate distribution(s) to use is part of *input analysis*, which is covered in Section 6.1. For now, we'll assume that the given distributions are appropriate. To change the processing-time distribution, simply change the Processing Time property to *Random.Triangular*(0.25, 1, 1.75) and leave the Units property as Minutes. By using the *Random* keyword, we can sample statistically independent observations from some 19 common distributions (as of the writing of this book) along with the continuous and discrete empirical distributions for cases where none of the standard distributions provides an adequate fit. These distributions, their required parameters, and plots of their density or probability-mass functions are discussed in detail in the "Distributions" subsection of the "Expressions Editor, Functions and Distributions" section in the "Modeling in Simio" part of the *Simio Reference Guide*. The computational methods that Simio uses to generate random numbers and random variates are discussed in Sections 6.3 and —refsec:RVG.

Now that we have completed Model 4-3, we must *verify* our model as discussed in Section 4.2.5. As noted, the verification process involves developing a set of expectations, developing and running the model, and making sure that the model results match our expectations. When there's a mismatch between our expectations and the model results, we must find and fix the problems with the model, the expectations, or both. For Model 4-1, the process of developing expectations was fairly simple — we were modeling an $M/M/1$ queueing system so we could calculate the exact values for the performance metrics. The process isn't quite so simple for Model 4-3, as we no longer have exponential processing times, and we've added entity-transfer times between the arrival and service, and between the service and the departure. Moreover, these two modifications will tend to counteract each other in terms of the queueing metrics. More specif-

Table 4.6: Model 4-3 (modified version) results.

Metric being estimated	Simulation
Utilization (ρ)	0.800 ± 0.004
Number in system (L)	4.232 ± 0.146
Number in queue (L_q)	3.205 ± 0.143
Time in system (W)	0.088 ± 0.003
Time in queue (W_q)	0.066 ± 0.003

Table 4.7: Model 4-3 results.

Metric being estimated	Simulation
Utilization (ρ)	0.801 ± 0.003
Number in system (L)	2.833 ± 0.069
Number in queue (L_q)	1.805 ± 0.066
Time in system (W)	0.059 ± 0.001
Time in queue (W_q)	0.037 ± 0.001

ically, we've reduced the variation in the processing times, so we'd expect the numbers of entities in system and in the queue as well as the time in system to go down (relative to Model 4-1), but we've also added the entity-transfer times, so we'd expect the number of entities in the system and the time that entities spend in the system to go up. As such, we don't have a set of expectations that we can test. This will be the case quite often as we develop more complex models. Yet we're still faced with the need for model verification.

One strategy is to develop a *modified model* for which we can easily develop a set of expectations and to use this modified model during verification. This is the approach we'll take with Model 4-3. There are two natural choices for modifying our model: Set the entity transfer times to 0 and use an $M/G/1$ queueing approximation (as described in Chapter 2), or change the processing-time distribution to exponential. We chose the latter option and changed the Processing Time property for the ATM to $Random.Exponential(1)$. Since we're simply adding 17 seconds of transfer time to each entity, we'd expect ρ, L_q, and W_q to match the $M/M/1$ values (Table 4.2), and W to be 17 seconds greater than the corresponding $M/M/1$ value. The results for running 500 replications of our model with replication length 30 hours and warm-up of 20 hours (the same conditions as we used in Section 4.2.5) are given in Table 4.6. These results appear to match our expectations (we could run each replication longer or run additional replications if we were concerned with the minor deviations between the average values and our expectations, but we'll leave this to you). So if we assume that we have appropriately verified the modified model, the only way that Model 4-3 wouldn't be similarly verified is if either we mistyped the Processing Time property, or if Simio's implementation of either the random-number generator or Triangular random-variate generator doesn't generate valid values. At this point we'll make sure that we've entered the property correctly, and we'll assume that Simio's random-number and random-variate generators work. Table 4.7 gives the results of our experiment for Model 4-3 (500 replications, 30 hours run length, 20 hours warm-up). As expected, the number of

entities in the queue and the entities' time in system have both gone down (this
was expected since we've reduced the variation in the service time). It's worth
reiterating a point we made in Section 4.2.5: We can't (in general) *prove* that
a model is verified. Instead, we can only collect evidence until we're convinced
(possibly finding and fixing errors in the process).

4.5   Beyond Means: Simio MORE (SMORE) Plots

We've emphasized already in several places that the results from stochastic
simulations are themselves random, so need to be analyzed with proper statistical methods. So far, we've tended to focus on means — using averages from
the simulation to estimate the unknown population means (or expected values
of the random variables and distributions of the output responses of interest).
Perhaps the most useful way to do that is via confidence intervals, and we've
shown how Simio provides them in its Experiment Pivot Grids and Reports.
Means (of anything, not just simulation-output data) are important, but seldom do they tell the whole tale[6] since they are, by definition, the average of an
infinite number of replications of the random variable of interest, such as a simulation output response like average time in system, maximum queue length, or
a resource utilization, so don't tell you anything about spread or what values
are likely and unlikely. This is among the points made by Sam Savage in his
engagingly-titled book, *The Flaw of Averages: Why We Underestimate Risk in
the Face of Uncertainty* [50]. In the single-period static inventory simulation
of Section 3.2.3, and especially in Figure 3.3, we discussed histograms of the
results, in addition to means, to see what kind of upside and downside risk
there might be concerning profit, and in particular the risk that there might be
a loss rather than a profit. Neither of these can be addressed by averages or
means since, for example, if we order 5000 hats, the *mean* profit seemed likely
to be positive (95% confidence interval $8734.20 \pm $3442.36), yet there was a
30% *risk* of incurring a loss (negative profit).

So in addition to the Experiment Pivot Grid and Report, which contain
confidence-interval information to estimate means, Simio includes a new type
of chart for reporting output statistics. *Simio MORE (SMORE)* plots are a
combination of an enhanced *box plot*, first described by John Tukey in 1977
[58], a histogram, and a simple dot plot of the individual-replication summary
responses. SMORE plots are based on the *Measure of Risk and Error (MORE)*
plots developed by Barry Nelson in [42], and Figure 4.21 shows a schematic
defining some of their elements. A SMORE plot is a graphical representation of
the run results for a summary output performance measure (*response*), such as
average time in system, maximum number in queue, or a resource utilization,
across multiple replications. Similar to a box plot in its default configuration, it
displays the minimum and maximum observed values, the sample mean, sample
median, and "lower" and "upper" percentile values (points at or below which are
that percent of the summary responses across the replications). The "sample"
here is composed of the summary measures *across* replications, not observations
from *within* replications, so this is primarily intended for terminating simula-

---

[6]or, for that matter, tail(s)

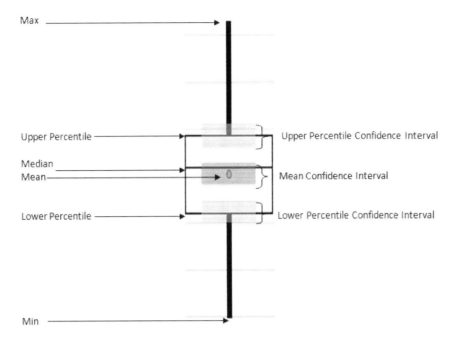

Figure 4.21: SMORE plot components (from the *Simio Reference Guide*).

tions that are replicated multiple times, or for steady-state simulations in which appropriate warm-up periods have been identified and the model is replicated with this warmup in effect for each replication. A SMORE plot can optionally display confidence intervals on the mean and both lower/upper percentile values, a histogram of observed values, and the responses from each individual replication.

SMORE plots are generated automatically based on experiment *Responses*. For Model 4-3, maybe we're interested in the average time, over a replication, that customers spend in the system (the time interval between when a customer arrives to the ATM and when that customer leaves the ATM). Simio tracks this statistic automatically and we can access within-replication average values as the expression ATMCustomer.Population.TimeInSystem.Average. To add this as a Response, click the Add Response icon in the Design ribbon (from the Experiment window) and specify the Name (AvgTimeInSystem) to label the output, and Expression (ATMCustomer.Population.TimeInSystem.Average) properties (see Figure 4.22). A reasonable question you might be asking yourself right about now is, "How do I know to type in ATMCustomer.Population. TimeInSystem.Average there for the Expression?" Good question, but Simio provides a lot of on-the-spot, smart, context-sensitive help.

When you click in the Expression field you'll see a down arrow on the right; clicking it brings up another field with a red X and green check mark on the right — this is Simio's *expression builder*, discussed more fully in Section 5.1.7, and shown in Figure 4.23. For now, just experiment with it, starting by clicking in the blank expression field and then tapping the down-arrow *key* on your

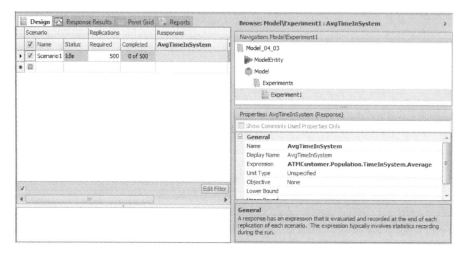

Figure 4.22: Defining the experiment Response for average time in system in Model 4-3.

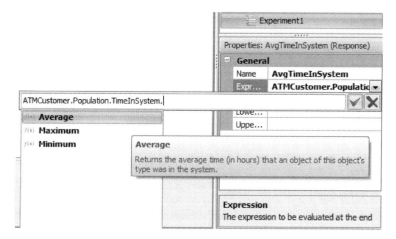

Figure 4.23: Using the Simio expression builder for the average-time-in-system Response for a SMORE plot.

keyboard to open up a menu of possible ways to get started, at the left edge of the expression. In the list that appears below, find ATMCustomer (since that's the name of the entities, and we want something about entities here, to wit, their time in system) and *double*-click on it; note that that gets copied up into the expression field. Next, type a period just to the right of ATMCustomer in the expression field, and notice that another list drops down with valid possibilities for what comes next. In this case we are looking for a statistic for the entire population of ATMCustomer entities, not just a particular entity, so double-click on Population. Again you are provided a list of choices; double-click on TimeInSystem at the bottom of the list (since that's what we want to know about our ATMCustomer entities). If at any point you lose your drop-down list, just type the down arrow again. As before, type a period on the right of the expression that's gradually getting built in the field, and double-click on Average in the list that appears (since we want the average time in system of ATMCustomer entities, rather than the maximum or minimum — though the latter two would be valid choices if you wanted to know about them rather than the average). That's the end of it, as you can verify by trying another period and down arrow but nothing happens, so click on the green check mark on the right to establish this as your expression.

You can add more Responses in this way, repeatedly clicking on the Add Response icon and filling in the Properties as above, and when viewing the SMORE plots you can rotate among them via a drop-down showing the Names you chose for them. Go ahead and add two more: the first with Name = MaxQueueLength and Expression = ATM1.AllocationQueue. MaximumNumberWaiting, and the second with Name = ResourceUtilization and Expression = ATM1.ResourceState.PercentTime(1). We invite you to poke through the expression builder to discover these, and in particular the (1) in the last one (note that as you hover over selections in the drop-downs from the expression builder, helpful notes pop up, as in Figure 4.23, describing what the entries are, including the (1)). The percents desired for the Lower and Upper Percentiles, and the Confidence Level for the confidence intervals can be set using the corresponding experiment properties (see Figure 4.24); by default, the lower and upper percentiles are set for 25% and 75% (i.e., the lower and upper quartiles) as in traditional box plots, though you may want to spread them out more than that since the "box" to which your eye is drawn contains the results from only the middle *half* of the replications in the default traditional settings, and maybe you'd like your eye to be drawn to something that represents more than just half (e.g., setting them at 10% and 90% would result in a box containing 80% of the replications' summary results).

To view your SMORE plots, select the Response Chart tab in the Experiment window. Figure 4.25 shows the SMORE plot for the average time in system from a 500-replication run of Model 4-3 described above (30-hour run length, 20-hour warmup on each replication), with the Confidence Intervals and Histogram showing, but not the individual replication-by-replication Observations; we left the percentiles at their defaults of 75% for upper and 25% for lower. The Rotate Plot button allows you to view the plot horizontally rather than vertically, if you prefer. The numerical values used to generate the SMORE plot, like the confidence-interval endpoints, are also available by

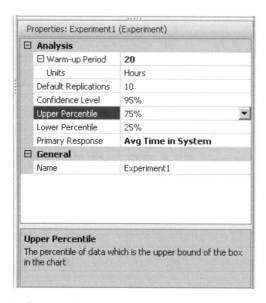

Figure 4.24: Setting the percentile and confidence interval levels.

clicking on the Raw Data tab at the bottom of the SMORE plot window, so you can see what they actually are rather than eyeballing them from the graph. We see from Figure 4.25 that the expected average time in system is just under 0.058 hour (3.5 minutes), and the median is a bit lower, consistent with the histogram shape's being skewed to the right. Further, the upper end of the box plot (75th percentile) is about 0.064 hour (3.8 minutes), so there's a 25% chance that the average time in system over a replication will be more than this. And the confidence-intervals seem reasonably tight, indicating that the 500 replications we made are enough to form reasonably precise conclusions.

Figure 4.26 shows the SMORE plots (with the same elements showing) for the maximum queue length; we temporarily moved the Upper Percentile from its default 75% up to 90%. Since this is for the *maximum* (not average) queue length across each replication, this tells us how much space we might need to hold this queue over a whole replication, and we see that if we provide space for 14 or 15 people in the ATM lobby, we'll always (remember, we're looking at the maximum queue length) have room to hold the queue in about 90% of the replications. Play around a bit with the Upper Percentile and Lower Percentile settings in the Experiment Design window; of course, as these percentiles move out toward the extreme edges of 0% and 100%, the edges of the box move out too, but what's interesting is that the Confidence Intervals on them become wider, i.e., less precise. This is because these more extreme percentiles are inherently more variable, being based on only the paucity of points out in the tails, and are thus more difficult to estimate, so the wider confidence intervals keep you honest about what you know (or, more to the point, what you don't know) concerning where the true underlying percentiles really are.

The distribution of the observed server utilization in Figure 4.27 shows that it will be between 77% and 83% about half the time, which agrees with the

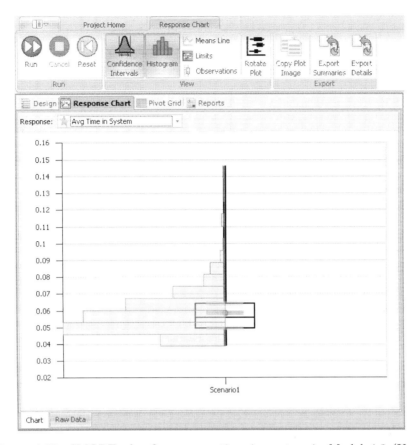

Figure 4.25: SMORE plot for average time in system in Model 4-3 (Upper Percentile = 75%, Lower Percentile = 25% for box boundaries).

queueing-theoretic expected utilization of

$$\frac{\text{E(service time)}}{\text{E(interarrival time)}} = \frac{(0.15 + 1.00 + 1.75)/3}{1.25} = 0.80$$

(the expected value of a triangular distribution with parameters *min, mode, max* is (*min + mode + max*)/3, as given in the *Simio Reference Guide*). However, there's a chance that the utilization could be as heavy as 90% since the histogram extends up that high (the maximum utilization across the 500 replications was 90.87%, as you can see in the Raw Data tab at the bottom).

As originally described in [42], SMORE plots provide an easy-to-interpret graphical representation of a system's *risk* and the sampling *error* associated with the simulation — far more information than just the average over the replications, or even a confidence interval around that average. The confidence intervals in a SMORE plot depict the sampling error in estimation of the percentiles and mean — we can reduce the width of the confidence intervals (and, hence, the sampling error) by increasing the number of replications. Visually, if the confidence-interval bands on the SMORE plot are too wide to suit you,

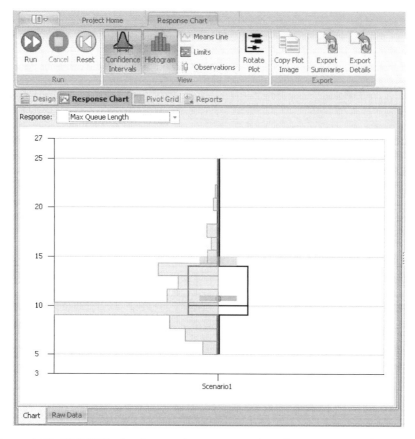

Figure 4.26: SMORE plot for maximum queue length in system in Model 4-3 (Upper Percentile = 90%, Lower Percentile = 25% for box boundaries).

then you need to run more replications. Once the confidence intervals are sufficiently narrow (i.e., we're comfortable with the sampling error), we can use the upper and lower percentile values to get a feeling for the variability associated with the response. We can also use the histogram to get an idea about the distribution shape of the response (e.g., in Figures 4.25-4.27 it's apparent that the distributions of average time in system and maximum queue length are skewed right or high, but the distribution of utilizations is fairly symmetric). As we'll see in Chapter 5, SMORE plots are quite useful to see what the differences might be across multiple alternative scenarios in an output performance measure, not only in terms of just their means, but also their relative spread and distribution.

4.6 Exporting Output Data for Further Analysis

Simio is a simulation package, not a statistical-analysis package. While it does provide some statistical capabilities, like the confidence intervals and SMORE plots that we've seen (and a few more that we'll see in future chapters), Simio

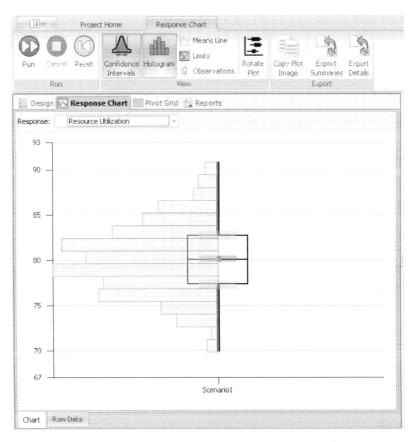

Figure 4.27: SMORE plot for server utilization in Model 4-3 (Upper Percentile = 75%, Lower Percentile = 25% for box boundaries).

makes it easy to export the results of your simulation to a CSV file that can then be easily read into a variety of dedicated statistical-analysis packages like SAS, JMP, SPSS, Stata, S-PLUS, or R, among many, many others, for post-processing your output data after your simulations have run. For relatively simple statistical analysis, the CSV file that you can ask Simio to export for you can be read directly into Excel and you could then use its built-in functions (like =AVERAGE, =STDEV, etc.) or the Data Analysis Toolbar that comes with Excel, or perhaps a better and more powerful third-party statistical-analysis add-in like StatTools from Palisade Corporation. With your responses from each replication exported in a convenient format like this, you're then free to use your favorite statistics package to do any analysis you'd like, such as hypothesis tests, analysis of variance, or regressions. Remember, from each replication you get a single summary value (e.g., average time in system, maximum queue length, or server utilization over the replication), not individual-entity results[7] from

---

[7]You can collect individual-entity observations manually with a Write step in add-on process logic or with simple customization of a standard library object as described in Section 10.4. Simio Version 6, entering Beta as this book went to press, introduces logs to collect

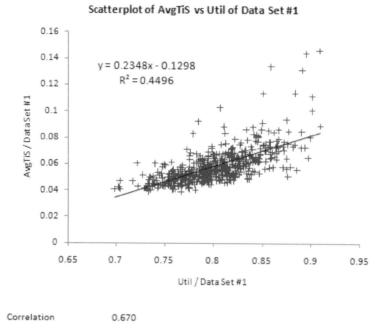

Figure 4.28: StatTools scatterplot of average time in system vs. server utilization in Model 4-3.

within replications, and those summary values are independent and identically distributed observations to which standard statistical methods will apply; your "sample size," in statistics parlance, is the number of replications you ran.

You've already seen how this export to a CSV file can be done, in Section 4.2.3, from the Experiment window via the Export Details icon in the Pivot Grid window. Depending on the size of your model, and the number of replications you made, this exported CSV file could be fairly large, and you may need to do some rearranging of it, or extracting from it, to get at the results you want. But once you have the numbers you want saved in this convenient format, you can do any sort of statistical analysis you'd like, including perhaps data mining to try to discover patterns and relationships from the output data themselves.

For example, in the 500 replications we used to make the SMORE plots in Section 4.5, we exported the data and extracted the 500 average times in system to one column in an Excel spreadsheet, and the 500 utilizations in a second column. Thus, in each of the 500 rows, the first-column value is the average time in system on that replication, and the second-column value is the server utilization *in that same replication*. We then used the StatTools statistical-analysis add-in to produce the scatterplot and correlation in Figure 4.28. We see that there's some tendency for average time in system to be higher when the server utilization is higher, but there are plenty of exceptions to this since there is a lot of random "noise" in these results. We also used the built-in Excel Chart

---

tally and state statistic observations and Dashboard Reports to display them.

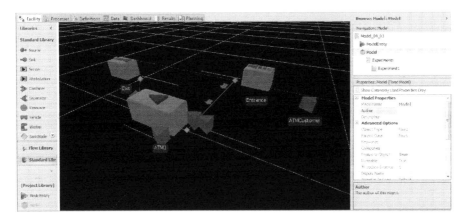

Figure 4.29: Model 4-3 in 3D view.

Tools to superpose a linear regression line, and show its equation and R^2 value, confirming the positive relationship between average time in system and server utilization, with about 45% of the variation in average time in system's being explained by variation in server utilization. This is just one very small example of what you can do by exporting the Simio results to a CSV file, extracting from it appropriately (maybe using custom macros or a scripting language), and then reading your simulation output data into powerful statistical packages to learn a lot about your models' behavior and relationships.

4.7   Basic Model Animation

Up until now we've barely mentioned animation, but we've already been creating useful animations. In this section we'll introduce a few highlights of animation. We'll cover animation in much greater detail in Chapter 8. Animation generally takes place in the Facility Window. If you click in the Facility Window and hit the "H" key, it will toggle on and off some tips about using the keyboard and mouse to move around the animation. You might want to leave this enabled as a reminder until you get familiar with the interface.

One of Simio's strengths is that when you build models using the Standard Library, you're building an animation as you build the model. The models in Figures 4.8 and 4.20 are displayed in two-dimensional (2D) animation mode. Another of Simio's strengths is that models are automatically created in 3D as well, even though the 2D view is commonly used during model building. To switch between 2D and 3D view modes, just tap the 2 and 3 keys on the keyboard, or select the View ribbon and click on the 2D or 3D options. Figure 4.29 shows Model 4-3 in 3D mode. In 3D mode, the mouse buttons can be used to pan, zoom, and rotate the 3D view. The model shown in Figure 4.29 shows one customer entity at the server (shown in the Processing.Contents queue attached to the ATM1 object), five customer entities waiting in line for the ATM (shown in the InputBuffer.Contents queue for the ATM1 server object), and two customers on the path from the entrance to the ATM queue.

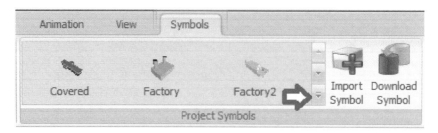

Figure 4.30: Navigating to Simio symbol library.

Let's enhance our animation by modifying Model 4-3 into what we'll call Model 4-4. Of course, you should start by saving `Model_04_03.spfx` to a file of a new name, say `Model_04_04.spfx`, and maybe to a different directory on your system so you won't overwrite our file of the same name that you might have downloaded; Simio's Save As capability is via the yellow pull-down tab just to the left of the Project Home tab on the ribbon.

If you click on any symbol or object, the Symbols ribbon will come to the front. This ribbon provides options to change the color or texture applied to the symbol, add additional symbols, and several ways to select a symbol to replace the default. We'll start with the easiest of these tasks — selecting a new symbol from Simio's built-in symbol library. In particular, let's change our entity picture from the default green triangle to a more realistic-looking person.

Start by clicking on the Entities object we named ATM Customer. In the Symbols ribbon now displayed, look for the Project Symbols category as illustrated in Figure 4.30. The wide section displays the top of the built-in library. If you click the expand button as indicated by the red arrow you'll see the entire library consisting of about 200 symbols organized into 13 categories. Scroll down to the "People" category. If you hover the mouse (don't click yet) over any symbol, you'll see an enlarged view to assist you with selection. Click on the symbol named "Woman1" to select it and apply it as the new symbol to use for your highlighted object — the entity. The entity in your model should now look like that in Figure 4.31. Note that under the People folder there is an folder named Animated that contains symbols of people with built-in animation like Walking, Running, and Talking. Using these provides even more realistic animation, but that is probably overkill for our little ATM model.

You may notice that Figure 4.31 also contains a symbol for an ATM machine. Unfortunately this was not one of the symbols in the library. If you happened to have an ATM symbol readily available you could import it. Or if you're particularly artistic you could draw one. But for the rest of us Simio provides a much easier solution — download it from Trimble 3D Warehouse[8]. This is a huge repository of symbols that are available free, and Simio provides a direct link to it. Let's change the picture of our server to that of an ATM machine.

Start by clicking on the Server object we'd previously named ATM1. Now go to the Symbol ribbon and click on the Download Symbol icon. Type **ATM** into the Search box and click on the Search button. You'll see the first 12 of several

---

[8]Formerly known as Google 3D Warehouse

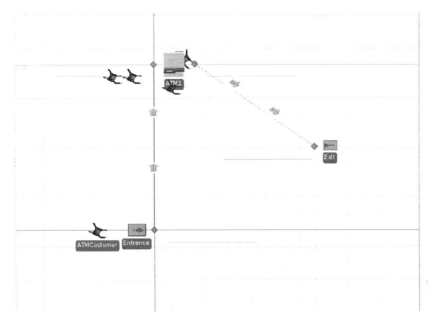

Figure 4.31: Model 4-4 with woman and ATM symbols in 2D view.

hundred symbols that have ATM in their name or description, something like Figure 4.32. Note that many of these don't involve automated teller machines, but many do, and in fact you should find at least one on the first screen that meets our needs. You can click on an interesting object (we chose one by Compo) and look at it in more detail and even rotate it to view in 3D at other angles. When you're satisfied with your selection, click on Download Model. After a brief delay for downloading, the symbol will be displayed in an editing window where you can change the name (to ATM1), add a description, and change the orientation. One of the most important things is to verify is that the size (in meters) is correct. You cannot change the ratio of the dimensions, but you can change any one value if it was sized wrong. In our case our ATM is about 1 meter wide and 1.5 meters high, which seems about right. Click OK and we're done applying a new symbol to our ATM server.

Now your new Model 4-4 should look something like to Figure 4.31 when running in the 2D view. If you change to the 3D view ("3" key) you should see that those new symbols you selected also look good in 3D, as in Figure 4.33.

Of course we've barely scratched the surface of animation possibilities. You could draw walls (see the Drawing ribbon), or add features like doorways and plants. You can even import a schematic or other background to make your model look even more realistic. Feel free to experiment with such things now if you wish, but we'll defer formal discussion of these topics until Chapter 8.

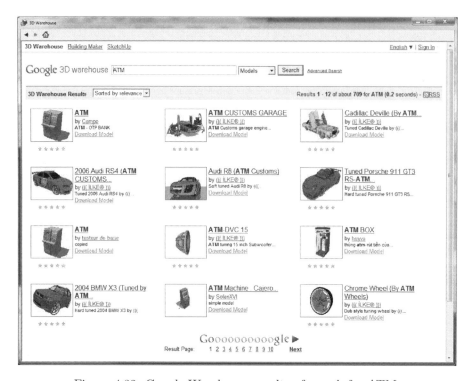

Figure 4.32: Google Warehouse results of search for ATM.

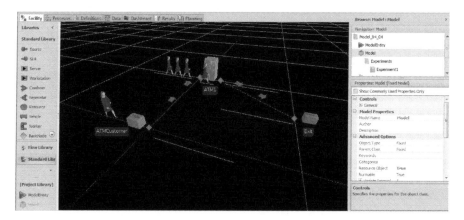

Figure 4.33: Model 4-4 with woman and ATM symbols in 3D view.

4.8 Model Debugging

As hard as it may be to believe, sometimes people make mistakes. When those mistakes occur in software they are often referred to as "bugs." Many things can cause a bug including a typo (typing a 2 when you meant to type a 3), a misunderstanding of how the system works, a misunderstanding of how the simulation software works, or a problem in the software. Even the most experienced simulationist will encounter bugs. In fact *a significant part of most modeling efforts is often spent resolving bugs* — it is a natural outcome of using complex software to model complex systems accurately. It is fairly certain that you will have at least a few bugs in the first real model that you do. How effectively you recognize and dispatch bugs can determine your effectiveness as a modeler. In this section we will give you some additional insight to improve your debugging abilities.

The best way to minimize the impact of bugs is to follow proper iterative development techniques (see Section 1.5.3). If you work for several hours without stopping to verify that your model is running correctly, you should expect a complex and hard-to-find bug. Instead, pause frequently to verify your model. When you find problems you will have a much better idea of what caused the problem and you will be able to find and fix it much more quickly.

The most common initial reaction to a bug is to assume it is a software bug. Although it is certainly true that most complex software, regardless of how well-written and well-tested it is, has bugs, it is also true that the vast majority of problems are user errors. Own the problem. Assume that it is your error until proven otherwise and you can immediately start down the path to fixing it.

4.8.1 Discovering Subtle Problems

How do you even know that you have a problem? Many problems are obvious — you press Run and either nothing happens or something dramatic happens. But the worst problems are the subtle ones — you have to work at it to discover if there even is a problem.

- In Section 4.2.2 we discussed the importance of developing expectations before running the model. Comparing the model results to our expectations is the first and best way to discover problems. In Section 4.2.5 we also discussed a basic model-verification process. The following steps extend that verification a bit deeper.

- Watch the animation. Are things moving where you think they should move? If not, why not?

- Enhance the animation to be more informative. Use floating labels and floor labels to add diagnostic information to entities and other objects.

- Examine the output statistics carefully. Are the results and the relationships between results reasonable? For example is it reasonable that you have a very large queue in front of a resource with low utilization? Add custom statistics to provide more information when needed.

- Finally, the same debugging tools described below to help resolve a problem can be used to determine if any problem even exists.

4.8.2   The Debugging Process

Okay, you are convinced that you have a bug. And you have taken ownership by assuming (for now) that the bug is due to some error that you have introduced. Good start. Now what? There are many different actions that you can try, depending on the problem.

- Look through all of your objects, especially the ones that you have added or changed most recently.

  - Look at all properties that *have* been changed from their defaults (in Simio these are all bold and their categories are all expanded). Ensure that you actually meant to change each of these and that you made the correct change.

  - Look at all properties that *have not* been changed from their defaults. Ensure that the default value is meaningful; often they are not.

- Minimize entity flow. Limit your model to just a single entity and see if you can reproduce the problem. If not, add a second entity. It is amazing how many problems can be reproduced and isolated with just one or two entities. A minimal number of entities help all of the other debugging processes and tools work better. In Simio, this is most easily done by setting Maximum Arrivals on each source to 0, 1, or 2.

- Minimize the model. Save a copy of your model, then start deleting model components that you think should have no impact. If you delete too much, simply undo, then delete something else. The smaller your model is, the easier it will be to find and solve the problem.

- If you encountered a warning or error, go back and look at it again carefully. Sometimes messages are somewhat obscure, but there is often valuable information embedded in there. Try to decode it.

- Follow your entity(ies) step by step. Understand exactly why they are doing what they are doing. If they are not going the way they should, did you accidentally misdirect them? Or perhaps not direct them at all? Examine the output results for more clues.

- Change your perspective. Try to look at the problem from a totally different direction. If you are looking at properties, start from the bottom instead of the top. If you are looking at objects, start with the one you would normally look at last. This technique often opens up new pathways of thought and vision and you might well see something you didn't see the first time[9].

---

[9]In banking, people often do verification by having one person read the digits in a number from left to right and a second person read those same digits from right to left. This breaks the pattern-recognition cycle that sometimes allows people to see what they expect or want to see rather than what is really there.

- Enlist a friend. If you have the luxury of an associate who is knowledgeable in modeling in your domain, he or she might be of great help solving your problem. But you can also get help from someone with no simulation or domain expertise — just explain aloud the process in detail to them. In fact, you can use this technique even if you are alone — explain it to your goldfish or your pet rock. While it may sound silly, it actually works. Explaining your problem *out loud* forces you to think about it from a different perspective and quite often lets you find and solve your own problem!

- RTFM - Read The (um, er) Friendly Manual. Okay, no one likes to read manuals. But sometimes if all else fails, it might be time to crack the textbook, reference guide, or interactive help and look up how something is really supposed to work.

You don't necessarily need to do the above steps in order. In fact you might get better results if you start at the bottom or skip around. But definitely use the debugging tools discussed below to facilitate this debugging process.

4.8.3 Debugging Tools

Although animation and numerical output provide a start for debugging, better simulation products provide a set of tools to help modelers understand what is happening in their models. The basic tools include Trace, Break, Watch, Step, and Profiler.

Trace provides a detailed description of what is happening as the model executes. It generally describes entity flow as well as the events and their side-effects that take place. Simio's trace is at the Process Step level — each Step in a process generates one or more trace statements. Until you learn about processes, Simio trace may seem hard to read, but once you understand Steps, you will begin to appreciate the rich detail made available to you. The Simio Trace can be filtered for easier use as well as exported to an external file for post-run analysis.

Break provides a way to pause the simulation at a predetermined point. The most basic capability is to pause at a specified time. Much more useful is the ability to pause when an entity reaches a specified point (like arrival to a server). More sophisticated capability allows conditional breaks such as for "the third entity that reaches point A" or "the first entity to arrive after time 125." Basic break functionality in Simio is found by right-clicking on an object or step. More sophisticated break behavior is available in Simio via the Break Window.

Watch provides a way to explore the system state in a model. Typically when a simulation is paused you can look at model and object-level states to get an improved understanding of how and why model decisions and actions are being taken and their side effects. In Simio, watch capability is found by right-clicking on any object. Simio watch provides access to the properties, states, functions, and other aspects of each object as well as the ability to "drill down" into the hierarchy of an object.

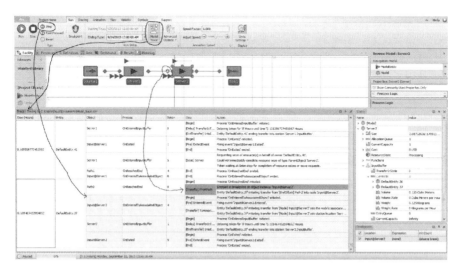

Figure 4.34: Using trace, watch, and break windows in custom layout.

Step allows you to control model execution by moving time forward by a small amount of activity called a step. This allows you to examine the actions more carefully and the side effects of each action. Simio provides two step modes. When you are viewing the facility view, the Step button moves the active entity forward to its next time advance. When you are viewing the process window the Step button moves the entity (token) forward one process step.

Profiler is useful when your problem is related to execution speed . It provides an internal analysis of what is consuming your execution speed. Identification of a particular step as processor intensive might indicate a model problem or an opportunity to improve execution speed by using a different modeling approach.

Trace, Break, Watch, and Step can all be used simultaneously for a very powerful debugging tool set. Combining these tools with the debugging process described above provides a good mechanism for better understanding your model and producing the best results.

Figure 4.34 illustrates these windows in a typical use. The black circle indicates the button used to display the Trace window and turn on the generation of model trace. You can see the trace from the running model until execution was automatically paused (a break) when the Break point set on the Server2 entry node (red circle) is reached. At that point the Step button (blue circle) was pushed and that resulted in an additional 11 lines of trace being generated as the entity moves forward until its next time advance (yellow background). The Watch window on the right side illustrates using a watch on Server2 to explore its input buffer and each entity in that buffer.

In the default arrangement, these debugging windows display as multiple tabs on the same window. You can drag and drop the individual windows to reproduce the window arrangement in Figure 4.34, or any window arrangement that meets your needs as discussed in Section 4.1.

4.9 Summary

In this chapter we've introduced Simio and developed several simple Simio models using the Standard Library and using Simio processes. Along the way, we integrated statistical analysis of simulation output, which is just as important as modeling in actual simulation projects, via topics like replications, run length, warmup, model verification, and the analysis capabilities made possible by the powerful SMORE plots. We started out with an abstract queueing model, and added some interesting context in order to model a somewhat realistic queueing system. In the process, we also discussed use of Simio *Paths* to model entity movement and basics of animation with Simio. All of these Simio and simulation-related topics will be covered in more detail in the subsequent chapters, with more interesting models.

4.10 Problems

1. Create a model similar to Model 4-1 except use an arrival rate, λ, of 120 entities per hour and a service rate, μ, of 190 entities per hour. Run your model for 100 hours and report the number of entities that were created, the number that completed service, and the average time entities spend in the system.

2. Develop a queueing model for the Simio model from Problem 1 and compute the exact values for the steady state time entities spend in the system and the expected number of entities processed in 100 hours.

3. Using the model from Problem 1, create an experiment that includes 100 replications. Run the experiment and observe the SMORE plot for the time entities spend in the system. Experiment with the various SMORE plot settings — viewing the histogram, rotating the plot, changing the upper and lower percentile values.

4. If you run the experiment from Problem 3 five (or any number of) times, you will always get the exact same results even though the interarrival and service times are supposed to be random. Why is this?

5. You develop a model of a system. As part of your verification, you also develop some expectation about the results that you should get. When you run the model, however, the results do not match your expectations. What are the three possible explanations for this mismatch?

6. In the context of simulation modeling, what is a *replication* and how, in general, do you determine how many replications to run for a given model?

7. What is the difference between a *steady-state* simulation and a *terminating* simulation?

8. What are the *initial transient period* and the *warm-up period* for a steady-state simulation?

9. Replicate the model from Problem 1 using Simio processes (i.e., not using objects from the Standard Library). Compare the run times for this model and the model from Problem 1 for 50 replications of length 100 hours.

10. Run the ATM model (Model 4-3) for 10 replications of length 240 hours (10 days). What are the maximum number of customers in the system and the maximum average number of customers in the system (recall that we mentioned that our model would not consider the physical space in the ATM). Was our assumption reasonable (that we did not need to consider the physical space, that is)?

11. Describe how SMORE plots give a quick view of a system's *risk* and the *sampling error* associated with a run of the model.

12. Animate your model from Problem 1 assuming that you are modeling a cashier at a fast food restaurant — the entities represent customers and the server represents the cashier at the cash register. Use Simio's standard symbols for your animation.

13. Modify your model from Problem 1 assuming that you are modeling a manufacturing process that involves drilling holes in a steel plate. The drilling machine has capacity for up to 3 parts at a time ($c = 3$ in queueing terms). The arrival rate should be 120 parts per hour and the processing rate should be 50 parts per hour. Use Google 3D Warehouse to find appropriate symbols for the entities (steel plates) and the server (a drill press or other hole-making device). Add a label to your animation to show how many parts are being processed as the model runs.

14. Build Simio models to confirm and cross-check the steady-state queueing-theoretic results for the four specific queueing models whose exact steady-state output performance metrics are given in Section 2.3. Remember that your Simio models are initialized empty and idle, and that they produce results that are subject to statistical variation, so design and run Simio Experiments to deal with both of these issues; make your own decisions about things like run length, number of replications, and Warm-up Period, possibly after some trial and error. In each case, first compute numerical values for the queueing-theoretic steady-state output performance metrics W_q, W, L_q, L, and ρ from the results in Section 2.3, and then compare these with your simulation estimates and confidence intervals. All time units are in minutes, and use minutes as well throughout your Simio models.

   a) $M/M/1$ queue with arrival rate $\lambda = 1$ per minute and service rate $\mu = 1/0.9$ per minute.

   b) $M/M/4$ queue with arrival rate $\lambda = 2.4$ per minute and service rate $\mu = 0.7$ per minute for each of the four individual servers (the same parameters used in the mmc.exe command-line program shown in Figure 2.2).

c) $M/G/1$ queue with arrival rate $\lambda = 1$ per minute and service-time distribution's being gamma(2.00, 0.45) (shape and scale parameters, respectively). You may need to do some investigation about properties of the gamma distribution, perhaps via some of the web links in Section 6.1.3.

d) $G/M/1$ queue with interarrival-time distribution's being continuous uniform between 1 and 5, and service rate $\mu = 0.4$ per minute (the same situation shown in Figure 2.3).

15. Build a Simio model to confirm and cross-check the steady-state queueing-theoretic results from your solutions to the $M/D/1$ queue of Problem 9 in Chapter 2. Remember that your Simio model is initialized empty and idle, and that it produces results that are subject to statistical variation, so design and run a Simio Experiment to deal with both of these issues; make your own decisions about things like run length, number of replications, and Warm-up Period, possibly after some trial and error. For each of the five steady-state queueing metrics, first compute numerical values for the queueing-theoretic steady-state output performance metrics W_q, W, L_q, L, and ρ from your solutions to Problem 9 in Chapter 2, and then compare these with your simulation estimates and confidence intervals. All time units are in minutes, and use minutes as well throughout your Simio model. Take the arrival rate to be $\lambda = 1$ per minute, and the service rate to be $\mu = 1/0.9$ per minute.

16. Repeat Problem 15, except use the $D/D/1$ queueing model from Problem 10 in Chapter 2.

17. In the processes-based model we developed in Section 4.3, we used the standard Token.TimeCreated token state to determine the time in system. Develop a similar model where you manually *mark* the arrival time (as illustrated in Figure 4.13 and used that value to record the time in system. Hint: You will need to create a custom token with a state variable to hold the value and use an Assign step to store the current simulation time when the token is created.

Chapter 5

Intermediate Modeling With Simio

The goal of this chapter is to build on the basic Simio modeling and analysis concepts presented in Chapter 4 so that we can start developing and experimenting with models of more realistic systems. We'll start by discussing a bit more about how Simio works and its general framework. Then we'll continue with the single-server queueing models developed in the previous chapter and will successively add additional features, including modeling multiple processes, conditional branching and merging, etc. As we develop these models, we'll continue to introduce and use new Simio features. We'll also resume our investigation of how to set up and analyze sound statistical simulation experiments, this time by considering the common goal of comparing multiple alternative scenarios. By the end of this chapter, you should have a good understanding of how to model and analyze systems of intermediate complexity with Simio.

5.1 Simio Framework

We've touched on quite a few Simio-specific concepts so far. In this section we'll provide additional detail on many of those concepts, as well as fill in some missing pieces that you'll need to know in coming sections. While parts of this section might seem a bit detailed (especially to the non-programmers in the audience), we'll hopefully connect the dots later in this chapter and through the remainder of the book.

5.1.1 Introduction to Objects

In the computer-programming world many professionals believe that *object-oriented programming* (OOP) is the de facto standard for modern software development — common languages such as C++, Java, and C# all support an object-oriented approach. Discrete-event simulation (DES) is moving down the same path. Many DES products are being developed using OOP but, more important to simulationists, a few also bring the benefits of true objects to the simulation model-building process. Simio (www.simio.com), Any-Logic (www.xjtek.com), Flexsim (www.flexsim.com), and Plant Simulation (www.plm.automation.siemens.com) are four popular DES products that provide a true OOP toolkit to modelers.

You might wonder why you should care. For many of the same reasons that OOP is revolutionizing the software-development industry, it is revolutionizing the simulation industry. The use of objects allows you to reduce large problems to smaller, more manageable problems. Objects help improve your models' reliability, robustness, reusability, extensibility, and maintainability. As a result, overall modeling flexibility and power are dramatically improved, and in some cases the modeling expertise required is lower.

In the rest of this section we will explore what Simio objects are and some important terminology. Then we will continue by discussing Simio object characteristics in some detail. While some of this may seem foreign if this is your first exposure to using objects, command of this material is important to understanding and effectively using the object-oriented paradigm.

What is an Object?

Simio employs an object approach to modeling, whereby models are built by combining objects that represent the physical components of the systems. An *object* is a self-contained modeling construct that defines that construct's characteristics, data, behavior, user interface, and animation. Objects are the most common constructs used to build models. You've already used objects each time you built a model using the Standard Library — the general-purpose set of objects that comes standard with Simio. Models 5-1, 5-3, and 5-4 demonstrated the use of the Source, Server, Sink, Connector, and Path objects from Simio's Standard Library.

An object has its own custom behavior that responds to *events* in the system as defined by its internal model. For example, a production-line model is built by placing objects that represent machines, conveyors, forklift trucks, and aisles, while a hospital might be modeled using objects that represent staff, patient rooms, beds, treatment devices, and operating rooms. In addition to building your models using the Standard Object Library objects, you can also build your *own* libraries of objects that are customized for specific application areas. And as you'll see shortly, you can modify and extend the Standard Library object behavior using process logic. The methods for building your own objects will be discussed in Chapter 10.

Object Terminology

So far we've used the word "object" somewhat casually. Let's add a little more precision to our object-related terminology, starting with the three tiers of Simio Object Hierarchy:

Object Definition An Object Definition defines how an object looks, behaves, and interacts with other objects. It consists of its *properties*, *states*, *events*, *external view*, and *logic*. An object definition may be part of your project or part of a library. Server and ModelEntity are examples of Object Definitions. To edit an object definition, you must first select that object

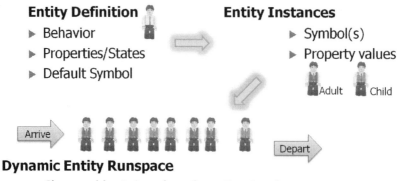

Figure 5.1: The three tiers of an entity object.

in the Navigation window, then all of the other model windows (e.g., the Processes and Definitions window) will correspond to that object.

Object Instance An Object Instance is created when you drag (or *instantiate*) an object into your model. An Object Instance includes the property values and may define one or more symbols, but refers back to the corresponding object definition for all the other aspects of the object definition. An Object Instance exists only in your model, but you can have many instances corresponding to a single Object Definition. `Server1`, `ATM1`, `DefaultEntity`, and `ATMCustomer` are all examples of object instances. To edit an object instance you must select that object instance in a facility view - then you can change its property values and animation.

Object Runspace An Object Runspace (also referred to as an *Object Realization*) is a third tier of objects that holds the current value of an object's states. Each has a unique ID and may reference a changeable symbol. Object Runspaces are created when you start a run, and in some cases can be dynamically created and destroyed during a run. All Object Instances are destroyed when a run ends. A state in the Runspace for a dynamic object (e.g., an entity) may be referenced using the terminology *InstanceName[InstanceID].StateName*. For example `LiftTruck[2].Priority` would reference the priority state of the second LiftTruck.

The three tiers of an Entity object are illustrated in Figure 5.1. We have a single entity definition, in this case a person. The entity instances are placed in the model, in this case an Adult person and a Child person. When the model is run, we might create many Adult and Child Runspaces that hold the states of the dynamic entities that are created.

We'd be remiss not to mention three other terms that are frequently used when we're discussing objects:

Model A Model is another name to describe an Object Definition. When you're building a model of your system or subsystem, you're building an object, and vice-versa. We tend to use the word Object when describing a component (such as the `Server` object), and use the word Model when referring to the highest-level object (such as the name `Model_05_03` on Model 5-3 from the previous chapter). But since even the highest-level object can later be used as a component in some other model, there's really little difference.

Project A Project is a collection of Object Definitions (or Models). Most often the objects in a project are related by being designed to work together or to work hierarchically. The file extension SPFX stands for Simio Project File. Because it's common to think of a project as "building a model," we often refer to "the model" as a short-hand way of referring what is technically "the project." Even our example projects are named in this fashion (e.g., Model 5-1) but we trust that you'll understand.

Library A Library is a collection of Object Definitions, or in fact, just another name for a Project. Any Project file can be loaded as a Library using the Load Library button on the Project Home ribbon.

These are key Simio concepts to understand for both using existing Simio objects as well as building your own custom objects. In the next several chapters, we'll be using existing Simio objects to construct models, and in Chapter 10 we'll cover the building of custom Simio objects. Technically, when we build a model, we're already developing a custom Simio object definition, but we'll refer to these objects as simply "models." If you're new to object-oriented approaches, the fact that a model is an object definition probably seems a bit strange, but hopefully once you gain some experience with Simio, it will begin to make sense.

Object Characteristics

As discussed above, an object is defined by its properties, states, events, external view, and logic. Let's discuss each in more detail.

Properties are input values that can be specified when you place (or instantiate) an object into your model. For example, an object representing a server might have a property that specifies the service time. When you place the server object into your facility model, you would also specify this property value. In Model 5-1, we used the Source object's Interarrival Time property to specify the interarrival-time distribution and the Server's Processing Time property to specify the server-processing-time distribution to form an M/M/1 queueing system.

States are dynamic values that may change as the model executes. For example, the busy and idle status of a server object could be maintained by a state variable named Status that's changed by the object each time it starts or ends service on a customer. In Model 5-2, we defined and used the model state `WIP` to track the number of entities in the system. When an entity arrives to the system, we increment the `WIP` state and when and entity departs, we decrement

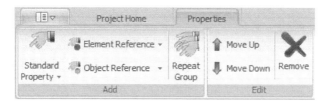

Figure 5.2: Properties ribbon on the Properties view of the Definitions window.

the `WIP` state. So, at any point in simulated time a specific state provides one component of the overall object's state. For example, in our single-server queueing system, the system state would be composed of the number of entities in the system (specified by the `WIP` model state) and the arrival time of each entity in the system (specified by the `Entity.TimeCreated` entity states).

Events are things that the object may "fire" at selected times. We first discussed the concept of dynamic-simulation events in Section 3.3.1. In our single-server queueing model, we defined entity arrival and service-completion events. In Simio, events are associated with objects. For example, a server object might have an event fire each time the server completes processing of a customer, or an oil-tank object might fire an event whenever it reaches full or empty. Events are useful for informing other objects that something important has happened and for incorporating custom logic.

The **External View** of an object is the 2D/3D graphical representation of the object. This is what you'll see when it's placed in your facility model. So, for example, when you use objects from the Standard Library, you're seeing the objects' external views when you add them to your model.

An object's **Logic** defines how the object responds to specific events that may occur. For example, a server object may specify what actions take place when a customer arrives to the server. This logic gives the object its unique behavior.

One of the powerful features of Simio is that whenever you build a model of a system, you can turn that model into an object definition by simply adding some input properties and an external view. Your model can then be placed as a sub-model within a higher-level model. Hence, hierarchical modeling is very natural in Simio. We'll expand on this in Chapter 10.

5.1.2   Properties and States

We briefly mentioned properties and states above, but because you'll likely be using them a great deal, they deserve some additional discussion.

Properties

Properties are defined within an object (see Figure 5.2) to collect information from the user[1] to customize that object's behavior. To create or view the properties for an object you must select the object in the Navigation window (*not*

---

[1] In this context the term *User* refers to the person who instantiates the object and completes its properties. In the case of Standard Library objects, *you* are the user of the

the Facility window), then select the Definitions tab and the Properties item. Properties specific to this object (if any) will appear at the top of this window; properties that are inherited from another object appear in an expandable area just below the other properties.

Properties can be of many different types including:

- Standard: Integer, Real, Expression, Boolean, DateTime, String, ...

- Element: Station, Network, Material, TallyStatistic, ...

- Object: Entity, Transporter, or other generic or specific object reference

- Repeat Group: A repeating set of any of the above

Tip: When you create a new property it will show up in the object's Property Window. Sometimes a new property will appear to get "lost." Make sure that you're looking in the properties window of the same object to which you added the property (e.g., if you add a property to the ModelEntity object, don't expect to see it appear on the Model object or a Server object). And if you haven't given the property a category, by default it will show up in the General category.

Properties can't change *definition* during a run. For example, if the user defined the value of 5.3 for a ProcessTime property, it will always return 5.3 and it can't be changed without stopping the run. However, some property definitions can contain random variables and dynamic states that could return different values even though the definition remains constant. For example, in Model 5-3, we used *Random.Triangular*(0.25, 1, 1.75) as the Process Time property for the Server object. In this case, the property value itself doesn't change throughout the model run (it's always the triangular distribution with parameters 0.25, 1, and 1.75), but an individual *sample* from the specified distribution is returned from each entity visit to the object. Similarly, we could set a property definition using the name of a state that changes over the model run so that each time the property is referenced, it returns the current value of the state.

The value of a property is unique to each object instance (e.g., the copy of the object that's placed into the facility window). So for example, you could place three instances of ModelEntity (let's call them PartA, PartB, and PartC) in your model by dragging three copies of the ModelEntity object from the Project Library onto the Facility view. Each of these will have a property named *Initial Desired Speed* but each one could have a different value of that property (e.g., The Initial Desired Speed of PartA doesn't necessarily equal that of PartB). However because properties can't change their values during a run, those property values are efficiently stored once, with the instance, regardless of how many dynamic entities (Runspaces) are created as the model runs.

States

States are defined within an object (see Figure 5.3) to hold the value of some-

---

objects created by Simio LLC employees. But you will learn in Chapter 10 that you can create your own objects and that the *user* of those objects may be someone other than yourself.

Figure 5.3: States ribbon on the States view of the Definitions window.

thing that might change while a model is running[2]. There are two categories of states: Discrete and Continuous. A **Discrete State** (or discrete-change state) may change via an assignment only at discrete points in time (e.g., it may change at time 1.2457, then change again at time 100.2). A **Continuous State** (or continuous-change state) will change continuously and automatically when it's rate or acceleration is non-zero.

States can be one of several types, including:

- Real: A discrete state that may take on any numeric value.

- Integer: A discrete state that may take on only integer (whole) numbers.

- Boolean: A discrete state that may take on only a value of *True* (1) or *False* (0).

- DateTime: A discrete state that may be assigned a value in a DateTime format.

- List: An integer value corresponding to one of several entries in a string list (discussed in Section 7.5), often used with statistics and animation.

- String: A discrete state that may be assigned a string value (e.g., "Hello") and may be manipulated using the *String Expression* functions such as String.Length.

- Element Reference: A discrete state that may be assigned an element reference value (e.g., indicates an element such as a TallyStatistic or a Material).

- Object Reference: A discrete state that may be assigned an object reference value (e.g., refers to either a generic Object or a specific type of object like an Entity).

- Level: A continuous state that has both a real value (the level) and a rate of change. The state will change continuously when the rate is non-zero. The rate may change only discretely.

- Level with Acceleration: A continuous state that has a real value (the level), a rate of change, and an acceleration. The state may change continuously over time based on the value of a rate of change and acceleration.

---

[2]Since state values can change, they are often referred to as "state variables" and you should consider the terms "state" and "state variables" synonymous in this context.

The rate may change continuously when the acceleration is non-zero. Both the rate and the acceleration may change discretely.

States can be associated with any object. A state on an entity is often thought of as an *attribute* of that entity. A state on a model can in some ways be thought of as a global variable on that model (e.g., the model state WIP that we used in Model 5-2). But you can have states on other objects to accomplish tasks like recording throughput, costs, or current contents.

States do *not* appear in the Properties Window, although a Property that's used to designate the initial value of a State often *does* appear in the Properties Window. To create or view the states for an object you must select the object in the Navigation window (*not* the Facility window). For example to add a state to the ModelEntity object, first click on ModelEntity in the navigation window. Once you have the proper object selected, you can select the Definitions tab and the States item. States specific to this object (if any) will appear at the bottom of this window; states that are inherited from another object appear in an expandable area at the top of this window.

The value of a state is unique to each object runspace (i.e. the copies of the object that are dynamically generated during a run) but *not* unique to the object instance. So for example, an instance of ModelEntity placed in your model (let's call it PartA) has a property named Initial Desired Speed that is common to all entities of type PartA. But all instances of ModelEntity also have a state named Speed. Because these states can be assigned as they move through the model, each entity created (the dynamic runspace) could have a different value for the state Speed. Hence the value of the Speed state must be stored with the dynamic entity runspaces. Because of the extra memory involved with a state, you should generally use a Property instead of a State unless you need the ability to change values during a run.

Editing Object Properties

Whenever you select an object in the Facility window (by clicking on it), the properties of the selected object are displayed for editing in the Properties window in the Browse panel on the right. For example, if you select a Server object, the properties for the Server will be displayed in the Properties window. The properties are organized into categories that can be expanded and collapsed. In the case of the Server in the Standard Library, the Process Logic category is initially expanded, and all others are initially collapsed. Clicking on +/- will expand and collapse a category. Whenever you select a property, a description of it appears at the bottom of the grid. As discussed above, the properties may be different types such as strings, numbers, selections from a list, and expressions; these are often reflected in the property. For example, the Ranking Rule for the Server is selected from a drop-down list and the Processing Time is specified as an expression. If you type an invalid entry into a property field, the field turns a salmon color and the error window will automatically open at the bottom. If you double-click on the error in the error window it will then automatically take you to the property field where the errors exist. Once you correct the error, the error window will automatically close.

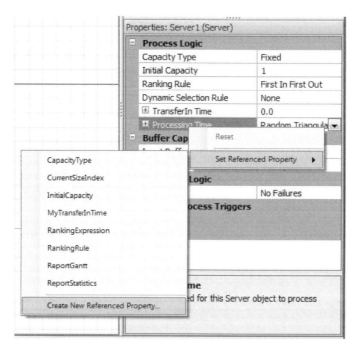

Figure 5.4: Defining a new referenced property.

Like much of the OOP world, Simio uses a "dot" notation for addressing an object's data. The general form is "xxx.yyy" where yyy is a component of xxx. This is sometimes repeated for several levels when you have sub-components and even sub-sub-components. For example `Server1.Capacity.Allocated.Average` provides the average allocated capacity for `Server1`. When you use the Property window this is mostly hidden from you, but you'll sometimes see this in a drop-down list and you'll see it as you use the expression builder (briefly discussed earlier in Section 4.5, and more fully below in Section 5.1.7) to construct and edit expressions. As you become more experienced, you may choose to type the dot notation directly rather than using the expression builder.

Referenced Properties

Referenced properties are a special application of properties that are generally used in the definition of other properties. For example, you could define a referenced property named `PaintingTime` and then use `PaintingTime` as the definition of the *ProcessingTime* property in a Server object. You can use any existing property in this way, or you can create a new referenced property by right clicking on a property as shown in Figure 5.4.

The are three common reasons to use referenced properties:

- When others are using your model, you might want to make some key parameters (referred to as *Controls*) easy to find and change. All referenced properties are displayed under a Controls category in the model Properties window (right click on Model in the Navigation Window, then

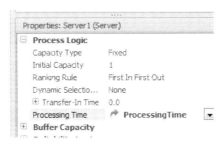

Figure 5.5: Example of using a referenced property.

select Properties). You can customize how they're displayed by changing Default Value, Display Name, Category Name and other aspects in the Definitions - Properties window.

- Referenced properties provide an easy way to share properties (e.g., you can specify the value once in a property and use (or reference) it in multiple places). This simplifies experimentation with different property values.

- When you do model experimentation from the Experiment window, referenced properties automatically[3] show up as experiment *controls*. These controls are the items that you may want to vary to define multiple alternative scenarios, such as altering server capacities, which we'll do in Section 5.5.

Referenced properties are displayed with a green arrow marker when used as property values (see Figure 5.5). When running a model interactively, the values of the controls are set in the model Properties window. When running experiments the values specified for controls in the model properties window are ignored and instead the values that are associated with each scenario in the Experiment Design window are used. You'll see examples of using referenced properties in the models in the current and subsequent chapters.

Properties and States Summary

Table 5.1 summarizes some of the differences between properties and states. Note that new users are very often confused about properties and states and, in particular, when one is used rather than the other. Hopefully we can clear up the natural confusion as we develop models with more and more features/components and discuss the use of their properties and states.

5.1.3   Tokens and Entities

Many simulation packages have the concept of an Entity — generally the physical "things" like parts and people that move around in a system. But something else is often required to do lower-level logic where an entity may not be involved

---

[3]You can prevent a property from being displayed in the Experiment by editing the property definition and changing its `Visible` value to `False`.

Table 5.1: Key differences between Properties and States.

	Properties	States
Basic Data Types	17	8
Runtime Change	No	Yes
Where Stored	Object instance	Object runspace
Server Example	Processing time	Number processed
Entity Example	Initial speed	Current speed
Cost Example	Cost per hour	Accrued cost
Failure Example	Failure rate	Last failure time
Batching Example	Desired batch size	Current batch size

(like system control logic) or where an entity may actually be doing multiple actions at one time (like waiting for an event during a processing delay). Simio provides Tokens for this extra flexibility.

Tokens

A *Token* is a delegate of an object that executes the steps in a process. Typically, our use of a Token is as a delegate of an Entity, but in fact *any* object that executes a process will do so using a token as a delegate. A token is created at the beginning of a process and is destroyed at the end of that same process. As the token moves through the process it executes the actions as specified by each step. A single process may have many active tokens moving through it in parallel. And a single object can have many tokens representing it.

A token may also carry its own properties and states. For example, a token might have a state that tracks the number of times the token has passed through a specific point in the logic. In most cases you can simply use the default token in Simio, but if you require a token with its own properties and states you can create one in the Token panel within the Definitions window.

A token carries a reference to both its parent object and its associated object (Figure 5.6). The *parent object* is an instance of the object in which the process is defined. For example a process defined inside the Server object definition would indicate a server (perhaps `Server1`) as its parent. The *associated object* is the related object (separate from the parent object) that triggered this process to execute. For example, a process that's triggered by an entity arriving to an object will have that entity as the associated object. The process logic can refer to properties, states, and functions of both the parent object and the associated object. Hence, the token could reference the arriving entity to specify the delay time on the Delay step. To reference the associated object it's necessary to precede the property or state name with the object-class name. For example `ModelEntity.TimeCreated` would return the value of the TimeCreated function for the associated object of type ModelEntity.

Entities

In Simio, Entities are part of an object model and can have their own intelligent behavior. They can make decisions, reject requests, decide to take a rest, etc.

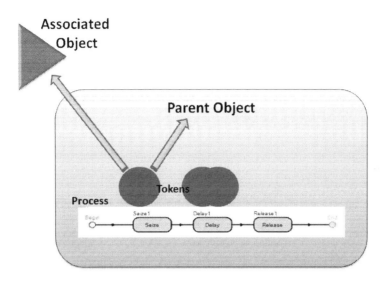

Figure 5.6: Relationships between Tokens, Objects, and Processes.

Entities have object definitions just like the other objects in the model. Entity objects can be dynamically created and destroyed, move across a network of links and nodes, move through 3D space, and move into and out of fixed objects.

Examples of entity objects include customers, patients, parts, and work pieces. Entities don't flow through processes as tokens do. Instead a token is created as a delegate of the entity to execute the process. The movement of entities into and out of objects may trigger an event, which might execute a process. When the process is executed, a Token is created that flows through the steps of a process.

Entities have a physical location within the Facility window and they can reside in either Free Space, in a Station, on a Link, or at a Node (discussed further in Chapter 8). When talking about the location type of an Entity, we're referring to the location of the Entity's leading edge. Examples of Stations within Simio are: the parking station of a Node (`TransferNode1.ParkingStation`), the input buffer of a Server object (`Server1.InputBuffer`), and the processing station of a Workstation object (`Workstation1.Processing`).

5.1.4 Processes

In Section 4.3 we built an entire model using nothing but Simio Processes. That was an extreme case for using processes. In fact, every model uses processes because the logic for all objects is specified using processes. We'll explore that more in Chapter 10. But there's yet another common use for processes, Add-on Processes, that we'll explore here.

A *process* is a set of actions that take place over (simulated) time, which may change the state of the system. In Simio a process is defined as a flowchart using *steps* that are executed by tokens and may change the state of one or more elements. Referring again to Figure 5.6, you'll note an object (Parent Object) that contains a process consisting of three steps. The arrows on the first blue token indicate that it has references to both its parent object (e.g., a server) and its associated object (e.g., an entity that initiated this process execution). Note that the other blue tokens in the figure also have similar references.

Processes can be triggered in several ways:

- *Simio-defined processes* are automatically executed by the Simio engine. For example, the OnInitialized process is executed by Simio for each object on initialization.

- *Event-triggered processes* are user-defined processes that are triggered by an event that fires within the model.

- *Add-on processes* are incorporated into an object definition to allow the user of that object to insert customized logic into the "standard" behavior of the object.

For now, let's just discuss add-on processes a bit more.

Add-on Processes

While object-based tools are well-known to provide ease of use, they generally have a major disadvantage. If an object isn't built to match your application perfectly (and they seldom are), then you either ignore the discrepancy (risking invalid or unusable results), or you change the object or build your own (either of which can be very difficult in some software). To eliminate that problem, Simio introduced the concept of *add-on processes*. Add-on processes allow you to supplement the standard object logic with additional behavior. Add-on process *triggers* are part of every standard library object. They exist to provide a high degree of flexibility to modelers.

An add-on process is similar to other processes but is triggered by another object. For example, a Server may have an add-on process that is triggered when an Entity that enters the Server. The object (the Server, in this case) provides a set of Add-on Process Triggers where you can enter the name of a process you want to execute. These processes are often quite simple, but provide incredible flexibility. For example, the Simio Server supports several types of failures, but none of them accounts for the availability of an external resource (e.g., an electrician) before the repair can be started. You could ignore that resource constraint (probably unwise), or create your own object (perhaps tedious), or simply add two processes with one step each to implement this capability. Server has a Repairing add-on process trigger where you'd add a process with a Seize step to obtain an electrician before starting the repair time. Server also has a Repaired add-on process trigger where you'd add a process with a Release step to free the electrician after the repair is complete.

It's actually quite easy to create these processes. If a process is to be shared or referenced from several different objects, you can go to the Processes tab and create the processes, then complete the logic. But if you're using a process only once, there's an even-easier way. We'll illustrate this by continuing the above example.

Create a new model and place a Source, Server (`Server1`), and Sink, connected by paths and add a Worker (`Worker1`) in it. For Server1 change the Failure type to `Calendar Time Based`, set the Uptime Between Failures to 10 `Minutes` and set the Time to Repair to 1 `Minute`. Leave all other properties on all objects at their defaults. At this point we have a simple model using the Standard Library objects that incorporates failures. Now let's customize the Server behavior to make it use the Worker during repairs.

1. In the `Server1` properties under Add-on Process Triggers, double click Failed. Simio will automatically create the add-on process, name it appropriately (Server1_Failed), categorize it with similar processes (Server1 Add-on Processes category), and then leave you in the process window ready to edit the process.

2. Drag a Seize step into the process.

3. Complete the Seize step properties: Double click on the Rows property and then click the Add button to get the resource dialog. Select `Worker1` for the object name to seize. Then click close.

4. Back in the Facility window in the `Server1` properties, under Add-on Process Triggers, double-click Repaired. Again note all the automatic behavior that will leave you in the Process window.

5. Drag a Release step into the Server1_Repaired process.

6. Complete the Release step properties: Double click on the Rows property and then click the Add button to get the resource dialog. Select `Worker1` for the object name to release. Then click close.

Your model should look something like Figure 5.7. If you run the model and look at the interactive results you should see the server failed for 10% of the time and likewise the Worker utilized for a corresponding 10% of the time. Of course, you could make this model more interesting by having the worker do other things when not busy. And a bit later we'll learn how to move the worker between tasks and consider the worker's current location in the processing. But for now, you've hopefully learned that using add-on processes can provide a lot of flexibility and isn't all that difficult to do.

5.1.5 Objects as Resources

In earlier sections we discussed the concept of a *resource* that represents a constraint in the system. In a previous model (Section 4.3), we specifically used the Resource object from the Standard Library. That Resource object was provided in the library for your convenience, but you're not limited to that particular object. In fact, *any Simio object can be used as a resource* (Note that the object builder may specifically disable that feature by setting the property Resource Object in the object definition to "False" or hiding some or all of the resource-related properties.).

Being able to be used as a resource includes the capability to:

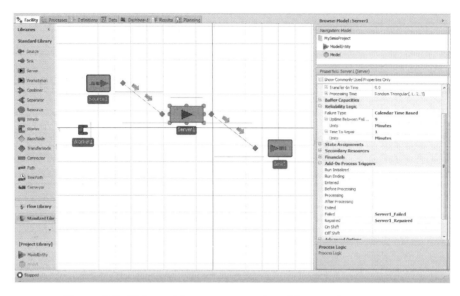

Figure 5.7: Using add-on processes to add a repair resource.

- Have a capacity constraint.

- Have a queue where tokens wait for capacity to become available.

- Have automatic statistics on scheduled and allocated capacity.

- Intelligently interact with the object attempting to seize it.

Let's explore each of these in a bit more detail.

Resources have a capacity constraint. The capacity can be 0 (no availability), Infinity (no constraint), or any integer in between. The capacity may also follow a schedule (see Sections 5.3.3 and 7.2). Calendar-based schedules may be defined by selecting the Schedules window under the Data Tab. Here you can configure recurring (such as daily and weekly) schedules and also configure Exceptions to those schedules, such as for holidays or scheduled overtime.

For ultimate flexibility the capacity of a resource can be controlled using process logic and assign steps. This approach allows you to adjust capacity dynamically in reaction to model situations and also to account for any desired transition rules (e.g., what happens to a busy resource when it goes off-shift).

Resources have a queue where tokens wait for capacity to become available. Entities don't directly take resource capacity. Rather, an entity has a token execute a Seize step inside a process to request that a resource allocate capacity to the entity. All tokens wait in a single allocation queue for a specific resource. If the requested resource isn't available, the token (and associated entity) must wait. An entity that wants to seize more than one resource, or one of a list of resources, may generate a token waiting in the allocation queue for each resource involved.

The ranking rule for allocation queues may be specified with either static or dynamic ranking. Static ranking rules include First In First Out, Last In First

Out, Smallest Value First, and Largest Value First. The latter two rules require a Ranking Expression that's often a state like Entity.Priority. This expression is evaluated when the queue is entered to determine proper rank. While static ranking is very efficient[4], dynamic ranking allows superior flexibility when it's necessary to make a real-time determination of the "best" allocation strategy. Dynamic Ranking examines every waiting entity each time an allocation is attempted. In addition to Smallest Value First and Largest Value First, it includes more sophisticated rules such as Campaign[5] Up, Campaign Down, and Campaign Cycle. Many rules such as Critical Ratio can be implemented by simply using the Ranking Expression and Filter Expression dynamic selection properties, but advanced users also have the capability to create their own dynamic allocation rules (see Chapter 10).

Resources have automatic statistics on scheduled and allocated capacity. These may also be supplemented with state statistics (e.g., Idle, Busy, Blocked, Failed, ...). Every resource automatically tracks the average, minimum, and maximum number of capacity units scheduled and capacity units allocated. Many of the standard objects take advantage of the List State feature to track additional state-based detail. For example, the Resource Object tracks the following states:

Idle The resource isn't allocated to any task.

Busy The resource has capacity allocated to one or more tasks.

Failed The resource is failed and not allocated to any tasks.

OffShift The resource is off shift and not allocated to any tasks.

FailedBusy The resource is failed but allocated to one or more tasks and still assumed "Busy" though failed.

OffShiftBusy The resource is off shift but allocated to one or more tasks and still assumed "Busy" though off-shift.

In addition to output statistics, you can also display pie charts of these data in your model or dashboard.

A resource may intelligently interact with the objects attempting to seize it. This feature provides a great deal of modeling latitude. Since objects may have autonomous behavior, this allows the concept of "smart" resources. In Simio the process of seizing a resource is a negotiation between two objects — both objects must "agree" before the seize occurs. A resource may choose if, and when, it will allow itself to be seized. For example, a resource may prevent itself from being seized close to the end of a shift. Or maybe the resource will choose to remain idle because it's aware that a higher-priority request will arriving soon. Or it may only consider requests that are in the proximity of its current location.

---

[4]Because dynamic ranking rules must examine every entity each time a decision is made, when the queues are very large it might cause slower execution speed.

[5]*Campaigns* are often used in industries like painting and rolling mills where the exact sequence of processing is very important, like painting successively darker colors or rolling successively narrower strips of steel.

As you can see, significant intelligence can be built into a resource. Intelligent control of allocations is accomplished by adding logic to the OnEvaluatingAllocation standard process. Some Standard Library objects also have an Evaluating Seize Request add-on process trigger. You'll learn more about how to take advantage of these in Chapter 10.

5.1.6 Data Scope

The definition of an object can't know anything about its encompassing object or objects parallel to it in the encompassing object because it can't know at the time of definition how it will be used, or what object in which it will be placed. For example, if you have a state named DayOfWeek defined on the model, the objects in your model can't directly reference DayOfWeek. This can be worked around by passing in that item via a property to permit an indirect reference. In fact, that's exactly what most of the properties in the Standard Library object are for.

Only the information that an object chooses to make public is available outside that object. For example, if you have an object that's designed to track the calendar and it contains a state named DayOfYear, if DayOfYear is defined as public, then the model and other objects would be able to reference it. If DayOfYear isn't public then only the calendar object would be able to reference its value.

Add-on processes are at the scope of the encompassing object. This means that Steps in an add-on process can directly reference information in the encompassing model (e.g., the main model), as well as the parent object and the associated object. This also means that data-scope issues are somewhat rare until you get into building your own objects in Chapter 10.

5.1.7 Expression Builder

For expression fields (e.g., Processing Time), Simio provides an *expression builder* to simplify the process of entering complex expressions; we guided you through a few examples of its use just briefly in Section 4.5, but provide more detail here. When you click in an expression field a small button with a down arrow appears on the right. Clicking on this button opens the expression builder, with a red X and green check mark on the right. The expression builder is very similar to IntelliSense in the Microsoft family of products and tries to find matching names or keywords as you type using the dot notation discussed in Section 5.1.2. Many items in the expression list have required or optional sub-items. Items that have sub-items are identified with a >> at the right edge, indicating that more choices are available. An item that is *not* in bold indicates that it's not yet a valid choice — you can't stop there and have a valid expression, you must select a sub-item. If the item itself is in bold, that indicates that it *is* a valid choice — it either has no sub-items or it has only optional sub-items. For example in Figure 5.8 you'll see that Agent, DefaultEntity, and Entity are not valid choices, indicated by not being bolded — you'd have to select a sub-item to use one of these. The other five items, all in bold, are valid choices. Of those, the first two (AllocationQueue and Capacity) have

Figure 5.8: Excerpt of expression builder choices.

optional sub-items as indicated by the arrows. The other three bold items have no sub-items.

You can use math operators $+, -, *, /$ and $\hat{}$ to form expressions, and parentheses as needed to control the order of calculation, like $2*(3+4)\hat{}(4/3)$. Logical expressions can be built using $<, <=, >, >=, !=, ==, \&\&, ||$, and $!$. Logical expressions return a numerical value of 1 when true, and 0 when false. For example $10*(A > 2)$ returns a value of 10 if A is greater than 2; otherwise it returns a value of 0. Arrays (up to 10 dimensions) are 1-based (i.e., the beginning subscript is 1, not 0) and are indexed with square brackets (e.g., B[2,3,1] indexes into the three dimensional state array named B). Model properties can be referenced by property name. For example if Time1 and Time2 are properties that you've added to your model then you can enter an expression such as (Time1 + Time2)/2.

A common use of the expression builder is to enter probability distributions. These are specified in Simio in the format Random.DistributionName(Parameter1, Parameter2, ..., ParameterN), where the number and meaning of the parameters are distribution-dependent. For example Random.Uniform(2,4) will return a continuous uniform random sample between 2 and 4. To enter this in the expression builder begin typing "Random." As you type, the drop list will jump to the word Random. Typing a period will then complete that word and display all possible distribution names. Typing a "U" will then jump to Uniform(min, max). Pressing enter or tab will then add this to the expression. You can then type in the numerical values to replace the parameter names min and max. Note that highlighting a distribution name in the list will automatically bring up a description of that distribution. Although it is not listed in the expression-builder argument lists, all distributions have an optional final parameter that is the random-number stream. Most people will leave this defaulted, but in some cases you may want to specify the stream number to use with a particular distribution.

Math functions such as Sin, Cos, Log, etc., are accessed in a similar way using the keyword Math. Begin typing "Math" and enter a period to complete the word and provide a list of all available math functions. Highlighting a math function in the list will automatically bring up a description of the function.

Object functions may also be referenced by function name. A function is a value that's internally maintained or computed by Simio and can be used in an expression but can't be assigned. For example, all objects have a function named ID that returns a unique integer for all active objects in the model. Note that in the case of dynamic entities the ID numbers are reused to avoid generating very large numbers for an ID. Object and math functions are listed and described in detail in the Simio Help and in the Reference Guide (see the Support ribbon).

5.1.8   Costing

One important modeling objective is to predict expected cost or compare scenarios based on cost. In some cases this can be accomplished by comparing the capital investment associated with one scenario versus another. Such simple cases can often be modeled with a few states to capture costs, or even have the cost entirely accounted for external to the model. But often there is need to measure the operating cost of one scenario versus another. In these cases you may have direct and indirect costs and widely differing production in terms of product mix, throughput at the subsystem level, and total throughput. These more complex cases often benefit from the more comprehensive approach of *Activity-Based Costing (ABC)*. ABC is a method of assigning costs to products or services based on the resources consumed. This helps us recognize more costs as direct costs instead of indirect, which in turn provides a more accurate understanding of how individual products and services contribute to costs under various scenarios.

Simulation is an ideal tool for implementing ABC because the activities are usually modeled at the same detail at which costs must be measured. The model typically "knows" which resources are being used by each entity and the duration involved. In some simulation products, ABC must be implemented by adding logic and data items to track each cost as it is being incurred. More comprehensive simulation products have built-in ABC. Simio is in the latter category.

Most objects in the Simio Standard Library have a *Financials* category. The properties in the category vary with the complexity and features of the object. It could be as simple as just having a busy cost per time unit (Usage Cost Rate), but it can be much more complicated than that to accurately capture all costs. For example most (not all) workers still get paid while they are idle, waiting for the next activity (Idle Cost rate). Some resources incur a fixed cost with each entity handled (Cost Per Use). And some resources and entities (like an airplane) are expensive enough that they incur a significant cost when they are just sitting idle (Holding Cost Rate). In many cases it is not enough to simply accrue the costs, but they must also be allocated or "rolled up" to the proper department (Parent Cost Center).

In Figure 5.9 you will notice that both the Server and the Worker have similar properties for calculating resource costs (e.g., the cost of seizing and holding the Server or Worker and the costs for it to be idle while scheduled for work). The Server also has properties to calculate the cost accrued while

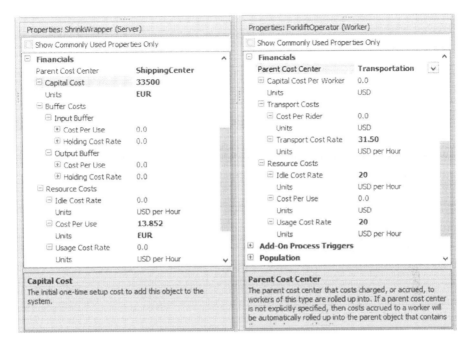

Figure 5.9: Financial properties on Server (left) and Worker.

waiting in its input and output buffers. The Worker object has no buffer costing. Instead it potentially has a cost each time it carries an entity, as well as a transportation cost while it is moving. It is rare for a single instance of any object to have all of these costs – typically, one resource may have a cost per use while another might have busy and idle cost rates.

The left side of Figure 5.9 illustrates an automated shrink wrapping machine (a Server). It requires a capital cost of 33,500 EUR, but the only cost of use is a fixed 13.852 EUR each time it is used. The shrink wrapper costs get rolled up to the Shipping Center cost center. The right side of Figure 5.9 illustrates a fork lift operator (a Worker) who gets paid 20 USD per hour whether busy or idle. But when the operator uses the fork lift to transport something it incurs an additional 31.50 USD per hour to account for the cost of the vehicle. The fork lift operator's costs get rolled up to the Transportation cost center.

One final note is that multinational companies and projects often involve work being done across the world and hence in multiple currencies. Simio recognizes most common world currencies (I don't think it yet handles goats or Bitcoins). The Financials category under the Run tab Advanced Options allows you to specify the exchange rate for the currencies you want to use as well as the default currency to be used in reports.

5.2 Model 5-1: PCB Assembly

Now that we have a better understanding of the Simio framework, we'll return our focus to modeling and analysis. The models in this chapter relate to

Figure 5.10: Example printed circuit board (Courtesy of the Center for Advanced Vehicle Electronics (CAVE) at Auburn University: cave.auburn.edu).

a simplified assembly operation for printed-circuit boards (PCBs). PCBs are principal components in virtually all electronics products. PCB assembly basically involves adding electronic components of various types to a special board designed to facilitate electronic signaling between the components. Figure 5.10 shows an example printed-circuit board. PCB assembly generally involves several sequential-processing steps, inspection and rework operations, and packaging operations. Through this chapter, we'll successively add these operations to our models, but for now, we'll start with a straightforward enhancement to our single-server queueing models from Chapter 4.

Model 5-1 focuses on a single operation where surface-mount components are placed on the board (the black rectangular shaped components on the board shown in Figure 5.10, for example) and the subsequent inspection operation. Boards arrive to the placement machine from an upstream process at the rate of 10 boards per hour (assume for the time being that the arrival process is Poisson, i.e., the interarrival times are exponentially distributed). The placement-machine processing times are triangularly distributed with parameters (3, 4, 5) minutes. After component placement, boards are inspected to make sure that all components were placed correctly. A human performs the inspection and the inspection times are uniformly distributed between 2 and 4 minutes. Historical records indicate that 92% of inspected boards are found to be "good" and the remaining 8% are found to be "bad." Since the interarrival, component-placement, and inspection times are all random, we expect some queueing and we'd like to estimate the queue lengths and utilizations for the placement machine and the inspector. We'll also estimate the time that boards spend in our small system and how much of this time is spent waiting, and we'll count the number of good and bad boards.

We could (and some of you likely will) jump into the model building at this

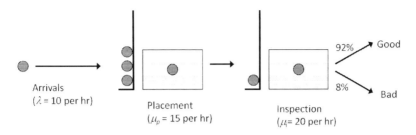

Figure 5.11: Example 5-1 queueing model.

point, but it makes sense to spend a little effort developing some *expectation* about our system, since (as described in Section 4.2.5) we'll need to verify our model before we can use it to estimate our performance metrics of interest. Recall that the basic verification process involves comparing your model results with your expectations. Figure 5.11 shows the basic queueing model of our initial PCB assembly system. Given the board arrival rate ($\lambda = 10$) and the respective service rates ($\mu_p = 15$ and $\mu_i = 20$), all in units of boards per hour, we can compute the exact server utilizations at steady state (the utilizations are independent of the specific inter-arrival and processing time distributions). Since $\rho_p = \lambda/\mu_p = 0.667 < 1$, we know that the arrival rate to the inspector is exactly 10 boards/hour (why?) and therefore, $\rho_i = \lambda/\mu_i = 0.500$. Since both of the server utilizations are strictly less than 1, our system is *stable* (we can process entities at a faster rate than they arrive, so the number of entities in the system will be strictly finite, i.e., the system won't "explode" with an ever-growing number of boards forever). If we assume that the placement and inspection times are exponentially distributed, we can also calculate the expected number of boards in our system, $L = 3.000$ and the expected time that a board spends in the system, $W = 0.3000$ hours (see Chapter 2 for details), both at steady-state, of course. While the times are not exponentially distributed, as we saw in Chapter 4 it's easy to make them so initially in our simulation model in order to verify it.

Armed with our expectation, we can now build the Simio model. Compared to our models in Chapter 4, Model 5-1 includes two basic enhancements: *sequential processes* (inspection follows component placement); and *branching* (after inspection, some boards are classified as "good" and others are classified as "bad"). Both of these enhancements are commonly used and are quite easy in Simio. Figure 5.12 shows the completed Simio model, consisting of an entity object (PCB), a source object (Source1), two server objects (Placement and Inspection), and two sink objects (GoodParts and BadParts). For our initial model, we used Connectors to connect the objects and specify the entity flow. Recall that a Connector transfers an entity from the output node of one object to the input node of another in zero simulated time. Our sequential processes are modeled by simply connecting the two server objects (boards leaving the Placement object are transferred to the Inspection object). Branching is modeled by having two connectors originating at the same output node (boards leaving inspection are transferred to either the GoodParts sink or the BadParts

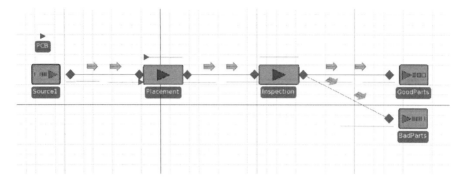

Figure 5.12: Model 5-1.

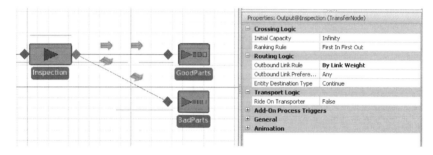

Figure 5.13: Output node Routing Logic.

sink). To tell Simio how to decide which branch to select for a given entity, we must specify the *Routing Logic* for the node, which Figure 5.13 illustrates for Model 5-1. Here, the output node for the Inspection object is selected and the Outbound Link Rule property in the Routing Logic group is set to By Link Weight. This tells Simio to choose one of the outbound links randomly using the connectors' Selection Weight properties to determine the relative probabilities of selecting each link[6]. Since our model specifications indicated that 92% of inspected boards are good, we set the *Selection Weight* property to 0.92 for the link to the GoodParts object (see Figure 5.14) and 0.08 for the link to the BadParts object. This link-weight mechanism allows us easily to model *n*-way random branching by simply assigning the relative link weight to each outgoing link. Note that we could have also used 92 and 8 or 9.2 and 0.8 or any other proportional values for the link weights — like a cake recipe ... 5 parts flour, 2 parts sugar, 1 part baking soda, etc. — Simio simply uses the specified link weights to determine the proportion of entities to route along each path. Simio includes several other methods for specifying basic routing logic — we'll explore these in later models — but for now, we'll just use the random selection. The final step to complete Model 5-1 is to specify the interarrival and service-time distributions using the Interarrival Time and Processing Time properties as we did in our previous models in Chapter 4. As a verification step, we initially set

---

[6]Note that By Link Weight is the default rule and it will be used if no entity destination is set even if the rule is set to Shortest Path

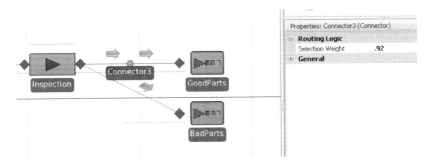

Figure 5.14: Selection weight for the link between the Inspection and GoodParts objects.

Table 5.2: Model 5-1 initial results (using exponential processing times).

Metric being estimated	Simulation
Placement Utilization (ρ_p)	0.667 ± 0.003
Inspection Utilization (ρ_i)	0.500 ± 0.003
Number in system (L)	2.995 ± 0.050
Time in system (W)	0.299 ± 0.004
Average number of "good" parts	9223.760 ± 37.768
Average number of "bad" parts	790.200 ± 7.614

Table 5.3: Model 5-1 initial results.

Metric being estimated	Simulation
Placement Utilization (ρ_p)	0.668 ± 0.002
Inspection Utilization (ρ_i)	0.501 ± 0.002
Number in system (L)	1.857 ± 0.020
Time in system (W)	0.185 ± 0.001
Average number of "good" parts	9207.600 ± 34.706
Average number of "bad" parts	813.720 ± 10.858

the processing times to be exponentially distributed so that we can compare to our queueing results. Table 5.2 gives the results from running 25 replications each with length 1200 hours and a 200 hour warm-up period. The results were all read from the Pivot Grid report as described in Section 4.2. These results are within the confidence intervals of our expectations, so we have strong evidence that our model behaves as expected (i.e., is verified). As such, we can change our processing time distributions to the specified values and have our completed model. Table 5.3 gives the results based on the same run conditions as our verification run. Note that, as expected, the utilizations and proportions of good and bad parts did not significantly change[7], but our simulation-generated estimates of the number of entities (L) and time in system (W) both went down. Note that we should have also expected these reductions as we reduced

---

[7]While the actual *numbers* did change, our *estimates* did not change in a probabilistic sense — remember that the averages and confidence-interval half widths are observations of random variables.

the variation in both of the processes when we switched from the exponential distribution to the triangular and uniform distributions (both of which, unlike the exponential distribution, are bounded on the right).

5.3   Model 5-2: Enhanced PCB Assembly

In this section, we'll enhance our PCB assembly model to add features commonly found in real systems. The specific additional features include:

- *Rework* — Where the inspected boards were previously classified as either Good or Bad, we'll now assume that 26% of inspected boards will require *rework* at a special rework station where an operator strips off the placed components. These rework boards all come out of what were previously deemed good boards, so that of the total coming out of inspection, 66% are now free-and-clear "good," 8% are hopelessly "bad," and 26% are rework. We'll assume that the rework process times are triangularly distributed with parameters (2, 4, 6) minutes. After rework, boards are sent back to component placement to be re-processed. Reworked boards are treated exactly like newly arriving boards at the placement operation, and have the same good, bad, and rework probabilities out of all subsequent inspections (note that there's no limit on the number of times a given board might be found in need of rework, and thus loop back through placement and inspection).

- *Worker Schedules* — The inspection and rework operations are both done by human operators, and we'll explicitly consider the hours of the work day that the operators work and will include "meal breaks." Inspection is done throughout the work day, but rework is done only during "second shift." We continue to assume that placement is automated and requires no human operator.

- *Machine Failures* — The placement machine is an automated process and is subject to random *failures* where it must be repaired.

Since we have introduced the concept of worker schedules into our model, we must now explicitly define the *work day*. For Model 5-2, we assume that a work day is 24 hours and we will be running our model continuously for 600 of these days.

5.3.1   Adding a Rework Station

Figure 5.15 shows the completed model. To add the rework station, we simply added an new server object (Rework), set the Processing Time and Name properties, and connected the input node of the server to the output node of the Inspection object, forming a 3-way branching process leaving inspection (be sure to update the link weights accordingly). Since reworked boards are sent back to the placement machine for processing, we connected the output node of the Rework object to the input node of the Placement object. This is our first example of a *merging* process. Here we have two separate "streams" of

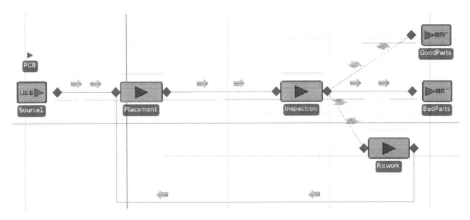

Figure 5.15: Model 5-2.

entities entering the Placement object — one from the Source1 object (newly arriving boards) and the other from the Rework object (boards that had previously been processed, failed inspection, and reworked). The Placement object (and its input node) will treat the arriving entities exactly the same, regardless of whether they arrived from the Source1 object or the Rework object. In a future model, we'll demonstrate how to distinguish between arriving entities and treat them differently depending on their state.

Given the structure of our system (and the model), it's possible for PCBs to be reworked multiple times since we don't distinguish between new and reworked boards at the placement or inspection processes. As such, we'd like to keep track of the number of times each board is processed by the placement machine and report statistics on this value (note that we could have just as easily chosen to track the number of times a board is reworked — the minimum would just be 0 rather than 1 in this case). This is an example of a *user-defined statistic* — a value in which we're interested, but Simio doesn't provide automatically. Adding this capability to our model involves the following steps:

1. Define an entity *state* that will keep track of the number of times the entity has been processed in the placement machine;

2. Add logic to increment that state value when an entity is finished being processed by the placement machine;

3. Create a *Tally Statistic* that will report the summary statistics over the run; and

4. Add logic to record the tally value when an entity departs the system (as either a good or bad part).

Note that this is similar to the *TimeInSystem* statistic that we modeled in Section 4.3. Both of these are examples of *observational statistics* (also called *discrete-time processes*), where each entity produces one observation — in the current case, the number of times the entity went through the placement process, and in Model 5-2, the time the entity spent in the system — and the

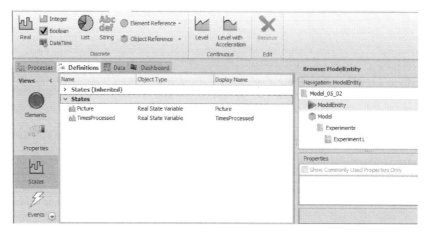

Figure 5.16: Creating the *TimesProcessed* entity state.

Tally Statistic reports summary statistics like average and maximum[8] over all observations (entities, in both cases).

A *state* is defined within an object and stores a value that changes over the simulation run (Section 5.1.2). This is exactly what we need to track the number of times an entity goes through the placement process — every time the entity goes through (exits) the process, we increment the value stored in the state. In this case, we need to define the state with respect to the entity so we select the ModelEntity in the Navigation window, select the Definitions tab, select States from the Panel, and click on Integer icon from the ribbon (see Figure 5.16). We named the state `TimesProcessed` for this model. Every entity object of type ModelEntity will now have a state named TimesProcessed and we can update the value of this state as the simulation runs. In our case, we need to increment the value (add 1 to the current value) at the end of the placement process. There are several ways that we could implement this logic in Simio. For this model, we'll use the *State Assignments* property associated with Server objects. This property allows us to make one or more state assignments when an entity either enters or exits the server; see Figure 5.17. Here we've selected the Placement object, opened the *Before Exiting* repeat group property in the *State Assignments* section, and added our state assignment to increment the value of the TimesProcessed state. This tells Simio to make the specified set of state assignments (in our case, the single assignment) just before an entity leaves the object. So at any point in simulated time an entity's TimesProcessed state indicates the number of times that entity has completed processing by the Placement object.

Next, we need to tell Simio that we'd like to track the value of the TimesProcessed state as an observational (tally) statistic just before entities leave the system. This is a two-step process:

1. Define the tally statistic; and

---

[8]But *not* standard deviation ... see Section 3.3.1 for why doing so would often be biased and thus invalid, even though it easily *could* be done.

Figure 5.17: Setting the State Assignments property for the Placement object.

Figure 5.18: Creating the NumTimesProcessed tally statistic.

2. Record the value (stored in the TimesProcessed entity state) at the appropriate location in the model.

First we'll define the tally statistic by selecting the model in the Navigation window, clicking on the Definitions tab, and selecting the Elements pane, and clicking on the Tally Statistic icon on the Ribbon to add the statistic (see Figure 5.18). Note that we named the tally statistic NumTimesProcessed by setting the Name property. Finally, we'll use *add-on processes* to record the value of the TimesProcessed state to the tally statistic.

Since we want to record the tally values just prior to entities' departing the system, we'll create add-on processes for the input nodes of the GoodParts and BadParts Sink objects. These processes will be executed when an entity enters the respective input nodes. To create the first add-on process, select the input node for the GoodParts Sink object and expand the Add-On Process Triggers group in the model properties panel. Three triggers are defined: Initialized, Entered, Exited. Since entities are simply exiting the system when they enter the Sink object, we could use any of these triggers for our purpose. We chose Entered. You can enter any name that you like for the add-on process, but there is a nice Simio shortcut for defining add-on processes. Simply double-click on the trigger property name (Entered, in this case) and Simio will define an appropriate name and open the processes panel with the newly created process selected. The add-on processes in this case are quite simple — just the single step of recording the tally value to the appropriate tally statistic. This is done using the *Tally* step (as shown in Figure 5.19). To add the step, simply select and drag the Tally step from the Common Steps panel onto the process

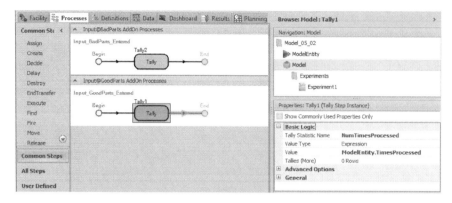

Figure 5.19: Add-on processes for the two nodes where entities depart the system.

and place it between the Begin and End nodes in the process. The important properties for the Tally step are the TallyStatistic Name, where we specify the previously defined Tally Statistic (`NumTimesProcessed`) and the Value, where we specify the value that should be recorded (the `TimesProcessed` entity state in this case). We also set the Value Type property to `Expression` to indicate that we'll be using an expression to be evaluated and recorded. Since we also want to record the tally for BadParts, we created a similar add-on process for the Entered trigger of the input node for the BadParts Sink object. Note that since the processes are identical, we could have defined a single add-on process and specified that process in the Entered triggers for both Sink objects. Now when we run the model, we'll see our Tally statistic reported in the standard Simio Pivot Grid and Reports and the values will be available in case we wish to use them in our process logic.

The process that we just described for creating a user-defined observational statistic is a specific example of a general process that we will frequently use. The general process for creating user-defined observational (tally) statistics is as follows:

1. Identify the *expression* to be observed. Often, this involves defining a new model or entity state where we can compute and store the expression to be observed (as it did with the previous example);

2. Add the logic to set the value of the expression for each observation, if it is not already part of the model;

3. Create the *Tally Statistic*; and

4. Tell Simio where to record the expression value using the *Tally* step in an add-on process at an appropriate location in the model.

We will follow this same general process wherever we incorporate user-defined observational statistics. Note that Standard Library node objects also provide Tally Statistics properties (see Figure 5.20) where the tallies can be specified

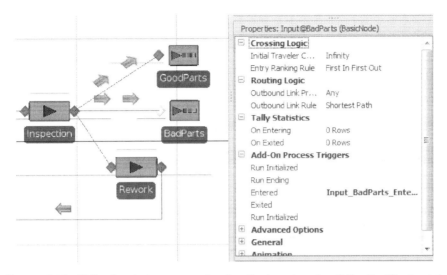

Figure 5.20: Tally Statistics properties for the input node of the BadParts sink object.

without using add-on processes. In this case, both methods produce the same results.

5.3.2 Using Expressions with Link Weights

Before continuing with Model 5-2, we would like to demonstrate using *expressions* for connector link weights. This is a very powerful, but often overlooked feature that can simplify the logic associated with link selection. Above, we specified fixed probabilities for the three possible inspection outcomes (good: 66%, rework: 26%, bad: 8%) and set the link weights accordingly. Suppose that we would like to limit the number of times that a board can be processed. For example, suppose that we had the following rules:

1. Boards can be processed a maximum of 2 times. Any board processed a third time should be rejected;

2. 95% of inspected board are good;

3. 5% of inspected boards are bad.

and we would like implement this logic in our model. In this case, the link selection is not totally probabilistic. Instead there is a condition that determines whether the board should be rejected (it's been processed 3 times) and, if that condition is not met, probabilities are used to determine between the remaining two alternative. Figure 5.21 shows a simple way to implement this logic using expressions for link weights. Note that the parenthetical components if the link weights will evaluate to either 0 or 1, so the probabilities are applied only if the board has been processed fewer than 3 times. Otherwise, the board will be rejected.

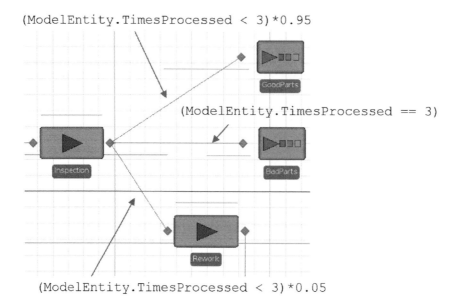

Figure 5.21: Example of using expressions with link weights.

This simple example illustrates the power of using expressions with the link-weight selection rule. New users will often unnecessarily resort to using more complex add-on processes in order to implement this type of logic. We encourage to you study this example carefully as these types of expressions can greatly simplify model development. Of course, if we were really implementing this system, we would like to reject the board *before* it is processed the third time – i.e., immediately after it fails inspection for the second time, but before it is reworked. We leave this as an exercise for the reader (see Problem 7 at the end if the chapter).

5.3.3   Resource Schedules

The rework and inspection operations of our PCB assembly system are both performed by human operators (the placement process is still automated). Human operators generally work in *shifts* and take breaks during the work day. Before describing how to implement worker schedules in Simio, we must first describe how Simio tracks the time of day in simulated time. Simio runs are based on a 24-hour clock with each "day" starting at 12 : 00 a.m. (midnight) by default. In our previous models, we've focused on the replication run length and warm-up period length, but have not discussed how runs are divided into 24-hour days.

In our PCB example, we'll assume that the facility operates three 8-hour work shifts per day and that each shift includes a 1-hour meal break (so a worker's day is actually 9 hours including the meal break). The shifts are as follows.

Table 5.4: Capacity schedule for the inspection resource.

Time period	Resource Capacity
12:00 a.m. – 4:00 a.m.	1
4:00 a.m. – 5:00 a.m.	0
5:00 a.m. – 12:00 p.m.	1
12:00 p.m. – 1:00 p.m.	0
1:00 p.m. – 8:00 p.m.	1
8:00 p.m. – 9:00 p.m.	0
9:00 p.m. – 12:00 a.m.	1

- First shift: 8:00 a.m. – 12:00 p.m., 12:00 p.m. – 1:00 p.m. (meal break), 1:00 p.m. – 5:00 p.m.

- Second shift: 4:00 p.m. – 8:00 p.m., 8:00 p.m. – 9:00 p.m. (meal break), 9:00 p.m. – 1:00 a.m.

- Third shift: 12:00 a.m. – 4:00 a.m., 4:00 a.m. – 5:00 a.m. (meal break), 5:00 a.m. – 9:00 a.m.

We'll also assume that, for multi-shift processes (inspection, in our case), a worker's first hour is spent doing paperwork and setting up for the shift, so that there's only one worker actually performing the inspection task during those periods when worker shifts overlap.

Given the work schedule that we described, the inspection operation operates 24-hours per day with three 1-hour meal breaks (12:00 noon – 1:00 p.m., 8:00 p.m. – 9:00 p.m., and 4:00 a.m. – 5:00 a.m.). This schedule would correspond to having three inspection operators, each of whom works 8 hours per day. Since Simio is based on a 24-hour clock that starts at 12:00 a.m., this translates to the inspection *resource capacity schedule* shown in Table 5.4. With this work schedule, we'd expect boards to queue in front of the inspection station during each of the three "meal periods" when there is no inspector working. Once we complete the model, we should be able to see this queueing behavior in the animation. There are multiple ways to implement this type of resource capacity schedule in Simio — for Model 5-2, we'll use Simio's built-in *Schedules* table. To use this table, we first define the schedule and then tell the Inspection object to use the schedule to determine the object's resource capacity.

To create our work schedule, select the model in the Navigation pane, select the Data tab, and the Schedules icon in the Data panel. You'll see two tabs — under the *Work Schedules* tab you'll see that a sample schedule named StandardWeek has already been created for you, and under the *Day Patterns* tab you'll see that a sample StandardDay has been created for you.

We'll start under the Day Patterns tab to define the daily work cycle by telling Simio the capacity values over the day. We can choose either to revise the sample pattern, or to replace it. Select StandardDay and then click on *Remove* to delete the existing pattern, then click on *Day Pattern* in the Create category of the ribbon to create a new one. Give the new pattern a *Name* of DayPattern1 and click the "+" to the left of it to expand the Work Periods definition. In the first work period row, type or use the arrows to enter 12:00:00

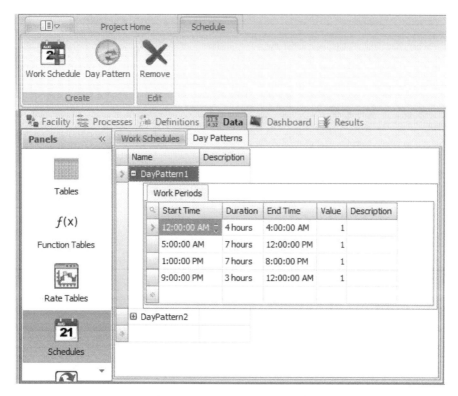

Figure 5.22: Definition of inspection day pattern.

AM for *Start Time* then type 4 for the *Duration*. You'll see that *End Time* is automatically calculated and that the *Value* (capacity of the resource) defaults to 1. Continue this process to add the three additional active work periods as indicated in Table 5.4. Your completed day pattern should look like that in Figure 5.22.

Now let's move back to the Work Schedules tab. Again, we could delete the sample schedule and create a new schedule, but this time we'll just change it to meet our needs. Change the Work Schedule *Name* to `InspectionSchedule` and change the *Description* to `Daily schedule for inspectors`. For our model, the work pattern is the same every day so we'll set the *Days* property value to 1 to indicate that the same pattern will repeat every (one) day. Finally, since we deleted the `StandardDay` pattern that was originally referenced here, the *Day 1* column indicates an error. To clear the error, just select the new `DayPattern1` from the pull-down list. Your completed work schedule should look like that in Figure 5.23.

Now that we have the schedule defined, we tell the Inspection object to use the schedule to specify the resource capacity. We can do this by selecting the Inspection object and setting the Capacity Type property to `WorkSchedule` and setting the Work Schedule property to `InspectionSchedule` (see Figure 5.24). Next, we need to specify the capacity schedule similarly for the rework operator. As specified above, the rework operation works only during the second shift,

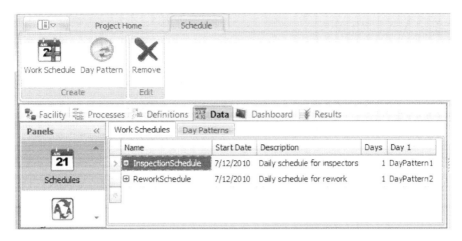

Figure 5.23: Inspection schedule.

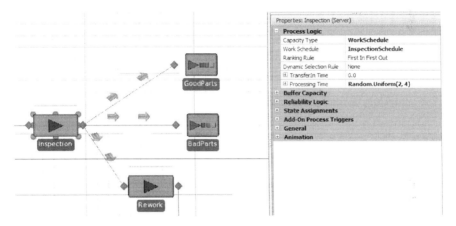

Figure 5.24: Specifying a WorkSchedule for the Inspection object.

which will be 4:00 p.m. – 8:00 p.m., 8:00 p.m. – 9:00 p.m. (meal break), and 9:00 p.m. – 1:00 a.m. So, as before, we created a new day pattern (DayPattern2) and referenced that day pattern in a new work schedule (ReworkSchedule) and then told the Rework object to use the schedule (see Figure 5.25). Schedules will be discussed in more detail in Section 7.2.

Before incorporating the placement-machine failures into our model, it's instructive to consider how incorporating resource capacity schedules affects our server-utilization statistics. In our previous models, we defined the utilization as the proportion of the run in which the server was busy (or the percentage resulting from multiplying the proportion by 100). This worked because the capacity of the server resources did not change throughout the run. Now that we have servers where the resource capacity changes over the model run, the utilization calculation is a bit more difficult — in fact, the *interpretation* of utilization itself is no longer straightforward. In particular, it's not clear how to account for the "off-shift" time for the server resources. As an example,

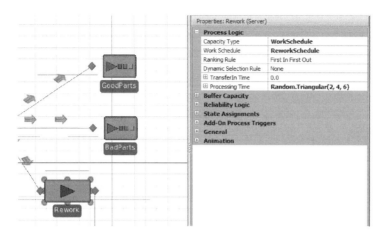

Figure 5.25: Specifying a WorkSchedule for the Rework object.

Table 5.5: Inspection utilization-related statistics.

Category	Statistic	Value
Capacity	ScheduledUtilization (Percent)	77.3324
	UnitsScheduled (Average)	0.8750
	UnitsUtilized (Average)	0.6767
Resource State	Offshift Time (Percent)	12.5000
	Processing Time (Percent)	67.6659
	Starved Time (Percent)	19.8341

assume that the rework operator is busy 6.28 hours for a particular day. One method of computing utilization involves dividing the busy time by the day length (24 hours), yielding a utilization of 26.17%. However, if you were the rework operator, you'd likely say "Wait a minute, I was way busier than that!" The problem (from the operator's perspective) is that the utilization calculation doesn't take into consideration that the rework operator is only *available* 8 hours of every 24-hour day (i.e., only during second shift). As a second calculation, we could divide the busy time by the 8 hours that the operator is actually available, yielding a utilization of 78.50% — a value with which the rework operator would likely agree. Note that neither of these interpretations is technically "incorrect," they are simply two different interpretations of the concept of resource utilization, and Simio provides both as part of its default output.

Figure 5.26 shows a portion of the standard Pivot Grid report from a run of 25 replications of Model 5-2 (without the placement-machine failures) of run length 125 days with a 25-day warm-up period. The portion of the Pivot Grid shows the Server Resource statistics for the Inspection object. The values in which we're particularly interested for our utilization discussion are extracted to Table 5.5. The statistics are defined as follows.

- ScheduledUtilization (Percent) — The percent utilization of the object's scheduled capacity during the run, calculated by returning the value of

[Resource]	Capacity	ScheduledUtilization	Percent	77.3324	76.3331	78.2596	0.2262
		UnitsAllocated	Total	32,468.7600	32,029.0000	32,917.0000	100.0532
		UnitsScheduled	Average	0.8750	0.8750	0.8750	0.0000
			Maximum	1.0000	1.0000	1.0000	0.0000
		UnitsUtilized	Average	0.6767	0.6679	0.6848	0.0020
			Maximum	1.0000	1.0000	1.0000	0.0000
	ResourceState	OffShift Time	Average (Ho...	1.0000	1.0000	1.0000	0.0000
			Occurrences	300.0000	300.0000	300.0000	0.0000
			Percent	12.5000	12.5000	12.5000	0.0000
			Total (Hours)	300.0000	300.0000	300.0000	0.0000
		Processing Time	Average (Ho...	0.0927	0.0913	0.0944	0.0003
			Occurrences	17,520.3600	17,220.0000	17,756.0000	43.6879
			Percent	67.6659	66.7915	68.4771	0.1979
			Total (Hours)	1,623.9805	1,602.9952	1,643.4507	4.7501
		Starved Time	Average (Ho...	0.0275	0.0264	0.0289	0.0003
			Occurrences	17,315.9600	17,019.0000	17,541.0000	42.4014
			Percent	19.8341	19.0229	20.7085	0.1979
			Total (Hours)	476.0195	456.5493	497.0048	4.7501

Figure 5.26: Resource section of the Pivot Grid for the Inspection object.

the expression $100 * (Capacity.Utilized.Average/Capacity.Average)$.

- UnitsScheduled (Average) — The average scheduled capacity of this object during the run.

- UnitsUtilized (Average) — The average number of capacity units of this object that have been utilized during the run.

- Offshift Time (Percent) — The percentage of the run during which the resource is in the Offshift state.

- Processing Time (Percent) — The percentage of the run during which the resource is in the Processing state.

- Starved Time (Percent) — The percentage of the run during which the resource is in the Starved state. Note that *starved* in this context means that the resource is available, but there are no entities available to process.

In our simple example above involving the rework operator, the 26.17% measure is equivalent to the $UnitsUtilized * 100$ and the 78.50% is equivalent to the $ScheduledUtilization$. Similarly, the values in Table 5.5 tell us that the inspection operators are busy roughly 77.3% of the time that they are working, and that there is an inspection operator busy roughly 67.7% of the entire day. The $UnitsScheduled$ value of 0.8750 is based on the fact that the three 1-hour meal breaks mean that inspection operators are scheduled to be available 21 out of the 24 hours each day (or 0.8750 of the day).

Note that there will not always be the same relationship between the Capacity values and the Resource State values. In particular, the following relationships hold because the resource capacity values in the schedule were either 0 or 1.

Table 5.6: Modified inspection utilization-related statistics.

Category	Statistic	Value
Capacity	ScheduledUtilization (Percent)	77.4332
	UnitsScheduled (Average)	1.7500
	UnitsUtilized (Average)	1.3551
Resource State	Offshift Time (Percent)	12.5000
	Processing Time (Percent)	82.5262
	Starved Time (Percent)	4.9738

$$Average\ UnitsScheduled = \frac{Processing\ Time\ \% + Starved\ Time\ \%}{100}$$

$$Average\ UnitsUtilized = \frac{Processing\ Time\ \%}{100}$$

If the resource capacities are greater than 1 during a schedule (e.g., if some shifts had 2 or 3 inspectors working), the Resource State values are independent of the number of units of a resource that are busy, whereas the Capacity values take the specific numbers of units of resource capacity into consideration. For example, if we halve the inspection processing rate (set the Processing Time property to *Random.Uniform*(4, 8)) and change the schedule so that there are 2 inspectors rather than 1 during the work cycles, we get the results shown in Table 5.6. In this model, the ScheduledUtilization and Offshift Time percentage did not change (as we'd expect — make sure you understand why this is true!), but the other values did. The ProcessingTime percentage went up because it measures the percentage of time that either or both of the units of capacity are busy. Similarly, the Starved Time percentage went down because now both units of capacity have to be starved in order for the inspection resource to be in the starved state.

5.3.4   Machine Failures

The final enhancement for Model 5-2 is to incorporate random machine failures for the placement machine (note that we also gave a machine-failure example in Section 5.1.4). As with most complex mechanical devices, the placement machine is subject to failures where it must be repaired before in can continue processing boards. Simio's Server object from the Standard Library supports three different types of failures:

- *Calendar Time Based* — the uptime is based on calendar time and you specify the uptime between failures;

- *Event Count Based* — the uptime is based on the number of times an event occurs and you specify the number of events; and

- *Processing Count Based* — the uptime is based on the number of times the server processes an entity and you specify the number of times that the server processes between failures.

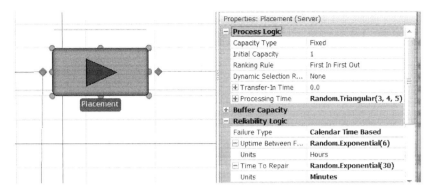

Figure 5.27: Defining the failures for the placement object.

ResourceState	Failed Time	Average (Mi...	30.1609	25.2118	31.4205	1,341.7422
		Occurrences	182.4000	172.0000	192.0000	5.0938
		Percent	7.6324	6.5831	9.6811	0.6448
		Total (Minutes)	5,495.3416	4,739.8099	5,592.8473	244,466.2059
	Processing Time	Average (Mi...	139.2685	133.3260	139.1878	6,195.5234
		Occurrences	476.4000	349.0000	574.0000	53.5915
		Percent	89.9698	88.6505	90.8264	0.5266
		Total (Minutes)	64,778.2442	65,329.7631	64,722.3171	,881,724.9950
	Starved Time	Average (Mi...	5.8229	6.2676	5.7113	259.0390
		Occurrences	298.3000	162.0000	403.0000	57.5357
		Percent	2.3978	1.4917	3.3770	0.4623
		Total (Minutes)	1,726.4143	1,930.4270	1,684.8356	76,801.9860

Figure 5.28: Placement resource state results after adding failures.

For our placement-machine failures, we'll use the Calendar Time Based failures. Historical records show that our placement machine fails after about 6 hours of uptime and that the time required to repair the machine after it fails is approximately 30 minutes. As both the uptime and repair durations are highly variable, we'll assume that both are exponentially distributed. To incorporate this failure model into Model 5-2, we need to set the *Failure Type* property to *Calendar Time Based*, the *Uptime Between Failures* property to *Random.exponential*(6) Hours, and the *Time to Repair* property to *Random.exponential*(30) Minutes (see Figure 5.27). Now when we run the model (using the same run parameters as in the previous section), we can see that the ResourceState category for the Placement object now includes a section for Failed Time (see Figure 5.28) and that our estimate of the percentage of time that the machine spends in the failed state is approximately 7.63%.

Calendar Time Based failures are most commonly used to model random mechanical failures where the time between failures is a function of time. Processing Count Based failures are most commonly used to model failures where the time between failures is a function of the number of times the process is used. Raw-material replenishment is an example of a situation that would call for this type of failure process. In this situation the machine has an inventory of raw materials that is exhausted by processing parts. When the raw-material in-

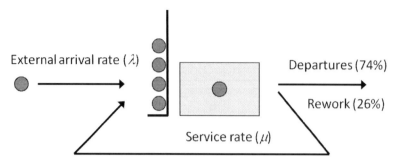

Figure 5.29: Re-entrant flow.

ventory is empty, the machine can't continue processing until the raw-material inventory is replenished. As such, the time between failures will be a function of the number of parts processed (or, more specifically, the Processing Count). Event Count Based failures are based on the number of occurrences of an event, which can be a standard Simio-defined event or a user-defined event.

5.3.5   Verification of Model 5-2

Before using our model to analyze our system, we need to have some confidence that our model is "correct." In the current context, this involves verifying that our model behaves as expected (i.e., that we've correctly translated the specifications into our working Simio model). When we verified Model 5-1 and the models in Chapter 4, our basic strategy was to develop some expectations to which we could compare the model output. However, in those models, the queueing analysis that we used to develop our expectation was very straightforward. With Model 5-2, we're beginning to add model components that are difficult to incorporate into standard queueing models even though Model 5-2 is still a very simple model. Of course we expected this — if we could completely analyze systems using queueing (or other analytical methods), there would be no need for simulation! The two components that complicate our queueing analysis of Model 5-2 are:

1. Re-entrant flow — Since the boards that leave the rework station are sent back for re-processing, the arrival rate to the Placement station will be greater than the external arrival rate of 10 boards/hour; and

2. Resources changing capacity during the run — With the incorporation of resource-capacity schedules and machine failures, the capacities of the Placement, Inspection, and Rework resources will vary over time.

We'll consider these issues one at a time. Since the arrival rate to the Placement station will be greater than the 10 boards/hour we had in Model 5-1, we expect the Placement resource utilization also to be higher. To estimate the new utilization, we need to determine (or estimate) the new arrival rate. Figure 5.29 illustrates the re-entrant flow. If we let λ' be the *effective arrival*

Table 5.7: Model 5-2 verification-run results.

Metric being estimated	Simulation
Placement Utilization (ρ_p)	0.9013 ± 0.0743
Inspection Utilization (ρ_i)	0.7778 ± 0.0647
Rework Utilization (ρ_4)	0.7029 ± 0.1219
Percent failed time (placement)	7.6662 ± 0.0707

rate (i.e., the arrival rate that the process actually sees), then:

$$\lambda' = \lambda + 0.26 \times \lambda'$$

Solving for λ', after a few steps of algebra, we get:

$$\lambda' = \lambda/(1 - 0.26)$$
$$= 1.3514 \times \lambda$$

Ignoring the machine failures for now, we'd expect the utilization of the placement machine to be $13.514/15 = 0.9019$. Since the utilization is strictly less than 1, the departure rate will equal the arrival rate and the expected utilization for the inspection station will be $13.514/20 = 0.6757$ (ignoring the resource-capacity schedule). Similarly, since the inspection utilization is also strictly less than 1, the expected utilization for the rework station will be $(13.514 \times 0.26)/15 = 0.2342$ (again, ignoring the resource-capacity schedule). If you run Model 5-2 ignoring the resource-capacity schedules and machine failures, you'll see that the model matches our expectation.

The resource-capacity schedules and the machine failures will have the effect of reducing the available capacities of the resources. For the resource-capacity schedules, it's instructive to consider the *daily* arrival and service rates for the inspection and rework operations. Since the inspection operators work a combined 21 hours/day (due to the three one-hour meal breaks), the daily processing rate is $21 \times 20 = 420$ and the expected utilization is $(24 \times 13.514)/420 = 0.7722$. Similarly, since the inspection operator works only during second shift, the daily processing rate is $8 \times 15 = 120$ and the expected utilization is $(24 \times 13.514 \times 0.26)/120 = 0.7027$.

We'll take a slightly different approach for the machine failures. Recall that the expected uptime for the placement machine is 6 hours and the expected time to repair it is 30 minutes. As such, the machine is failed for approximately 30 minutes out of every 390 minutes, or 7.692% of the time. Since the expected percentage of time that the placement machine is idle $((1 - 0.9010) \times 100 = 9.91\%)$ is greater than the expected percentage of failed time, we'd expect the system to remain stable.

Table 5.7 gives the pertinent results based on 50 replications of 600-day runs with 100-day warmup. As these values appear to match our expectation, we have strong evidence that our model is correct (verified) and we can now use the model to analyze the system.

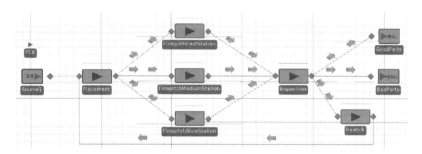

Figure 5.30: Model 5-3.

5.4 Model 5-3: PCB Model With Process Selection

The final enhancement that we'll add to our PCB model involves adding a second component-placement process immediately after our current placement process. The second placement process will add *fine-pitch* components to the PCB. Because of the placement-location accuracy required by these components, the placement process takes significantly longer than does the regular component-placement process. As such, we'll use three placement machines in *parallel* so that the fine-pitch placement process will be able to keep up with the regular placement, inspection, and rework processes. Figure 5.30 shows the completed model. Note that this is a common configuration for serial production systems that include processes with different speeds.

Just to make the model interesting, we'll also assume that the three fine-pitch placement machines are not identical. More specifically, we'll assume that we have a fast machine, a medium machine, and a slow machine where the processing times are random and are distributed as follows:

- Fast — Triangular(8, 9, 10);

- Medium — Triangular(10, 12, 14); and

- Slow — Triangular(12, 14, 16).

Finally, we'll also assume that the fast machine is subject to random failures with the up times' being exponentially distributed with mean 3 hours and repair times' being exponentially distributed with mean 30 minutes. Note that for this model, we assume that the medium and slow machines do not fail.

As shown in Figure 5.30, we removed the Connection object between the Placement object and the Inspection object and added three new server objects representing the three fine-pitch placement machines (FinepitchFastStation, FinepitchMediumStation, and FinepitchSlowStation). We set the Capacity Type property to Fixed and the Initial Capacity property to 1 for each station and we set the Processing Time properties to the specified processing-time distributions given above. For the FinepitchFastStation object, we also added the failures (see Figure 5.31). We next connected the output node of the Placement object to the input nodes of the three new server objects using Connector objects (*branching*) and similarly connected the output nodes of the new server objects to the input node of the Inspection object (*merging*).

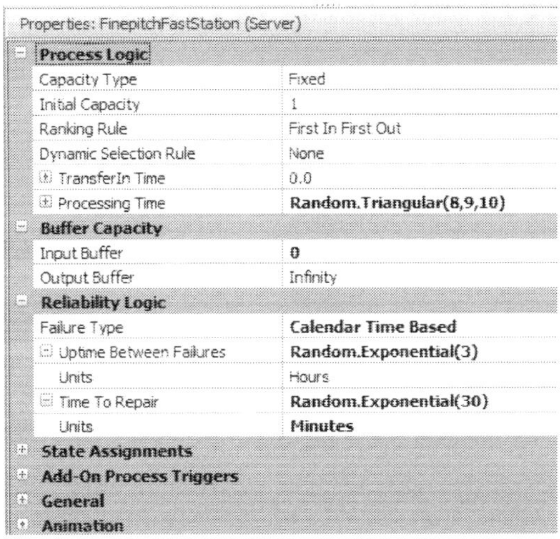

Figure 5.31: FinepitchFastStation object properties.

In our first example of branching in Section 5.2, we used *link weights* to select randomly between the alternative entity destinations. We could easily use this same logic to select the fine-pitch machine, but this is likely not what we want. Rather than selecting the machine randomly, we'd like to select the "first available" machine. We'll use a Simio *node list* to implement this logic. In order to use the node list, we must first define the list and then tell the Placement object's output node to use the list to determine the destination for the PCB objects. To define the list, select the Model in the Navigation pane, select the Definitions tab, and the Lists icon from the Definitions panel. Note that Simio supports String, Object, Node, and Transporter lists. Click on the Node icon in the Ribbon to create a new node list and then add the input nodes for the three fine-pitch objects (Input@FinepitchFastStation, Input@FinepitchMediumStation, and Input@FinepitchSlowStation) to the node list in the bottom pane of the main window (see Figure 5.32). Note that each of the Node list items in the bottom pane is a pull-down list that includes all of the nodes in the model.

Now that we have the list defined, we need to tell Simio how to select one of the list elements. In the real system, we'd like to send the PCB to the machine that will be available first. Since the processing times are random and the machines are subject to random failures, we won't know for sure which one will be available first, so we'll use a surrogate measure — we'll select the machine with the fewest boards already waiting for or en route to the machine. To implement this logic, select the output node for the Placement object and set the parameters to the values shown in Figure 5.33. The *Entity Destination Type* and *Node List Name* properties tell Simio to use the FinePitchStations list (Figure 5.32) and the *Selection Goal* property indicates that the smallest value of the *Selection Expression*. So, when an entity arrives at the node, Simio will evaluate the expression for each of the candidate nodes in the list, and will route the entity to the node with a minimum value of that expression.

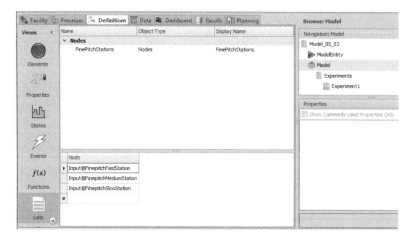

Figure 5.32: FinePitchStations node list.

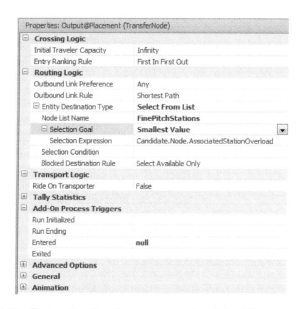

Figure 5.33: Properties for the output node of the Placement object.

The expression value, `Candidate.Node.AssociatedStationOverload` requires some explanation. The *Candidate.Node* keyword indicates that we want to consider each candidate node from the list. The *AssociatedStationOverload* function returns the *Overload* for each of the stations associated with the nodes (the fine-pitch machine objects, in our case). The *AssociatedStationOverload* property is defined in the *Simio Reference Guide* as:

> For an external input node, returns the current difference between the load and capacity values (a positive difference indicating an 'overload') for the station locations inside the node's associated object that may be entered using the node.

> The associated station 'load' is defined as the sum of entities currently en route to the node intending to enter the stations, plus the current number of entities arrived to the node but still waiting to enter the stations, plus the current number of entities occupying the stations. This function returns the difference between that load and the current station capacity (Overload = Load - Capacity).

So when an entity reaches the output node of the Placement object, Simio will select the input node for the fine-pitch machine that has minimum overload (a measure of the workload waiting for the machine, in our context) and then route the entity to that node. If there are multiple nodes with the same overload value (e.g, when the first board arrives, all three machines are empty and idle), Simio will select the node based on the list order (i.e., the first node with the minimum value)[9]. For our problem, we put the fast machine first so that it would have priority in the case of ties. In addition to AssociatedStationOverload, The AssociatedStation keyword can be used in combination with many values associated with the station. Common uses include (you can see and select from the alternatives by opening the value in the expression builder and using the down arrow to open the display):

1. *AssociatedStation.Capacity* — For an external input node, returns the current capacity of the station;

2. *AssociatedStation.Capacity.Remaining* — For an external input node, returns the current available (unused) capacity of the station;

3. *AssociatedStation.Contents* — For an external input node, returns the number of entities currently waiting in the station;

Now that the logic for the fine-pitch placement process has been incorporated into our model, we'll add one more set of user-defined statistics so that we can track the proportion of times that the fast, medium, and slow machines are selected, respectively. We do this for two reasons: first, for model verification — we expect approximately 38%, 33%, and 29% of the processing to be done by the fast, medium, and slow machines, respectively (see Problem 8 at the end

---

[9] Functions like `AssociatedStationOverload` and `AssociatedStationLoad` don't work well when *Input Buffer Capacity* is `Infinity`. You might recall from basic math that Infinity plus anything equals Infinity. So when infinite capacity queues are involved, a tie will result.

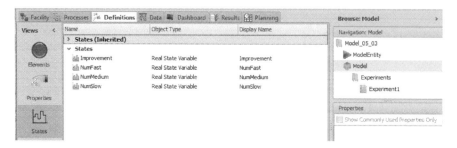

Figure 5.34: Defining the model states to track the fast-pitch machine process counts.

of the chapter); and second, to demonstrate how to set up another user-defined statistic. The steps for creating the user-defined statistics are:

1. Define a *model state* to record the number of times each machine is used to process a board;

2. Increment the corresponding model state when one of the fast-pitch machines completes processing; and

3. Tell Simio to report an *Output Statistic* for each proportion.

For the first step, select the Model in the Navigation window, select the Definitions tab, and the States icon from the Definitions panel. Then add the three Discrete States (NumFast, NumMedium, and NumSlow — see Figure 5.34). Note that when we defined the user statistic to track the numbers of times boards are processed by the placement machine (Section 5.3.1), we used an entity state rather than a model state. This was because we were tracking an *entity* characteristic in that case, whereas we're tracking a *model* characteristic in this case. This is a very important distinction when you're creating user-defined statistics. For the second step, we'll use the same method that we used in Section 5.3.1 — we'll use the *State Assignments* property associated with the Server objects and will increment the NumFast, NumMedium, and NumSlow states in the respective *BeforeExiting* repeat group. Finally, select the Elements icon from the model Definitions Panel and add the three *Output Statistics* to compute the proportion of processing steps completed by each machine. Recall that an Output Statistic reports the value of an expression at the end of a replication. The expression for the proportion of times the fast machine is used is:

$$\texttt{NumFast/(NumFast+NumMedium+NumSlow)}$$

This will tell Simio to compute the proportion at the end of the replication and to keep track of and report the statistics across replications. The output statistics for the medium and slow machine are similarly defined except we replace the NumFast in the numerator with NumMedium and NumSlow, respectively.

We could also have tracked the proportion statistics by defining the output statistics using the Simio-provided processing counts. In particular, the Simio property

FinepitchFastStation.Processing.NumberExited

gives the number of entities that exit the FinepitchFastStation object. As such, the same proportion we specified above could be specified using the expression

FinepitchFastStation.Processing.NumberExited /

(FinepitchFastStation.Processing.NumberExited+

FinepitchMediumStation.Processing.NumberExited+

FinepitchSlowStation.Processing.NumberExited)

Comparing these two methods, the benefit of using your own states is that you know *exactly* how the values are tracked and you can customize them as you see fit, but using the Simio-provided states is somewhat easier (since you don't have to define or increment the states). It's important to know how to use both methods — defining your own states and using Simio's built-in states.

5.5 Model 5-4: Comparing Multiple Alternative Scenarios

One of the primary uses of simulation is to evaluate alternative system designs. For example, suppose that we are considering purchasing an additional fine-pitch placement machine for our PCB assembly system and we would like to know whether to purchase an additional fast machine, medium machine, or slow machine (presumably, these would cost different amounts). In this section, we'll create Model 5-4 by modifying Model 5-3 so that we can compare the predicted performance of the system under each of these alternative designs. While we could easily modify the model to add capacity to one of the fine-pitch stations, run the model, save the results, repeat two additional times (changing the capacity of a different fine-pitch station each time), and manually compare the alternatives using the saved results, we'll take a different approach that demonstrates the power of Simio's experiment facilities. The first step is to define *Referenced Properties* (see Section 5.1.2) for the Initial Capacity properties for the three fine-pitch server objects so that we can easily manipulate the respective server capacities in our experiment.

Figure 5.35 shows the properties for the FinepitchFastStation server with the Initial Capacity property set to the referenced property InitialCapacity-Fast (note the green arrow indicating that the property value is a referenced property). To define a referenced property and set the Initial Capacity server property to this referenced property in a single step, simply right-click on the Initial Capacity property label in the Properties window (make sure you're on the label and not the property), select Set Referenced Property from the pop-up menu, and select Create New Referenced Property from the bottom of the menu. This will bring up a dialog box where you enter the Reference Property name (we used InitialCapacityFast). Next, follow the same procedure to define the referenced properties for the medium and slow fine-pitch servers.

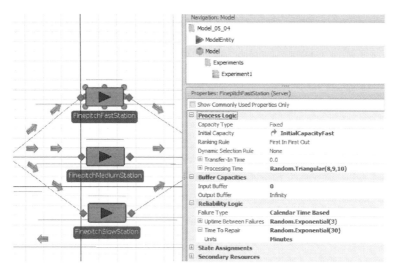

Figure 5.35: Setting the referenced property for the fast fine-pitch station.

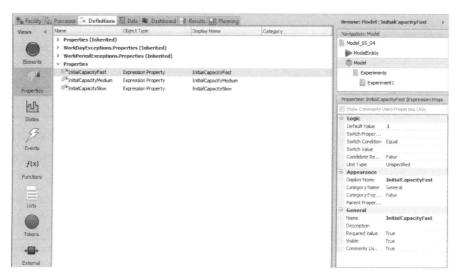

Figure 5.36: Referenced properties for Model 5-4.

After defining the three referenced properties, select the Model from the Navigation window, select the Definitions tab and the Properties the panel to display the newly defined Reference Properties (see Figure 5.36). This is where we set the Default Value property (set to 1 in Model 5-4). At this point, we have defined the new referenced properties and told Simio to use these properties to define the Initial Capacity properties for the three fine-pitch servers. Note that the referenced property values for interactive runs can be set in the Model Properties (right-click on the Model in the Navigation window and select Properties from the context menu). The next step is to define the experiment

	Design	Response Results		Pivot Grid	Reports					

Scenario			Replications		Controls			Responses		
✓	Name	Status	Required	Completed	InitialCapacityFast	InitialCapacityMedium	InitialCapacitySlow	TIS	WIP	NumTimes
✓	Add Fast	Completed	300	300 of 300	2	1	1	8.93971	89.4942	1.35112
✓	Add Medium	Completed	300	300 of 300	1	2	1	9.97657	99.9508	1.351
✓	Add Slow	Completed	300	300 of 300	1	1	2	13.0986	131.297	1.35055

Figure 5.37: Experiment Design for Model 5-4 showing the referenced properties and Subset Selection results.

that we'll use to compare our three alternatives. Before defining the experiment, we'll adjust the Processing Time properties for the three fine-pitch stations so that the performance differences will be exaggerated (just for demonstration purposes). The new processing-time distributions are `Triangular(8,9,10)`, `Triangular(18,20,22)`, and `Triangular(22,24,26)` for the fast, medium, and slow servers, respectively.

An Experiment Design for our newly updated model is in Figure 5.37. Three aspects of our experiment warrant discussion at this point — the *Controls* and *Responses* (columns in the experiment table) and the *Scenarios* (rows in the experiment table). We introduced experiment *Responses* in Section 4.5, where we demonstrated how responses are used in Simio MORE (SMORE) plots. In Model 5-4, we've defined the following responses:

1. TIS (Time in System): The time that an average entity (PCB) spends in the system. Simio tracks this statistic automatically for each entity type (`PCB.Population.TimeInSystem.Average` for the PCB objects).

2. WIP (Work in Process): The average number of entities (PCBs) in the system. Simio also tracks this statistic automatically for each entity type (`PCB.Population.NumberInSystem.Average` for the PCB objects).

3. Num Times: The average number of times that an entity is processed by the placement machine. We defined the corresponding tally statistic in Section 5.3.1 and use the expression `NumTimesProcessed.Average` to access the average value.

You can see in Figure 5.37 that the "current" response values are displayed as the experiment runs. The controls columns include the referenced properties that we created and used to specify the initial capacities of the three fine-pitch placement machine objects. In an experiment, the values for these properties can be set, and we can define a scenario by specifying the number of replications to run and values for each of the three referenced properties. As shown in Figure 5.37, we've defined three scenarios using the referenced properties to define the configurations — one for adding a fast machine, one for adding a medium machine, and one for adding a slow machine (and named accordingly). With multiple scenarios, you can explicitly control which ones Simio runs by checking/unchecking the boxes in the far left column of the experiment table. Depending on your computer processor type, Simio will run one or more scenarios simultaneously.

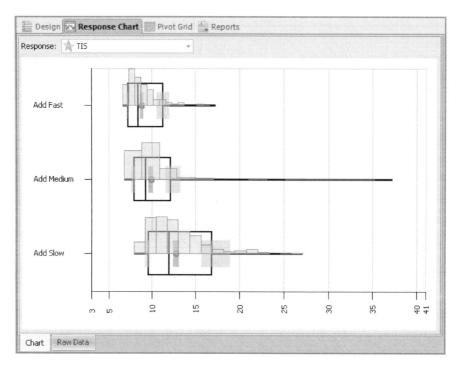

Figure 5.38: SMORE plot for the average-time-in-system Response in Model 5-4.

Figure 5.38 shows the Response Chart tab (the SMORE plot, introduced in Section 4.5) for our 3-scenario experiment with the TIS response selected. Note that the three SMORE plots are displayed on the same scale allowing a direct comparison. For these SMORE plots, we specified the Upper Percentile as 90% and the Lower Percentile as 10%, so the box in the box plot contains the middle 80% of the responses from the 300 replications; we also opted to display the plots with the value axes' being horizontal, via the Rotate Plot button in the Response Chart ribbon. As we might expect, the Add Fast scenario appears to be better (smaller average times in system) than the other two, and Add Medium appears better than Add Slow; you'll find a similar pattern with the WIP Responses selected from the pull-down near the top left of the chart (not shown). Figure 5.39 shows the SMORE plot for the average number of times that an entity is processed by the placement machine (the NumTimes Response). As we'd expect, the number of times that a board is processed does not appear to depend on whether we buy a fast, medium, or slow fine-pitch placement machine.

While some scenarios may "appear" better than others in the SMORE plots for TIS and WIP, we must be careful to use statistically valid comparisons before drawing formal conclusions. One route to this might be to use Simio's Export Details capability (discussed in Section 4.6) to create a CSV file with each response from each replication, and then import that into a dedicated statistical-analysis package that will provide several statistical-comparison methods. One

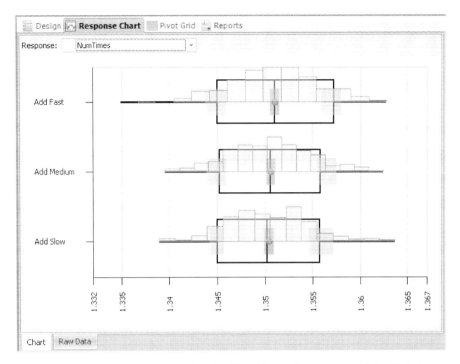

Figure 5.39: SMORE plot for the NumTimes Response in Model 5-4.

such method, if we're interested in comparing *means*, is the classical paired-*t* approach[10] to building a confidence interval on the difference between means of *two* scenarios, or doing hypothesis tests on the difference between *two* means. Note that this allows only *two* scenarios to be compared at a time, so in a situation like ours with more than two scenarios, we could perhaps do all pairwise comparisons (Fast vs. Medium, then Fast vs. Slow, then Medium vs. Slow), but we need to remember that the overall confidence level degrades when doing multiple statistical comparisons in this way. Alternatively, methods like analysis of variance (ANOVA) could be used to compare all the means at once, with standard post-test multiple-comparison methods (Tukey, Scheffé, Bonferroni, etc.) For details, see any standard statistics text.

Since this situation of comparing multiple alternative scenarios is quite common in simulation applications, special statistical methods have been developed for it. If you look back carefully at the TIS and WIP response columns in Figure 5.37, you'll notice that the Response values for the Add Medium and Add Slow scenarios are shaded light brown, but are shaded darker brown for the Add Fast scenario. This is a result of the Subset Selection Analysis available in Simio Experiments, which uses sound statistical methods from [6] to divide the scenarios into two subsets in terms of their estimated mean responses, "possible

---

[10]We'd use the *paired-t* approach, rather than the *two-sample-t* approach, since the same underlying random numbers are used to run all scenarios, so they are not independent. The paired approach can tolerate this non-independence, but the two-sample approach requires independent samples, so separate random numbers across all scenarios.

best" (shaded the darker brown) and "rejects" (shaded the light brown), based on the 300 replications that were already run for each scenario. The essential interpretation of this, per [43], is that "everything in the [possible-best] subset is statistically indistinguishable from the sample best" in terms of sample means. To make this Subset Selection happen, after completing your Experiment runs, first select the TIS response in the Experiment Design window, and in the resulting Properties grid under Objective, select Minimize (via the pulldown at the right) since smaller values for average time in system are better. Then, in the ribbon for the Experiment Design, click on the Subset Selection icon in the Analysis section, and the algorithm decides which scenarios are "possible best" (Add Fast, which is shaded the darker brown), and which are "rejects" (Add Medium and Add Slow, which are shaded the light brown). You should choose the overall statistical confidence level for this conclusion in the Experiment Properties under Confidence Level to be 90%, 95%, 98%, or 99%. You can repeat all this for the WIP Responses, where smaller is once again better, so you would again choose Minimize for the Objective property; we'd actually done this, and happened to get the same subsets of "possible best" and "rejects" as we did for the TIS responses, though such agreement for all responses is generally not to be expected. (We did not do Subset Selection for the NumTimes Responses since it's not clear whether we "want" this to be big or small.)

In Subset Selection, things don't always work out as conclusively as they did in this example, since the subset of "possible best" scenarios could contain more (maybe a lot more) than just a single scenario. In such a case, Simio provides an Add-In, "Select Best Scenario using KN," available under the Select Add-In icon on the Experiment ribbon, to narrow things down for you; again, this is based on means. Using algorithms in [26], this add-in commandeers your Experiment run, sequentially increasing the number of replications for some scenarios as called for by the algorithm, in order to isolate one scenario as very likely "the best," based on the Response you select (only one Response at a time can be considered). You must provide a confidence level for this selection, and an *indifference zone* (a value small enough so that you don't care if the mean of your selection is inferior to the mean of the true best scenario by this amount or less), as well as a replication limit that you might set based on your initial experimentation with the model. For more information on how this procedure works in Simio, see the Simio Reference Guide topic "Select Best Scenario Using KN Add-In," and of course [26] for the underlying methods; in Section 9.1.1 we use this KN Add-In to select the best scenario in a larger example.

5.6 Summary

In this chapter we've moved things along into intermediate modeling with several new Simio concepts, building on a firmer foundation of how the object orientation of Simio is defined. We also have discussed how to do a sound statistical analysis when comparing multiple alternate simulated scenarios. All of this provides a solid basis for further study of simulation with Simio, including model data, animation, and advanced modeling techniques, to be covered in the chapters that follow.

Table 5.8: Processing time parameters for Problem 4.

Process	Processing time distribution
Reception/application	triangular(5, 8, 11)
Vision exam	triangular(2, 4, 6)
Written exam	triangular(15, 15, 20)

5.7   Problems

1. What is the difference between an object *property* and an object *state*?

2. Consider a process associated with a Server object. What is the difference between a token's *parent* object and it's *associated* object?

3. Develop a queueing model that gives the steady-state values analogous to the values in Table 5.2 (Model 5-1 with exponential processing times at both stations).

4. Consider an office where people come to get their drivers' licenses. The process involves three steps for arriving customers — reception / application; a vision exam; and a written exam. Assume that customer arrivals are Poisson with a rate of 6 per hour (i.e., interarrival times are exponentially distributed with mean 10 minutes). The processing time distributions for the three processes are given in Table 5.8. The office has one person responsible for reception/application and one person responsible for administering the vision exams. The written exam is computer-based and there are three computer stations where customers can take the exam. Develop a Simio model of this system. Assume that the office opens at 9:00 a.m. and closes at 5:00 p.m. The performance metrics of interest include the time that customers spend in the system, the utilizations of the office employees and the computer stations, and the average and maximum numbers of customers in the reception/application queue, the vision exam queue, and the written exam queue. How many replications should be run in order to be confident about your results? Justify your answer.

5. Animate the model from Problem 4. If you did not already do so, specify a reasonable office layout and use Path object to incorporate customer movement through the office. Make sure that the distances between stations are reasonable for a drivers' license office and that the entity speed is appropriate for humans walking.

6. In Model 5-2, we assumed that the inspection workers did not overlap between shifts (i.e., there were never two inspectors working at the same time). Modify the model so that the inspectors *do* overlap for the first hour of one shift and the last hour of the other shift. Compare the results of the two models. Does having the inspectors overlap help?

7. In the description of Model 5-2, we noted that there is no limit on the number of times that a board may be found in need of rework. Modify Model 5-2 so that if a board fails the inspection more than two times, it

Table 5.9: Arrival and staffing data for Problem 9.

Time Period	C	P	λ_1	λ_2
8:00 a.m. - 11:00 a.m.	1	2	10	12
11:00 a.m. - 3:00 p.m.	2	3	10	20
3:00 p.m. - 7:00 p.m.	2	2	10	15
7:00 p.m. - 10:00 p.m.	1	1	5	12

is rejected as a bad part. Count the number of boards that are rejected because of 3 or more failures of inspection.

8. In the description of Model 5-3, we indicated that as part of our model verification, we had predicated the proportions of parts that would go to the fast, medium and slow fine-pitch placement machines (38%, 33%, and 29%, respectively). Develop a queueing model to estimate these proportions.

9. Consider a pharmacy where customers come to have prescriptions filled. Customers can either have their doctor fax their prescriptions ahead of time and come at a later time to pick up their prescriptions or they can walk in with the prescriptions and wait for them to be filled. Fax-in prescriptions are handled directly by a pharmacist who fills the prescriptions and leaves the filled prescriptions in a bin behind the counter.

Historical records show that approximately 59% of customers have faxed their prescriptions ahead of time and the times for a pharmicist to fill a prescription are triangularly distributed with parameters (2,5,8) minutes. If an arriving customer has faxed his/her prescription in already, a cashier retrieves the filled prescription from the bin and processes the customer's payment. Historical records also indicate that the times required for the cashier to retrieve the prescriptions and process payment are triangularly distributed with parameters (2,4,6) minutes. If an arriving customer has not faxed the prescription ahead of time, the cashier processes payment and sends the prescription to a pharmacist, who fills the prescription. The distributions of the cashier times and pharmacist times are the same as for the fax-in customers (triangular(2,4,6) and triangular(2,5,8), respectively). Fax-in and customer arrival rates vary during the day as do the pharmacy staffing levels. Table 5.9 gives the arrival and staffing data where C is the number of cashiers, P is the number of Pharmacists, λ_1 is the arrival rate for fax-in prescriptions and λ_2 is the arrival rate for customers.

Develop a Simio model of this pharmacy. Performance metrics of interest include the average time fax-in prescriptions take to be filled, the average time customers spend in the system, and the scheduled utilizations of the cashiers and pharmacists. Assume that the pharmacy opens at 8:00 a.m. and closes at 10:00 p.m. and you can ignore faxes and customers that are still in the system at closing time (probably not the best customer service!). Use 500 replications for your analysis and generate the SMORE plots for the performance metrics of interest.

Chapter 6

Input Analysis

In the simulation models of Chapters 3-5, there were many places where we needed to specify probability distributions for input, as part of the modeling process. In Model 3-3 we said that the demand for hats was a discrete uniform random variable on the integers $\{1000, 1001, ..., 5000\}$. In Models 3-5, 4-1, and 4-2 we said that interarrival times to the queueing system had an exponential distribution with mean 1.25 minutes, and that service times were exponentially distributed with mean 1 minute; in Models 4-3 and 4-4 we changed that service-time distribution to be triangularly distributed between 0.25 and 1.75 with a mode of 1.00, which is probably a more realistic shape for a service-time distribution than is the exponential (which has a mode of 0). The models in Chapter 5 used exponential, triangular, and uniform input distributions to represent interarrival, processing, inspection, and rework times. And in the models in the chapters to follow there will be many places where we need to specify probability distributions for a wide variety of inputs that are probably best modeled with some uncertainty, rather than as fixed constants.

This raises two questions:

1. How do you specify such input distributions in practice (as opposed to just making them up, which we authors get to do, after some experimentation since we're trying to demonstrate some ideas and make some points)?

2. Once you've somehow specified these input distributions, how do you "generate" values from them to make your simulation go?

We'll discuss both in this chapter, but mostly the first question since it's something that *you* will have to do as part of model building. The second question is fortunately pretty much covered within simulation software like Simio; still, you need to understand the basics of how those things work so you can properly design and analyze your simulation experiments.

We do assume in this chapter (and for this whole book in general) that you're already comfortable with the basics of probability and statistics. This includes:

- All the probability topics itemized at the beginning of Chapter 2.

- *Random sampling*, and *estimation* of means, variances, and standard deviations.

- *Independent and identically distributed* (IID) data observations and RVs.

- *Sampling distributions* of estimators of means, variances, and standard deviations.

- *Point estimation* of means, variances, and standard deviations, including the idea of *unbiasedness*.

- *Confidence intervals* for means and other population parameters, and how they're interpreted.

- *Hypothesis tests* for means and other population parameters (though we'll briefly discuss in this chapter a specific class of hypothesis tests we need, *goodness-of-fit* tests), including the concept of the *p-value* (a.k.a. observed significance level) of a hypothesis test.

If you're rusty on any of these concepts, we strongly suggest that you now dust off your old probability/statistics books and spend some quality time reviewing before you read on here. You don't need to know by heart lots of formulas, nor have memorized the normal or t tables, but you do need familiarity with the ideas. There's certainly a lot more to probability and statistics than the above list, like Markov and other kinds of stochastic processes, regression, causal path analysis, data mining, and many, many other topics, but what's on this list is all we'll really need.

In Section 6.1 we'll discuss methods to specify *univariate* input probability distributions, i.e., when we're concerned with only one scalar variate at a time, independently across the model. Section 6.2 surveys more generally the kinds of inputs to a simulation model, including correlated, multivariate, and process inputs (one important example of which is a *nonstationary Poisson process* to represent arrivals with a time-varying rate). In Section 6.3 we'll go over how to generate *random numbers* (continuous observations distributed uniformly between 0 and 1), which turns out to be a lot trickier than many people think, and the (excellent) random-number generator that's built into Simio. Finally, Section 6.4 describes how those random numbers are transformed into *observations* (or *realizations* or *draws*) from the probability distributions and processes you decided to use as inputs for your model.

6.1 Specifying Univariate Input Probability Distributions

In this section we'll discuss the common task of how to specify the distribution of a univariate random variable for input to a simulation. Section 6.1.1 describes our overall tactic, and Section 6.1.2 delineates options for using observed real-world data. Sections 6.1.3 and 6.1.4 go through choosing one or more candidate distributions, and then fitting these distributions to your data; Section 6.1.5 goes into a bit more detail on assessing whether fitted distributions really are a good fit. Section 6.1.6 discusses some general issues in distribution specification.

	A	B	C	D	E
1	34.2	47 observed real-world service times.			
2	28.4				
3	26.9				
4	23.5				
5	21.9				
6	21.5				
7	32.6				

Figure 6.1: Excerpt of `Data_06_01.xls` with 47 observed service times (in minutes).

6.1.1 General Approach

Most simulations will have several places where we need to specify probability distributions to represent random numerical inputs, like the demand for hats, interarrival times, service times, machine up/down times, travel times, or batch sizes, among many other examples. So for each of these inputs, just by themselves, we need to specify a *univariate* probability distribution (i.e., for just a one-dimensional RV, not multivariate or vector-valued RVs). We also typically assume that these input distributions are *independent* of each other across the simulation model (though we'll briefly discuss in Section 6.2.2 the possibility and importance of allowing for correlated or vector-valued random inputs to a simulation model).

For example, in most queueing-type simulations we need to specify distributions for service times, which could represent processing a part at a machine in a manufacturing system, or examining a patient in an urgent-care clinic. Typically we'll have real-world data on these times, either already available or collected as part of the simulation project. Figure 6.1 shows the first few rows of the Excel spreadsheet file `Data_06_01.xls` (downloadable from the "students" section of the book's website as described in the Preface), which contains a sample of 47 service times, in minutes, one per row in column A. We'll assume that our data are IID observations on the actual service times, and that they were taken during a stable and representative period of interest. We seek a probability distribution that well represents our observed data, in the sense that a fitted distribution will "pass" statistical goodness-of-fit hypothesis tests, illustrated below. Then, when we run the simulation, we'll "generate" random *variates* (observations or samples or realizations of the underlying service-time RV) from this fitted distribution to drive the simulation.

6.1.2 Options for Using Observed Real-World Data

You might wonder why we don't take what might seem to be the more natural approach of just using our observed real-world service times directly by reading them right into the simulation, rather than taking this more indirect approach of fitting a probability distribution, and then generating random variates from that fitted distribution to drive the simulation. There are several reasons:

- As you'll see in later chapters, we typically want to run simulations for a very long time and often repeat them for a large number of IID *replications* in order to be able to draw statistically valid and precise conclusions from the output, so we'd simply run out of real-world input data in short order.

- The observed data in Figure 6.1 and in `Data_06_01.xls` represent only what happened when the data were taken at that time. At other times, we would have observed different data, perhaps just as representative, and in particular could have gotten a different range (minimum and maximum) of data. If we were to use our observed data directly to drive the simulation, we'd be stuck with the values we happened to get, and could not generate anything outside our observed range. Especially when modeling service times, large values, even if infrequent, can have great impact on typical queueing congestion measures, like average time in system and maximum queue length, so by "truncating" the possible simulated service times on the right, we could be biasing our simulation results.

- Unless the sample size is quite large, the observed data could well leave gaps in regions that should be possible but in which we just didn't happen to get observations that time, so such values could never happen during the simulation.

So, there are several reasons why fitting a probability distribution, and then generating random variates from it to drive the simulation, is generally preferred over using the observed real-world data directly.[1] Our job, then, is to figure out *which* probability distribution best represents our observed data.

6.1.3   Choosing Probability Distributions

Of course, there are many probability distributions from which to choose in order to model random inputs for your model, and you're probably familiar with several. Common continuous distributions include the normal, exponential, continuous uniform, and triangular; on the discrete side you may be familiar with discrete uniform, binomial, Poisson, geometric, and negative binomial. We won't provide a complete reference on all these distributions, as that's readily available elsewhere. The *Simio Reference Guide* (in Simio itself, tap the F1 key or click the "?" icon near the upper right corner) provides basic information on the approximately 20 distributions supported (i.e., which you can specify in your Simio model and Simio will generate random variates from them), via the Simio Reference Guide Contents-tab path "Modeling in Simio" → "Expression Editor, Functions and Distributions" → "Distributions," to select the distribution you want. Figure 6.2 shows this path on the left, and the gamma-distribution entry on the right, which includes a plot of the PDF (the smooth continuous line since in this case it's a continuous distribution), a

---

[1]One situation where driving the simulation directly from the observed real-world data might be a good idea is in model *validation*, to assess whether your simulation model accurately reproduces the performance of the real-world system. Doing this, though, would require that you *have* the output data from the real-world system against which to match up your simulation outputs, which might not be the case if your data-collection efforts were directed at specifying *input* distributions.

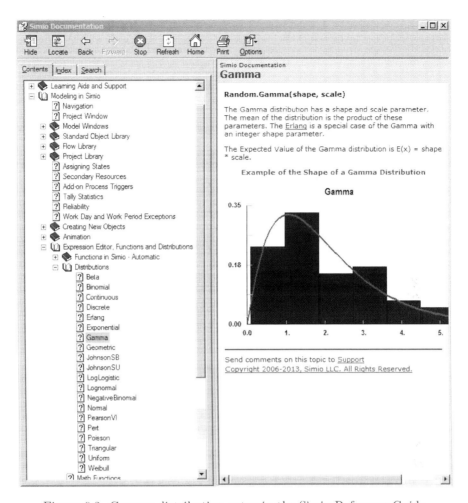

Figure 6.2: Gamma-distribution entry in the Simio Reference Guide.

histogram of possible data that might be well fitted by this distribution, and
some basic information about it at the top, including the syntax and parameterization for it in Simio. There are also extended discussions and chapters
describing probability distributions useful for simulation input modeling, such
as [53] and [30]; moreover, entire multi-volume books (e.g., [21], [22], [20], [11])
have been written describing in great depth many, many probability distributions. Of course, numerous websites offer compendia of distributions. A
search on "probability distributions" returned over 7 million results, such as
en.wikipedia.org/wiki/List_of_probability_distributions; this page in
turn has links to web pages on well over 100 specific univariate distributions,
divided into categories based on range (or *support*) for both discrete and continuous distributions, like en.wikipedia.org/wiki/Gamma_distribution, to
take the same gamma-distribution case as in Figure 6.2. Be aware that distributions are not always parameterized the same way, so you need to take care
with what matches up with Simio; for instance, even the simple exponential

distribution is sometimes parameterized by its *mean* $\beta > 0$, as it is in Simio, but sometimes with the *rate* $\lambda = 1/\beta$ of the associated Poisson process with events happening at mean rate λ, so inter-event times are exponential with mean $\beta = 1/\lambda$.

With such a bewildering array of possibly hundreds of probability distributions, how do you choose *one*? First of all, it should make *qualitative* sense. By that we mean several things:

- It should match whether the input quantity is inherently discrete or continuous. A batch size should not be modeled as a continuous RV (unless perhaps you generate it and then round it to the nearest integer), and a time duration should generally not be modeled as a discrete RV.

- Pay attention to the range of possible values the RV can take on, in particular whether it's finite or infinite on the right and left, and whether that's appropriate for a particular input. As mentioned above, whether the right tail of a distribution representing service times is finite or infinite can matter quite a lot for queueing-model results like time (or number) in queue or system.

- Related to the preceding point, you should be wary of using distributions that have infinite tails both ways, in particular to the left. This of course includes the normal distribution (though you know it and love it) since its PDF always has an infinite left (and right) tail, so will always extend to the left of zero and will thus always have a positive probability of generating a negative value. This makes no sense for things like service times and other time-duration inputs. Yes, it's true that if the mean is three or four standard deviations above zero, then the probability of realizing a negative value is "small," as basic statistics books call it and sometimes say you can thus just ignore it. The actual probabilities of getting negative values are 0.00134990 and 0.00003167 for, respectively, the mean's being three and four standard deviations above zero. But remember, *this is computer simulation* where we could easily generate hundreds of thousands, even millions of random variates from our input probability distributions, so it could well happen that you'll eventually get a negative one, which, depending on how that's handled in your simulation when it doesn't make sense, could create undesired or even erroneous results (Simio actually terminates your run and generates an error message for you if this happens, which is the right thing for it to do). The above two probabilities are respectively about one in 741 and one in 31,574, hardly out of the realm of possibility in simulation. If you find yourself wanting to use the normal distribution due to its shape (and maybe being a good fit to your data), know that there are other distributions, notably the Weibull, that can match the shape of the normal very closely, but have *no* tail to the left of zero in their PDFs, so have *no* chance of generating negative values.

So getting discrete vs. continuous right, and the range right, is a first step and will narrow things down.

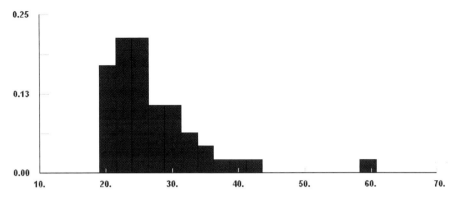

Figure 6.3: Histogram of the observed 47 service-times (in minutes) in `Data_06_01.xls`.

But still, there could be quite a few distributions remaining from which to choose, so now you need to look for one whose shape (PMF for discrete or PDF for continuous) resembles, at least roughly, the shape of a *histogram* of your observed real-world data. The reason for this is that the histogram is an empirical graphical estimate of the true underlying PMF or PDF of the data. Figure 6.3 shows a histogram (made with part of the Stat::Fit® distribution-fitting software, discussed more below) of the 47 observed service times from Figure 6.1 and the file `Data_06_01.xls`. Since this is a service time, we should consider a continuous distribution, and one with a finite left tail (to avoid generating negative values), and perhaps an infinite right tail, in the absence of information placing an upper bound on how large service times can possibly be. Browsing through a list of PDF plots of distributions (such as the *Simio Reference Guide* cited above and in Simio via the F1 key or the "?" icon near the upper right corner), possibilities might be Erlang, gamma (a generalization of Erlang), lognormal, Pearson VI, or Weibull. But each of these has parameters (like shape and scale, for the gamma and Weibull) that we need to estimate; we also need to *test* whether such distributions, with their parameters estimated from the data, provide an acceptable fit, i.e., an adequate representation of our data. This estimating and goodness-of-fit testing is what we mean by *fitting* a distribution to observed data.

6.1.4 Fitting Distributions to Observed Real-World Data

Since there are several stand-alone third-party packages available that do such distribution fitting and subsequent goodness-of-fit testing, Simio does not provide this capability internally. One such package is Stat::Fit® from Geer Mountain Software Corporation (`www.geerms.com`), which we chose to discuss in part since it has a free "textbook" version available for download from the "students" area of this book's web site as described in the Preface. Another reason we chose Stat::Fit is that it can export the specification of the fitted distribution in the proper parameterization and syntax for simple copy/paste directly into Simio. A complete Stat::Fit manual comes with the textbook-version download as a .pdf file, so we won't attempt to give a complete description of Stat::Fit's capabilities, but rather just demonstrate some of its basics. Other packages include the Distribution Fitting tool in @RISK® (Palisade Corporation, www.palisade.com), and ExpertFit® (Averill M. Law & Associates, www.averill-law.com).

Here's how to get the textbook version of Stat::Fit running on your system. Download the .zip file from the "students" area of the book's website and unzip it to a convenient folder to which you have access (like C:\StatFit). You can run Stat::Fit by double-clicking on the statfit.exe file you just unzipped, or you can make it easier to run by creating a shortcut to it on your desktop: Right-click on your desktop, select New ▷ Shortcut, and specify the location of the statfit.exe file (e.g., C:\StatFit\statfit.exe), or browse to it. Currently, there is no formal "installation" procedure for Stat::Fit, like you may be familiar with from other software.

The Stat:Fit help menu is based on Windows Help, an older system no longer directly supported in Windows Vista or Windows 7. The Microsoft web site support.microsoft.com/kb/917607 has a full explanation as well as links to download an operating-system-specific version of WinHlp32.exe that should solve the problem and allow you to gain access to Stat::Fit's help system. Remember, there's an extensive .pdf manual included in the same .zip file for Stat::Fit that you can download from this book's web site.

Figure 6.4 shows Stat::Fit with our service-time data pasted into the Data Table on the left. You can just copy/paste your observed real-world data directly from Excel; only the first 19 of the 47 values present appear, but all 47 are there and will be used by Stat::Fit. In the upper right is the histogram of Figure 6.3, made via the menu path Input → Input Graph (or the Graph Input button on the toolbar); we changed the number of intervals from the default 7 to 17 (Input → Options, or the Input Options button labeled just "OPS" on the toolbar). In the bottom right window are some basic descriptive statistics of our observed data, made via the menu path Statistics → Descriptive.

To view PMFs and PDFs of the supported distributions, follow the menu path Utilities → Distribution Viewer and then select among them via the pull-down field in the upper right of that window. Note that the free textbook version of Stat::Fit includes only seven distributions (binomial and Poisson for discrete; exponential, lognormal, normal, triangular, and uniform for continuous) and is limited to 100 data points, but the full commercial version supports 33 distributions and allows up to 8000 observed data points.

We want a continuous distribution to model our service times, and don't want to allow even a "small" chance of generating a negative value, so the qualitatively sensible choices from this list are exponential, lognormal, triangular, and uniform (from among those in the free textbook version — the full version, of course, would include many other possible distributions). Though the histogram shape certainly would seem to exclude uniform, we'll include it anyway just to demonstrate what happens if you fit a distribution that doesn't fit very well. To select these four distributions for fitting, follow the menu path Fit → Setup, or click the Setup Calculations button (labeled just "SETUP") on the toolbar, to bring up the Setup Calculations window. In the Distributions tab (Figure 6.5), click one at a time on the distributions you want to fit in the

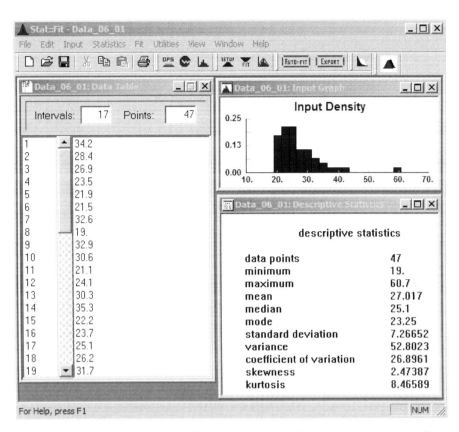

Figure 6.4: Stat::Fit with data file, histogram, and descriptive statistics (data are in minutes).

Distribution List on the left, which copies them one by one to the Distributions Selected list on the right; if you want to remove one from the right, just click on it there.

The Calculations tab (Figure 6.6) has choices about the method to form the Estimates (MLE, for Maximum Likelihood Estimation, is recommended), and the Lower Bound for the distributions (which you can select as "unknown" and allow Stat::Fit to choose one that best fits your data, or specify a "fixed" lower bound if you want to force that). The Calculations tab also allows you to pick the kinds of goodness-of-fit Tests to be done: Chi Squared, Kolmogorov-Smirnov, or Anderson-Darling,[2] and, if you selected the Chi Squared test, the kind of intervals that will be used (Equal Probability is usually best).

The Lower Bound specification requires a bit more explanation. Many distributions, like exponential and lognormal on our list, usually have zero as their customary lower bound, but it may be possible to achieve a better fit to your data with a non-zero lower bound, essentially sliding the PDF left or right (usually right for positive data like service times) to match up better with

---

[2]See Section 6.1.5 for a bit more on assessing goodness of fit; for more detail see [3] or [30].

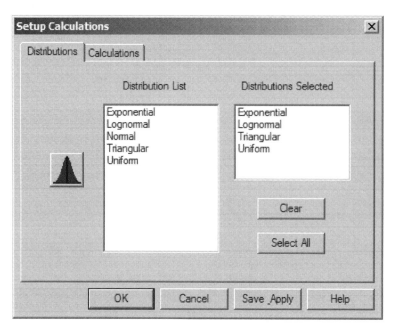

Figure 6.5: Stat::Fit Setup Calculations window, Distributions tab.

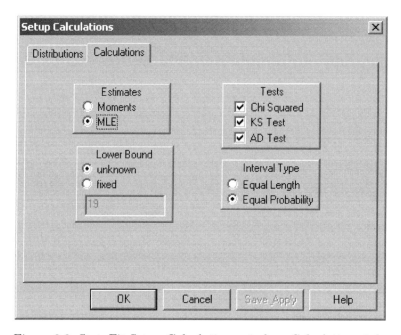

Figure 6.6: Stat::Fit Setup Calculations window, Calculations tab.

the histogram of your data. If you select "fixed" here you can decide yourself where you'd like the lower end of your distribution to start (in which case the field below it becomes active so you can enter your value — the default is the minimum value in your data set, and entering 0 results in the customary lower bound). But if you select "unknown" here the numerical field becomes inactive since this selection lets Stat::Fit make this determination, which will usually be a bit less than the minimum observation in your data set to get a good fit. Click Save Apply to save and apply all your selections.

To fit the four distributions you chose to represent the service-time data, follow the menu path Fit → Goodness of Fit, or click the Fit Data button (labeled just "FIT") in the toolbar, to produce a detailed window of results, the first part of which is in Figure 6.7. A brief summary is at the top, with the test statistics for all three tests applied to each distribution; for all of these tests, a smaller value of the test statistic indicates a better fit (the values in parentheses after the Chi Squared test statistics are the degrees of freedom). The lognormal is clearly the best fit, followed by exponential, then triangular, and uniform is the worst (largest test statistics).

In the "detail" section further down, you can find, well, details of the fit for each distribution in turn; Figure 6.7 shows these for only the exponential distribution, and the other three distributions' fit details can be seen by scrolling down in this window. The most important parts of the detail report are the *p*-values for each of the tests, which for the fit of the exponential distribution is 0.769 for the Chi Squared test, 0.511 for the Kolmogorov-Smirnov test, and 0.24 for the Anderson-Darling test, with the "DO NOT REJECT" conclusion just under each of them. Recall that the *p*-value (for any hypothesis test) is the probability that you would get a sample more in favor of the alternate hypothesis than the sample that you actually got, if the null hypothesis is really true. For goodness-of-fit tests, the null hypothesis is that the candidate distribution adequately fits the data. So large *p*-values like these (recall that *p*-values are probabilities, so are always between 0 and 1) indicate that it would be quite easy to be more in favor of the alternate hypothesis than we are; in other words, we're not very much in favor of the alternate hypothesis with our data, so that the null hypothesis of an adequate fit by the exponential distribution appears quite reasonable.

Another way of looking at the *p*-value is in the context of the more traditional hypothesis-testing setup, where we pre-choose α (typically small, between 0.01 and 0.10) as the probability of a Type I error (erroneously rejecting a true null hypothesis), and we reject the null hypothesis if and only if the *p*-value is less than α; the *p*-values for all three of our tests for goodness of the exponential fit are well above any reasonable value of α, so we're not even close to rejecting the null hypothesis that the exponential fits well. If you're following along in Stat::Fit (which you should be), and you scroll down in this window you'll see that for the lognormal distribution the *p*-values of these three tests are even bigger — they show up as 1, though of course they're really a bit less than 1 before roundoff error, but in any case provide no evidence at all against the lognormal distribution's providing a good fit. But, if you scroll further on down to the triangular and uniform fit details, the *p*-values are tiny, indicating that the triangular and uniform distributions provide terrible fits to the data (as we

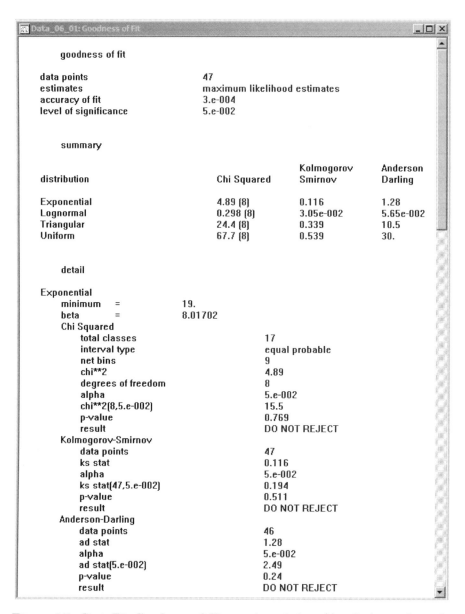

Figure 6.7: Stat::Fit Goodness of Fit results window (detail shown for only the exponential distribution; similar results for the lognormal, triangular, and uniform distributions are below these when scrolling down).

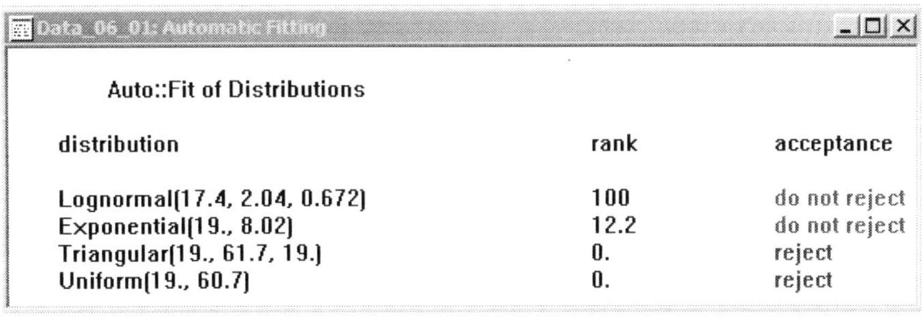

Data_06_01: Automatic Fitting

Auto::Fit of Distributions

distribution	rank	acceptance
Lognormal(17.4, 2.04, 0.672)	100	do not reject
Exponential(19., 8.02)	12.2	do not reject
Triangular(19., 61.7, 19.)	0.	reject
Uniform(19., 60.7)	0.	reject

Figure 6.8: Stat::Fit Auto::Fit results.

already knew from glancing at the histogram in the case of the uniform, but maybe not so much for the triangular).

If all you want is a quick summary of which distributions might fit your data and which ones probably won't, you could just do Fit → Auto::Fit, or click the Auto::fit button in the toolbar, to get first the Auto::Fit dialogue (not shown) where you should select the "continuous distributions" and "lower bound" buttons for our service times (since we don't want an infinite left tail), then OK to get the Automatic Fitting results window in Figure 6.8. This gives the overall "acceptance" (or not) conclusion for each distribution, referring to the null hypothesis of an adequate fit, without the p-values, and in addition the parameters of the fitted distributions (with Stat::Fit's parameterization conventions, which, as noted earlier, are not universally agreed to, and so in particular may not agree with Simio's conventions in every case). The "rank" column is an internal score given by Stat::Fit, with larger ranks indicating a better fit (so lognormal is the best, consistent with the p-value results from the detailed report).

A graphical comparison of the fitted densities against the histogram is available via Fit → Result Graphs → Density, or the Graph Fit button on the toolbar, shown in Figure 6.9. By clicking on the distributions in the upper right you can overlay them all (in different colors per the legend at the bottom); you can get rid of them by clicking in the lower-right window. This provides our favorite goodness-of-fit "test," the *eyeball test*, and provides a quick visual check (so we see just how ridiculous the uniform distribution is for our service-time data, and the triangular isn't much better).

The final step is to translate the results into the proper syntax for copying and direct pasting into Simio expression fields. This is on File → Export → Export fit, or the Export toolbar button, and brings up the EXPORT FIT dialog shown in Figure 6.10. The top left drop-down contains a list of several simulation-modeling packages, Simio among them, and the top right drop-down shows the distributions you've fit. Choose the distribution you want, let's say the fitted lognormal, then select the Clipboard radio button just below and see the valid Simio expression `17.4+Random.Lognormal(2.04, 0.672)` in the panel below. This expression now occupies the Windows clipboard, so if you go into the Simio application at this point and paste (CTRL-V) into an expression field that accepts a probability distribution, you're done. Note that Stat::Fit is

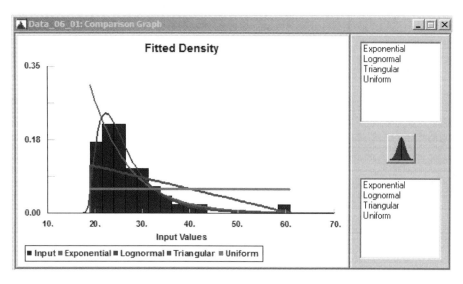

Figure 6.9: Stat::Fit overlays of the fitted densities over the data histogram.

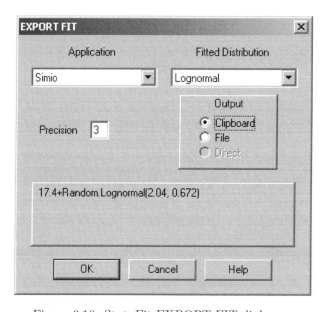

Figure 6.10: Stat::Fit EXPORT FIT dialogue.

recommending that we shift the lognormal distribution up (to the right) by 17.4 in order to get a better fit to our data, rather than stay with the customary lower end of zero for the lognormal distribution; this shift value of 17.4 is a little less than the minimum value of 19.0 in the data set, so that the fitted lognormal density function will "hit zero" on the left just a bit to the left of the minimum data value (so when generating variates, the smallest possible value will be 17.4).

You can save all your Stat::Tools work, including the data, distribution selection, results, and graphics. As usual, do File → Save or Save As, or the usual Save button. The default filename extension is .sfp (for Stat::Fit Project); we chose the file name Data_06_01.sfp.

6.1.5   More on Assessing Goodness of Fit

In Section 6.1.4 we gave an overview of fitting distributions to data with Stat::Fit, and testing how well the various candidate distributions do in terms of representing our observed data set. There are many ways to assess goodness of fit, far more than we can cover in depth; see [3] or [30] for more detail. But in this section we'll describe what's behind some of the main methods we described above, including formal hypothesis tests as well as informal visual heuristics. We won't treat the Anderson-Darling test; see [3] or [30].

Chi-Squared Goodness-of-Fit Test

In Figure 6.9, what we're looking for in a good fit is good (or at least reasonable) agreement between the histogram (the blue bars) and the PDFs of the fitted distributions (the lines of various colors in the legend at the bottom); for this discussion we'll assume we're fitting a continuous distribution. While this desire is intuitive, there's also a mathematical reason for it.

Let n be the number of real-world observed data points (so $n = 47$ in our example), and let k be the number of intervals defining the histogram along the horizontal axis (so $k = 17$ in our example); for now, we'll assume that the intervals are all of equal width w as they would be in a histogram, though they don't need to be for the chi-squared test, and in many cases they shouldn't be ... more on that below. Let the left endpoints of the k intervals be denoted $x_0, x_1, x_2, \ldots, x_{k-1}$ and let x_k be the right endpoint of the kth interval, so the jth interval is $[x_{j-1}, x_j)$ for $j = 1, 2, \ldots, k$. If O_j of our n observations fall in the jth interval (the *observed* frequency of the data in that interval), then O_j/n is the *proportion* of the data falling into the jth interval.

Now, let $\widehat{f}(x)$ be the PDF of the fitted distribution under consideration. If this is really the right PDF to represent the data, then the probability that a data point in our observed sample falls in the jth interval $[x_{j-1}, x_j)$ would be

$$p_j = \int_{x_{j-1}}^{x_j} \widehat{f}(x) dx,$$

and this should be (approximately) equal to the observed proportion O_j/n. So if this is actually a good fit, we would expect $O_j/n \approx p_j$ for all intervals j. Multiplying this through by n, we would expect that $O_j \approx np_j$ for all intervals

j, i.e., the *observed* and *expected* (if $\widehat{f}$ really is the right density function for our data) frequencies within the intervals should be "close" to each other; if they're in substantial disagreement for too many intervals, then we would suspect a poor fit. This is really what our earlier "eyeball" test is doing.

But even in the case of a really good fit, we wouldn't demand $O_j = np_j$ (exactly) for all intervals j, simply due to natural fluctuation in random sampling. To formalize this idea, we form the chi-squared test statistic

$$\chi^2_{k-1} = \sum_{j=1}^{k} \frac{(O_j - np_j)^2}{np_j},$$

and under the null hypothesis H_0 that the observed data follow the fitted distribution $\widehat{f}$, χ^2_{k-1} has (approximately[3]) a chi-squared distribution with $k-1$ degrees of freedom. So we'd reject the null hypothesis H_0 of a good fit if the value of the test statistic χ^2_{k-1} is "too big" (or "inflated" as it's sometimes called). How big is "too big" is determined by the chi-squared tables: for a given size α of the test, we'd reject H_0 if $\chi^2_{k-1} > \chi^2_{k-1,1-\alpha}$ where $\chi^2_{k-1,1-\alpha}$ is the point above which is probability α in the chi-squared distribution with $k-1$ degrees of freedom. The other way of stating the conclusion is to give the p-value of the test, which is the probability above the test statistic χ^2_{k-1} in the chi-squared distribution with $k-1$ degrees of freedom; as usual with p-values, we reject H_0 at level α if and only if the p-value is less that α.

Note that the value of the test statistic χ^2_{k-1}, and perhaps even the conclusion of the test, depends on the choice of the intervals. How to choose these intervals is a question that has received considerable research, but there's no generally accepted "right" answer. If there is consensus, it's that (1) the intervals should be chosen *equiprobably*, i.e., so that the probability values p_j of the integrals are equal to each other (or at least approximately so), and (2) so that np_j is at least (about) 5 for all intervals j (the latter condition is basically to discourage use of the chi-squared goodness-of-fit test if the sample size is quite small). One way to find the endpoints of equiprobable intervals is to set $x_j = \widehat{F}^{-1}(j/k)$, where $\widehat{F}$ is the CDF of the fitted distribution, and the superscript -1 denotes the functional inverse (not the arithmetic reciprocal); this entails solving the equation $\widehat{F}(x_j) = j/k$ for x_j, which may or may not be straightforward, depending on the form of $\widehat{F}$, and so may require use of a numerical-approximation root-finding algorithm like the secant method or Newton's method.

The chi-squared goodness-of-fit test can also be applied to fitting a discrete distribution to a data set whose values must be discrete (like batch sizes). In the foregoing discussion, p_j is just replaced by the *sum* of the fitted PMF values within the jth interval, and the procedure is the same. Note that for discrete distributions it will generally not be possible to attain exact equiprobability on the choice of the intervals.

---

[3]More precisely, *asymptotically*, i.e., as $n \to \infty$.

Kolmogorov-Smirnov Goodness-of-Fit Test

While chi-squared tests amount to comparing the fitted PDF (or PMF in the discrete case) to an "empirical" PDF or PMF (a histogram), Kolmogorov-Smirnov (K-S) tests compare the fitted CDF to a certain empirical CDF defined directly from the data. There are different ways to define empirical CDFs, but for this purpose we'll use $F_{\text{emp}}(x)$ = the proportion of the observed data that are $\leq x$, for all x; note that this is a step function that is continuous from the right, with a step of height $1/n$ occurring at each of the (ordered) observed data values. As before, let $\widehat{F}(x)$ be the CDF of a particular fitted distribution. The K-S test statistic is then the largest vertical discrepancy between $F_{\text{emp}}(x)$ and $\widehat{F}(x)$ along the entire range of possible values of x; expressed mathematically, this is

$$V_n = \sup_x \left| F_{\text{emp}}(x) - \widehat{F}(x) \right|,$$

where "sup" is the *supremum*, or the least upper bound across all values of x. The reason for not using the more familiar max (maximum) operator is that the largest vertical discrepancy might occur just before a "jump" of $F_{\text{emp}}(x)$, in which case the supremum won't actually be attained exactly at any particular value of x.

A finite (i.e., computable) algorithm to evaluate the K-S test statistic V_n is given in [30]. Let $X_1, X_2, \ldots, X_n$ denote the observed sample, and for $i = 1, 2, \ldots, n$ let $X_{(i)}$ denote the ith smallest of the data values (so $X_{(1)}$ is the smallest observation and $X_{(n)}$ is the largest observation); $X_{(i)}$ is called the *ith order statistic* of the observed data. Then the K-S test statistic is

$$V_n = \max \left\{ \max_{i=1,2,\ldots,n} \left[\frac{i}{n} - \widehat{F}(X_{(i)}) \right], \quad \max_{i=1,2,\ldots,n} \left[\widehat{F}(X_{(i)}) - \frac{i-1}{n} \right] \right\}.$$

It's intuitive that the larger V_n, the worse the fit. To decide how large is too large, we need tables or reference distributions for critical values of the test for given test sizes α, or to determine p-values of the test. A disadvantage of the K-S test is that, unlike the chi-squared test, the K-S test needs different tables (or reference distributions) for different hypothesized distributions and different sample sizes n (which is why we include n in the notation V_n for the K-S test statistic); see [14] and [30] for more on this point. Distribution-fitting packages like Stat::Fit include these reference tables as built-in capabilities that can provide p-values for K-S tests. Advantages of the K-S test over the chi-squared test are that it does not rely on a somewhat arbitrary choice of intervals, and it is accurate for small sample sizes n (not just asymptotically, as the sample size $n \to \infty$).

In Stat::Fit, the menu path Fit $\to$ Result Graphs $\to$ Distribution produces the plot in Figure 6.11 with the empirical CDF in blue, and the CDFs of various fitted distributions in different colors per the legend at the bottom; adding/deleting fitted distributions works as in Figure 6.9. While the K-S test statistic is not shown in Figure 6.11 (for each fitted distribution, imagine it as the height of vertical bar at the worst discrepancy between the empirical CDF and that fitted distribution), it's easy to see that the uniform and triangular CDFs are very poor matches to the empirical CDF in terms of the largest

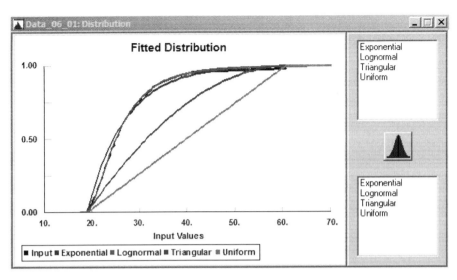

Figure 6.11: Stat::Fit overlays of the fitted CDFs over the empirical distribution.

vertical discrepancy, and that the lognormal is a much better fit (in fact, it's difficult to distinguish it from the empirical CDF in Figure 6.11).

P-P Plots

Let's assume for the moment that the CDF $\widehat{F}$ of a fitted distribution really *is* a good fit to the true underlying unknown CDF of the observed data. If so, then for each $i = 1, 2, \ldots, n$, $\widehat{F}(X_{(i)})$ should be close to the empirical proportion of data points that are at or below $X_{(i)}$. That proportion is i/n, but just for computational convenience we'd rather stay away from both 0 and 1 in that proportion (for fitted distributions with an infinite tail), so we'll make the small adjustment and use $(i-1)/n$ instead for this empirical proportion. To check out heuristically (not a formal hypothesis test) whether $\widehat{F}$ actually is a good fit to the data, we'll gauge whether $(i-1)/n \approx \widehat{F}(X_{(i)})$ for $i = 1, 2, \ldots, n$, by plotting the points $\left((i-1)/n, \widehat{F}(X_{(i)})\right)$ for $i = 1, 2, \ldots, n$; if we indeed have a good fit then these points should fall close to a diagonal straight line from $(0,0)$ to $(1,1)$ in the plot. Since both the x and y coordinates of these points are probabilities (empirical and fitted, respectively), this is called a *probability-probability plot*, or a *P-P plot*.

In Stat::Fit, P-P plots are available via the menu path Fit $\rightarrow$ Result Graphs $\rightarrow$ PP Plot, to produce Figure 6.12 for our 47-point data set and our four trial fitted distributions. As in Figures 6.9 and 6.11, you can add fitted distributions by clicking them in the upper right box, and remove them by clicking them in the lower right box. The P-P plot for the lognormal fit appears quite close to the diagonal line, signaling a good fit, and the P-P plots for the triangular and uniform fits are very far from the diagonal, once again signaling miserable fits

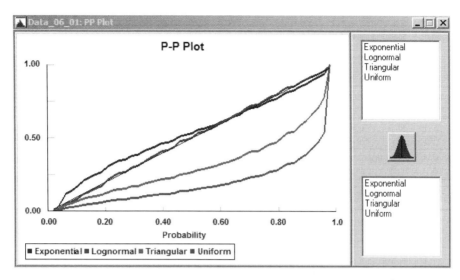

Figure 6.12: Stat::Fit P-P plots.

for those distributions. The exponential fit appears not unreasonable, though not as good as the lognormal.

Q-Q Plots

The idea in P-P plots was to see whether $(i-1)/n \approx \widehat{F}(X_{(i)})$ for $i = 1, 2, \ldots, n$. If we apply $\widehat{F}^{-1}$ (the functional inverse of the CDF $\widehat{F}$ of the fitted distribution) across this approximate equality, we get $\widehat{F}^{-1}\left((i-1)/n\right) \approx X_{(i)}$ for $i = 1, 2, \ldots, n$, since $\widehat{F}^{-1}$ and $\widehat{F}$ "undo" each other. Here, the left-hand side is a *quantile* of the fitted distribution (value below which is a probability or proportion, in this case $(i-1)/n$). Note that a closed-form formula for $\widehat{F}^{-1}$ may not be available, mandating a numerical approximation. So if $\widehat{F}$ really *is* a good fit to the data, and we plot the points $\left(\widehat{F}^{-1}\left((i-1)/n\right), X_{(i)}\right)$ for $i = 1, 2, \ldots, n$, we should again get an approximately straight diagonal line, but not between $(0,0)$ and $(1,1)$, but rather between $\left(X_{(1)}, X_{(1)}\right)$ and $\left(X_{(n)}, X_{(n)}\right)$. (Actually, Stat::Fit reverses the order and plots the points $\left(X_{(i)}, \widehat{F}^{-1}\left((i-1)/n\right)\right)$, but that doesn't change the fact that we're looking for a straight line.) Now, both the x and y coordinates of these points are quantiles (fitted and empirical, respectively), so this is called a *quantile-quantile plot*, or a *Q-Q plot*.

You can make Q-Q plots in Stat::Fit via Fit $\rightarrow$ Result Graphs $\rightarrow$ QQ Plot, to produce Figure 6.13 for our data and our fitted distributions. As in Figures 6.9, 6.11, and 6.12, you can choose which fitted distributions to display. The lognormal and exponential Q-Q plots appear to be the closest to the diagonal line, signaling a good fit, except in the right tail where neither does well. The Q-Q plots for the triangular and uniform distributions indicate very poor fits, consistent with our previous findings. According to [30], Q-Q plots tend to be sensitive to discrepancies between the data and fitted distributions

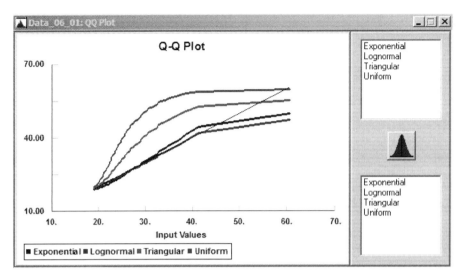

Figure 6.13: Stat::Fit Q-Q plots.

in the tails, whereas P-P plots are more sensitive to discrepancies through the interiors of the distributions.

6.1.6 Distribution-Fitting Issues

In this section we'll briefly discuss a few issues and questions that often arise when trying to fit distributions to observed data.

What If Nothing Fits?

The process just described in Section 6.1.4 is a kind of best-case scenario, and things don't always go so smoothly. It can definitely happen that, despite your (and Stat::Fit's) best efforts to find a standard distribution that fits your data, alas, all your p-values are small and you reject all your distributions' fits. Does this mean that your data are unacceptable or have somehow failed? No, on the contrary, it means that the "standard" list of distributions fails to have an entry that accommodates your data (which, after all, are "The Truth" about the phenomenon being observed, like service times — the burden is on the distributions to mold themselves to your data, not the other way around).

In this case, Stat::Fit can produce an empirical distribution, which is basically just a version of the histogram, and variates can be generated from it in the Simio simulation. On the Stat::Fit menus, do File → Export → Export Empirical, and select the Cumulative radio button (rather than the default Density button) for compatibility with Simio. This will copy onto the Windows clipboard a sequence of pairs v_i, c_i where v_i is the ith smallest of your data values, and c_i is the *cumulative* probability of generating a variate that is *less than or equal to* the corresponding v_i. Exactly what happens after you copy this information into Simio depends on whether you want a discrete or continuous distribution. In the Simio Reference Guide (F1 or "?" icon from within Simio), follow the Contents-tab path "Modeling in Simio" → "Expression Editor, Functions and Distributions" → "Distributions," then select either Discrete or Continuous according to which you want:

- If Discrete, the proper Simio expression is `Random.Discrete(v1, c1, v2, c2, ...)` where the sequence of pairs can go on as long as needed according to your data set, and you'll generate each value v_i with cumulative probability c_i (e.g., the probability is c_4 that you'll generate a value *less than or equal to* v_4, so you'll generate a value *equal to* v_4 with probability $c_4 - c_3$).

- If Continuous, the Simio expression is `Random.Continuous(v1, c1, v2, c2, ...)`, and the CDF from which you'll generate will pass through each (v_i, c_i) and will connect them with straight lines, so it will be a piecewise-linear "connect-the-dots" CDF rising from 0 at v_1 to 1 at the largest (last) v_i.

Note that in both the discrete and continuous cases, you'll end up with a distribution that's bounded both below (by your minimum data point) and above (by your maximum data point), i.e., it does not have an infinite tail in either direction.

What If I Have No Data?

Obviously, this is not a good situation. What people usually do is ask an "expert" familiar with the system (or similar systems) for, say, lowest and highest values that are reasonably possible, which then specify a uniform distribution. If you feel that uniform gives too much weight toward the extremes, you could instead use a triangular distribution with a mode (peak of the PDF, not necessarily the mean) that may or may not be in the middle of the range, depending on the situation. Once you've done such things and have a model working, you really ought to use your model as a sensitivity-analysis tool (see "What's the Right Amount of Data?" just below) to try to identify which inputs matter most to the output, and then try very hard to get some data on at least those important inputs, and try to fit distributions.

What If I Have "Too Much" Data?

Another extreme situation is when you have a very large sample size of observed real-world data of maybe thousands. Usually we're happy to have a lot of data, and really, we are here too. We just have to realize that, with a very large sample size, the goodness-of-fit hypothesis tests will have high statistical *power* (probability of rejecting the null hypothesis when it's really false, which strictly speaking, it always is), so it's likely that we'll reject the fits of all distributions, even though they may look perfectly reasonable in terms of the eyeball test. In such cases, we should remember that goodness-of-fit tests, like all hypothesis tests, are far from perfect, and we may want to go ahead and use a fitted distribution anyway even if goodness-of-fit tests reject it with a large sample size, provided that it passes the eyeball test.

What's the Right Amount of Data?

Speaking of sample size, people often wonder how much real-world data they need to fit a distribution. Of course, there's no universal answer possible to such a question. One way to address it, though, is to use your simulation model itself as a sensitivity-analysis tool to gauge how sensitive some key outputs are to changes in your input distributions. Now you're no doubt thinking "but I can't even build my model if I don't know what the input distributions are," and strictly speaking, you're right. However, you could build your model first, before even collecting real-world data to which to fit input probability distributions, and initially just use made-up input distributions — not arbitrary or crazy distributions, but in most cases you can make a reasonable guess using, say, simple uniform or triangular input distributions and someone's general familiarity with the system. Then vary these input distributions to see which ones have the most impact on the output — those are your critical input distributions, so you might want to focus your data collection there, rather than on other input distributions that seem not to affect the output as much.

What's the Right Answer?

A final comment is that distribution specification is not an exact science. Two people can take the same data set and come up with different distributions, both of which are perfectly reasonable, i.e., provide adequate fits to the data, but are different distributions. In such cases you might consider secondary criteria, such as ease of parameter manipulation to effect changes in the distributions' means. If that's easier with one distribution than the other, you might go with the easier one in case you want to try different input-distribution means (e.g., what if you had a server that was 20% faster on average?).

6.2 Types of Inputs

Having discussed univariate distribution fitting in Section 6.1, we should now take a step back and think more generally about all the different kinds of numerical inputs that go into a simulation model. We might classify these along two dimensions in a 2×3 classification: deterministic vs. stochastic, and scalar vs. multivariate vs. stochastic processes.

6.2.1 Deterministic vs. Stochastic

Deterministic inputs are just constants, like the number of automated check-in kiosks a particular airline has at a particular airport. This won't change during the simulation run — unless, of course, the kiosks are subject to breakdowns at random points in time, and then have to undergo repair that lasts a random amount of time. Another example of deterministic input might be the pre-scheduled arrival times of patients to a dental office.

But wait, have you *never* been late (or early) to a dental appointment? So arrival times might be more realistically modeled as the (deterministic) scheduled time, plus a "deviation" RV that could be positive for a late arrival and negative for an early arrival[4] (and maybe with expected value zero if we assume that patients are, on average, right on time, even if not in every case). This would be an example of a *stochastic* input, which involves (or just is) an RV. Actually, for this sort of a situation it's more common to specify a distribution for the interarrival times, as we did in the spreadsheet-imprisoned queueing Model 3-3.

Often, a given input to a simulation might arguably be either deterministic or stochastic:

- The walking time of a passenger in an airport from the check-in kiosk to security. The distances are the same for everybody, but walking speeds clearly vary.

- The number of items actually in a shipment, as opposed to the (deterministic) number ordered.

- The time to stamp and cut an item at a machine in a stamping plant. This could be close to deterministic if the sheet metal off the roll is pulled through at a constant rate, and the raising/dropping of the stamping die is at a constant rate. Whether to try to model small variations in this using RVs would be part of the question of level of detail in model building (by the way, more detail doesn't always lead to a "better" model).

- The time to do a "routine" maintenance on a military vehicle. While what's planned could be deterministic, we might want to model the extra (and random) time needed if more problems are uncovered.

Whether to model an input as deterministic or stochastic is a modeling decision.[5] It's obvious that you should do whichever matches the real system, but there still could be questions about whether it matters to the simulation output.

6.2.2 Scalar vs. Multivariate vs. Stochastic Processes

If an input is just a single number, be it deterministic or stochastic, it's a *scalar* value. Another term for this, used especially if the scalar is stochastic, is *univariate*. This is most commonly how we model inputs to simulations — one scalar number (or RV) at a time, typically with several such inputs across the model — and typically assumed to be statistically independent of each other. Our discussion in Section 6.1 tacitly assumed that our model is set up this way, with all stochastic inputs being univariate and independent of each other across the model. In a manufacturing model, such scalar univariate inputs could include the time for processing a part, and the time for subsequent inspection of it. And in an urgent-care clinic, such scalar univariate inputs might include interarrival times between successive arriving patients, their age, gender, insurance status, time taken in an exam room, diagnosis code, and

---

[4]Simio provides exactly this feature with a Source object using an Arrival Type of `Arrival Table`

[5]In Appendix A we will see how a given input might be deterministic during the planning phase and stochastic during the risk analysis phase.

disposition code (like go home, go to a hospital emergency room in a private car, or call for an ambulance ride to a hospital).

But there could be relationships between the different inputs across a simulation model, in which case we really should view them as components (or coordinates) of an input *vector*, rather than being independent; if some of these components are random, then this is called a *random vector*, or a *multivariate distribution*. Importantly, this allows for dependence and correlation across the coordinates of the input random vector, making it a more realistic input than if the coordinates were assumed to be independent — and this can affect the simulation's output. In the manufacturing example of the preceding paragraph, we could allow, say, positive correlation between a given part's processing and inspection times, reflecting the reality that some parts are bigger or more problematic than are other parts. It would also allow us to prevent generating absurdities in the urgent-care clinic like a young child suffering from arthritis (unlikely), or an elderly man with complications from pregnancy (beyond unlikely), both of which would be possible if all these inputs were generated independently, which is typically what we do. Another example is a fire-department simulation, where the number of fire trucks and ambulances sent out on a given call should be positively correlated (large fires could require several of each, but small fires perhaps just one of each); see [39] for how this was modeled and implemented in one project. While some such situations can be modeled just logically (e.g., for the health-care clinic, first generate gender and then do the obvious check before allowing a pregnancy-complications diagnosis), in other situations we can model the relationships statistically, with correlations or joint probability distributions. If such non-independence is in fact present in the real-world system, it can affect the simulation output results, so ignoring it and just generating all inputs independently across your model can lead to erroneous results.

One way to specify an input random vector is first to fit distributions to each of the component univariate random variables one at a time, as in Section 6.1; these are called the *marginal distributions* of the input random vector (since in the case of a two-dimensional discrete random vector, they could be tabulated on the margins of the joint-distribution table). Then use the data to estimate the cross correlations via the usual sample-correlation estimator discussed in any statistics book. Note that specifying the marginal univariate distributions and the correlation matrix does not necessarily completely specify the joint probability distribution of the random vector, except in the case of jointly distributed normal random variables.

Stepping up the dimensionality of the input random vector to an infinite number of dimensions, we could think of a (realization of) a *stochastic process* as being an input driving the simulation model. Models of telecommunications systems sometimes do this, with the input stochastic process representing a stream of packets, each arriving at a specific time, and each being of a specific size; see [4] for a robust method to fit a very general time-series model for use as an input stream to a simulation.

For more on such "nonstandard" simulation input modeling, see [23] and [24], for example.

6.2.3 Time-Varying Arrival Rate

In many queueing systems, the arrival rate from the outside varies markedly over time. Examples come to mind like fast-food restaurants over a day, emergency rooms over a year (flu season), and urban freeway exchanges over a day. Just as ignoring correlation across inputs can lead to errors in the output results, so too can ignoring nonstationarity in arrivals. Imagine how (mis-)modeling the freeway exchange as a flat "average" arrival rate over a 24-hour day, including the arrival rate around 3:00 a.m., would likely badly understate congestion during rush hours (see [17] for a numerical example of the substantial error that this produces).

The most common way of representing a time-varying arrival rate is via a *nonstationary Poisson process*, (also called a *non-homogeneous Poisson process*). Here, the arrival rate is a function $\lambda(t)$ of simulated time t, instead of constant flat rate λ. The number of arrivals during any time interval $[a, b]$ follows a (discrete) Poisson distribution with mean $\int_a^b \lambda(t)dt$. Thus, the mean number of arrivals is higher during time intervals where the rate function $\lambda(t)$ is higher, as desired (assuming equal-duration time intervals, of course). Note that if the arrival rate actually *is* constant at λ, this specializes to a stationary Poisson process at rate λ; this is equivalent to an arrival process with interarrival times that are IID exponential RVs with mean $1/\lambda$.

Of course, to model a nonstationary Poisson process in a simulation, we need to decide how to use observed data to specify an estimate of the rate function $\lambda(t)$. This is a topic that's received considerable research, for example [29] and the references there. One straightforward way to estimate $\lambda(t)$ is via a *piecewise-constant* function. Here we assume that, for time intervals of a certain duration (let's arbitrarily say ten minutes in the freeway-exchange example to make the discussion concrete), the arrival rate actually *is* constant, but it can jump up or down to a possibly-different level at the end of each ten-minute period. You'd need to have knowledge of the system to know that it's reasonable to assume a constant rate within each ten-minute period. To specify the level of the rate function over each ten-minute period, just count arrivals during that period, and hopefully average over multiple weekdays for each period separately, for better precision (the arrival rate within an interval need not be an integer). While this piecewise-constant rate-estimation method is relatively simple, it has good theoretical backup, as shown in [33].

Simio supports generating arrivals from such a process in its Source object, where entities arrive to the system, by specifying its "Arrival Mode" to be "Time Varying Arrival Rate." The arrival-rate function is specified separately in a Simio *Rate Table*. Section 7.3 provides a complete example of implementing a nonstationary Poisson arrival process with a piecewise-constant arrival rate in this way, for a hospital emergency department. Note that in Simio, all rates must be in per-*hour* units before entering them into the corresponding Rate Table; so if your arrival-rate data are on the number of arrivals during each ten-minute period, you'd first need to multiply these rate estimates for each period by 6 to convert them to per-*hour* arrival rates.

6.3 Random-Number Generators

Every stochastic simulation must start at root with a method to "generate" *random numbers*, a term that in simulation specifically means observations uniformly and continuously distributed between 0 and 1; the random numbers also need to be independent of each other. That's the ideal, and cannot be literally attained. Instead, people have developed numerical algorithms to produce a stream of values between 0 and 1 that *appear* to be independent and uniformly distributed. By "appear" we mean that the generated random numbers satisfy certain provable theoretical conditions (like they'll not repeat themselves for a very, very long time), as well as pass batteries of tests, both statistical and theoretical, for uniformity and independence. These algorithms are known as *random-number generators* (RNGs).

You can't just think up something strange and expect it to work well as a random-number generator. In fact, there's been a lot of research on building good RNGs, which is much harder than most people think. One classical (though outmoded) method is called the *linear congruential generator* (LCG), which generates a sequence of integers Z_i based on the recurrence relation

$$Z_i = (aZ_{i-1} + c)(\text{mod } m)$$

where a, c, and m are non-negative integer constants (a and m must be > 0) that need to be carefully chosen, and we start things off by specifying a *seed* value $Z_0 \in \{0, 1, 2, \ldots, m-1\}$. Note that mod m here means to divide $(aZ_{i-1}+c)$ by m and set Z_i to be the *remainder* of this division (you may need to think back to before your age hit double digits to remember long division and remainders). Because it's a remainder of a division by m, each Z_i will be an integer between 0 and $m-1$, so since we need our random numbers U_i to be between 0 and 1, we let $U_i = Z_i/m$; you could divide by $m-1$ instead, but in practice m will be quite large so it doesn't much matter. Figure 6.14 shows part of the Excel spreadsheet `Model_06_01.xls` (which is available for download as described in the book's Preface) that implements this for the first 100 random numbers; Excel has a built-in function =MOD that returns the remainder of long division in column F, as desired.

The LCG's parameters a, c, m, and Z_0 are in cells B3..B6, and the spreadsheet is set up so that you can change them to other values if you'd like, and the whole spreadsheet will update itself automatically. Looking down at the generated U_i values in column F, it might appear at first that they seem pretty, well, "random" (whatever that means), but looking a little closer should disturb you. Our seed was $Z_0 = 7$, and it turns out that $Z_{22} = 7$ as well, and after that, the sequence of generated numbers just repeats itself as from the beginning, and in exactly the same order. Try changing Z_0 to other values (we dare you) to try to prevent your seed from reappearing so quickly. As you'll find out, you can't. The reason is, there are only so many integer remainders possible with the mod m operation (m, in fact) so you're bound to repeat by the time you generate your mth random number (and it could be sooner than that depending on the values of a, c, and m); this is called *cycling* of RNGs, and the length of a cycle is called its *period*. Other, less obvious issues with LCGs, even if they had long periods, is the uniformity and independence appearance we need, which

E7	▼	f_x	=MOD($B$3*E6+$B$4, $B$5)

	A	B	C	D	E	F	G	H
1	**Linear Congruential Random-Number Generator**							
2								
3	a:	17		i		Z_i	U_i	
4	c:	8		0		7	n/a	
5	m:	23		1		12	0.5217	
6	Z_0:	7		2		5	0.2174	
7				3		1	0.0435	
8				4		2	0.0870	
9				5		19	0.8261	
10				6		9	0.3913	
11				7		0	0.0000	
12				8		8	0.3478	
13				9		6	0.2609	
14				10		18	0.7826	
15				11		15	0.6522	
16				12		10	0.4348	
17				13		17	0.7391	
18				14		21	0.9130	
19				15		20	0.8696	
20				16		3	0.1304	
21				17		13	0.5652	
22				18		22	0.9565	
23				19		14	0.6087	
24				20		16	0.6957	
25				21		4	0.1739	
26				22		7	0.3043	
27				23		12	0.5217	
28				24		5	0.2174	
29				25		1	0.0435	

Figure 6.14: A linear congruential random-number generator implemented in Model_06_01.xls, with the Excel function at the top for cell E7 (which contains the value for Z_3.

are actually more difficult to achieve and require some fairly deep mathematics involving prime numbers, relatively prime numbers, etc. (*number theory*).

Several relatively good LCGs were created (i.e., acceptable values of a, c, and m were found) and used successfully for many years following their 1951 development in [34], but obviously with far larger values of m (often $m = 2^{31} - 1 = 2,147,483,547$, about 2.1 billion or on the order of 10^9). However, computer speed has come a long way since 1951, and LCGs are no longer really serious candidates for high-quality RNGs; for one thing, an LCG with a period even as high as 10^9 can be run through its whole cycle in just a few minutes on common personal computers today. So other, different methods have been developed, though many still using the modulo remainder operation internally at points, with far longer periods and much better statistical properties (independence and uniformity). We can't begin to describe them here; see [31] for a survey of these methods, and [32] for testing RNGs.

Simio's RNG is the *Mersenne twister* (see [38], or the developer's website www.math.sci.hiroshima-u.ac.jp/~m-mat/MT/emt.html), and has a truly astronomical cycle length (10^{6001}, and by comparison, it's estimated that the observable universe contains about 10^{80} atoms). And just as importantly, it has excellent statistical properties (tested independence and uniformity up to 632 dimensions). So in Simio, you at least have two fewer things to worry about — running out of random numbers, and generating low-quality random numbers.

As implemented in Simio, the Mersenne twister is divided into a huge number of hugely long *streams*, which are subsegments of the entire cycle, and it's essentially impossible for any two streams to overlap. While you can't access the seed (it's actually a seed *vector*), you don't need to since you can, if you wish, specify the stream to use, as an extra parameter in the distribution specification. For instance, if you wanted to use stream 28 (rather than the default stream 0) for the shifted lognormal that Stat::Fit fitted to our service-time data in Section 6.1, you'd enter 17.4+Random.Lognormal(2.04, 0.672, 28).

Why would you ever want to do this? One good reason is if you're comparing alternative scenarios (say, different plant layouts), you'd like to be more confident that output differences you see are due to the differences in the layouts, and not due to just having gotten different random numbers. If you dedicate a separate stream to each input distribution in your model, then when you simulate all the plant-layout scenarios you're doing a better job of *synchronizing* the random-number use for the various inputs across the different scenarios, so that there's a better chance that each layout scenario will "see" the same jobs coming at it at the same times, the processing requirements for the jobs will be the same across the scenarios, etc. This way, you've at least partially removed "random bounce" as an explanation of the different results across the alternative scenarios.

Re-using random numbers in this way is one kind of a *variance-reduction technique*, of which there are several, as discussed in general simulation texts like [3] or [30]; this particular variance-reduction technique is called *common random numbers* since we're trying to use the same (common) random numbers for the same purposes across different scenarios. In addition to being intuitively appealing (comparing like with like, or apples with apples), there's actually a probabilistic grounding for common random numbers. If Y_A and Y_B are the same output-performance RVs (e.g., total time in system) for scenarios A and B, then for comparison we're interested in $Y_A - Y_B$, and $Var(Y_A - Y_B) = Var(Y_A) + Var(Y_B) - 2Cov(Y_A, Y_B)$, where Cov denotes covariance. By using common random numbers, we hope to correlate Y_A and Y_B positively, making $Cov(Y_A, Y_B) > 0$, and thus reducing the variance of the difference $Y_A - Y_B$ in the outputs from what it would be if we ran the scenarios independently (in which case $Cov(Y_A, Y_B)$ would be 0. If you're making multiple replications of your different scenarios, Simio also arranges for each replication across each of the scenarios to start at the same point within all the streams you might be using, so that synchronizing the use of the common random numbers stays intact even after the first replication. Also, when running multiple scenarios in Simio, it will start each scenario at the beginning of each random-number stream you're using.

One important thing to realize about all random-number generators is that they aren't really random at all, in the sense of being unpredictable. If you have a fixed algorithm for the generator, and use the same seed value (or seed vector, as is the case with the Mersenne twister), you'll of course get exactly the same sequence of "random" numbers. For this reason, they're sometimes called *pseudo-random*, which is technically a more correct term. So in simulation software like Simio, if you simply re-run your model you'll get exactly the same numerical results, which often surprises people new to simulation since it seems that this shouldn't happen when using a "random"-number generator. However, if you *replicate* your model multiple times within the same run, you'll just keep marching through the random-number stream(s) from one replication to the next so will get different and independent results, which is what's needed for statistical analysis of the output data from simulation. Actually, the fact that you get the same results if you re-run the same model is quite useful, e.g. for debugging.

6.4 Generating Random Variates and Processes

In Sections 6.1 and 6.2 we discussed how to choose probability distributions for the stochastic inputs to your model, and in Section 6.3 we described how random numbers (continuously uniformly distributed between 0 and 1) are generated. In this section we'll combine all that and discuss how to transform the uniform random numbers between 0 and 1 into draws from the input distributions you want for your model. This is often called *random-variate generation*. When using Simio, this is taken care of for you internally, at least for the approximately 20 distributions it supports. But it's still important to understand the basic principles, since you may find that you need to generate variates from other distributions on occasion.

Actually, we've already discussed random-variate generation for a couple of special cases in Chapter 3. In Section 3.2.3 we needed to generate discrete random variates distributed uniformly on the integers $1000, 1001, \ldots, 5000$. And in Section 3.3.2 we needed to generate continuous random variates from exponential distributions. In both cases we devised distribution-specific methods, and in the discrete-uniform case, at least, there was reasonable intuition about why that method was correct (in the exponential case, we just referred you here to this section, so this is the place).

There are some general principles, though, that provide guidance in figuring out how to transform random numbers into random variates from the input distributions you want. Probably the most important one is called *inverse CDF* since it involves finding the functional inverse (not the algebraic inverse or reciprocal) of the CDF F_X of your desired input distribution (X is the corresponding RV; we often subscript CDFs, PDFs, and PMFs with the pertinent RV just to clarify). As a reminder, the CDF gives you the probability that the associated RV will be less than or equal to its argument, i.e., $F_X(x) = P(X \leq x)$ where the RV X has CDF F_X.

Let's first consider the case of a continuous RV X with CDF F_X that's a continuous function. The basic idea is to generate a random number U, set $U = F_X(X)$, and (try to) solve that equation for X; that solution value X will be a random variate with CDF F_X, as we'll show in the next paragraph. The reason we say "try to" solve is that, depending on the distribution, solving that equation may or may not be simple. We denote its solution (simple or not) as $X = F_X^{-1}(U)$, where F_X^{-1} is the functional inverse of F_X, i.e., F_X^{-1} "undoes" whatever it is that F_X does.

Why does this work, i.e., why does the solution $X = F_X^{-1}(U)$ have distribution with CDF F_X? The key is that U is continuously uniformly distributed between 0 and 1, and we do need to rely on the quality of the random-number generator to guarantee that that's at least very close to being true. If you take any subinterval of $[0, 1]$, let's say $[0.2, 0.6]$, the probability that the random number U will fall in the subinterval is the width of the subinterval, in this case $0.6 - 0.2 = 0.4$; to see this, just remember that the PDF of U is

$$f_U(x) = \begin{cases} 1 & \text{if } 0 \leq x \leq 1 \\ 0 & \text{otherwise} \end{cases}$$

and that the probability that an RV lands in an interval is the area under its PDF over that interval (see Figure 6.15). So in particular, for any value w between 0 and 1, $P(U \leq w) = w$. So, starting with the "generated" variate $X = F_X^{-1}(U)$, let's find the probability that it's less than or equal to any value x in the range of X:

$$
\begin{aligned}
P(F_X^{-1}(U) \leq x) &= P(F_X(F_X^{-1}(U)) \leq F_X(x)) && \text{(apply } F_X\text{, an increasing} \\
&&& \text{function, to both sides)} \\[2ex]
&= P(U \leq F_X(x)) && \text{(definition of functional} \\
&&& \text{inverse)} \\[2ex]
&= F_X(x). && (U \text{ is uniformly distributed} \\
&&& \text{between 0 and 1, and} \\
&&& F_X(x) \text{ is between 0 and 1)}
\end{aligned}
$$

This shows that the probability that the generated variate $X = F_X^{-1}(U)$ is less than or equal to x is $F_X(x)$, i.e., the generated variate has CDF $F_X(x)$, exactly as desired. Figure 6.16 illustrates the inverse CDF method in the continuous

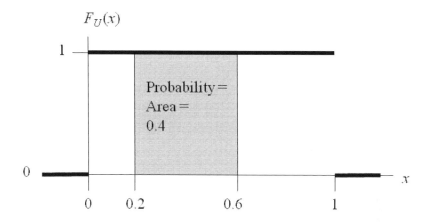

Figure 6.15: PDF of the continuous uniform distribution between 0 and 1 (i.e., of a random number U), and $P(0.2 \leq U \leq 0.6) = 1 \times (0.6 - 0.2) = 0.4$.

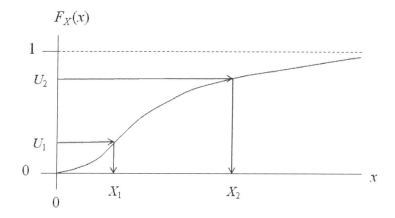

Figure 6.16: Illustration of the inverse CDF method for continuous random-variate generation.

case, with two random numbers U_1 and U_2 plotted on the vertical axis (they'll always be between 0 and 1, so will always "fit" within the vertical range of any CDF plot, since it too is between 0 and 1), and the corresponding generated variates X_1 and X_2 are on the x axis, in the range of the CDF $F_X(x)$ (this particular CDF appears to have a range of all positive numbers, like the gamma or lognormal distributions). So the inverse-CDF pictorially amounts to plotting random numbers on the vertical axis, then reading across to the CDF (which could be to either the left or the right, though it's always to the right in our example since the RV X is always positive), and then down to the x axis to get the generated variates. Note that bigger random numbers yield bigger random variates (since CDFs are nondecreasing functions, their functional inverses are too). Also, since the random numbers are distributed uniformly on the vertical axis between 0 and 1, they'll be more likely to "hit" the CDF where it's rising steeply, which is where its derivative (the PDF) is high — and this is exactly the outcome we want, i.e., the generated RVs are more dense where the PDF is high (which is why the PDF is called a *density* function). So the inverse-CDF method "deforms" the uniform distribution of the random numbers according to the distribution of the RV X desired.

To take a particular example, the same distribution we used in Section 3.3.2, suppose X has an exponential distribution with mean $\beta > 0$. As you can look up in any probability/statistics book, the exponential CDF is

$$F_X(x) = \begin{cases} 1 - e^{-x/\beta} & \text{if } x \geq 0 \\ 0 & \text{otherwise} \end{cases}$$

so setting $U = F_X(X) = 1 - e^{-X/\beta}$ and solving for U, a few lines of algebra yields $X = -\beta \ln(1 - U)$ as the variate-generation recipe, just as we got in Section 3.3.2. In the case of this exponential distribution, everything worked out well since we *had* a closed-form formula for the CDF in the first place, and furthermore, the exponential CDF was easily inverted (solving $U = F_X(X)$ for X) via simple algebra. With other distributions there might not even *be* a closed-form formula for the CDF (e.g., the normal) so we can't even start to try to invert it with algebra. And for yet other distributions, there might be a closed-form formula for the CDF, but inverting it isn't possible analytically (e.g., the beta distribution with shape parameters' being large integers). So, while the inverse CDF always works for the continuous case in principle, implementing it for some distributions could involve numerical methods like root-finding algorithms.

In the discrete case, the inverse-CDF *idea* is the same, except the CDF is not continuous — it's a piecewise-constant step function, with "riser" heights of the steps equal to the PMF values above the corresponding possible values of the discrete RV X. Figure 6.17 illustrates this, where the possible values of the RV X are x_1, x_2, x_3, x_4. So implementing this generally involves a search to find the right "riser" on the steps. If you think of projecting the step heights leftward onto the vertical axis (all between 0 and 1, of course), what you've done is divide the (vertical) interval $[0, 1]$ into subsegments of width equal to the PMF values. Then generate a random number U, search for which subinterval on the vertical axis contains it, and the corresponding x_i is the one you return as the generated variate X ($X = x_3$ in the example of Figure 6.17). Actually,

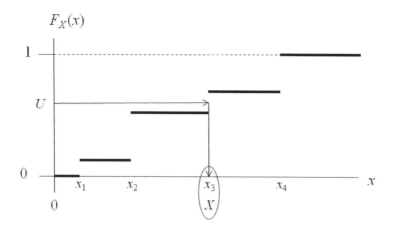

Figure 6.17: Illustration of the inverse CDF method for discrete random-variate generation.

the custom method we devised in Section 3.2.3 to generate the discrete uniform demands *is* the discrete inverse-CDF method, just implemented more efficiently as a formula rather than as a search. Such special "tricks" for some distributions are common, and sometimes are equivalent to the inverse CDF (and sometimes not).

Inverse CDF is, in a way, the best variate-generation method, but it is not the only one. There has been a lot of research on variate generation, focusing on speed, accuracy, and numerical stability. For overviews of this topic see, for example, [3] or [30]; a thorough encyclopedic treatment can be found in [8].

We briefly discussed *specifying* random vectors and process in Section 6.1.6, and for each one of those cases, we also need to think about how to *generate* (realizations of) them. Discussing these methods is well beyond our scope in this book, and the references in the preceding paragraph could be consulted for many situations. As mentioned in Section 6.2.3, Simio has a built-in method for an important one of these cases, generating nonstationary Poisson processes when the rate function is piecewise linear. However, in general, simulation-modeling software does not yet support general methods for generating correlated random variables, random vectors with multivariate distributions, or general random processes.

6.5 Problems

1. The Excel file `Problem_Dataset_06_01.xls`, available for download in the student area on the book's website as described in the Preface, contains 42 observations on interarrival times (in minutes) to a call center. Use Stat::Fit (or other software) to fit one or more probability distributions to these data, including goodness-of-fit testing and probability plots. What's your recommendation for a distribution to be used in the simulation model from which to generate these interarrival times? Provide the correct Simio

expression, heeding any parameterization-difference issues.

2. The Excel file `Problem_Dataset_06_02.xls`, available for download in the student area on the book's website as described in the Preface, contains 47 observations on call durations (in minutes) at the call center of Problem 1. Use Stat::Fit (or other software) to fit one or more probability distributions to these data, including goodness-of-fit testing and probability plots. What's your recommendation for a distribution to be used in the simulation model from which to generate these call-duration times? Provide the correct Simio expression, heeding any parameterization-difference issues.

3. The Excel file `Problem_Dataset_06_03.xls`, available for download in the student area on the book's website as described in the Preface, contains 45 observations on the number of extra tech-support people (beyond the initial person who originally took the call) needed to resolve the problem on a call to the call center of Problems 1 and 2. Use Stat::Fit (or other software) to fit one or more probability distributions to these data, including goodness-of-fit testing and probability plots. What's your recommendation for a distribution to be used in the simulation model from which to generate the number of extra tech-support people needed for a call? Provide the correct Simio expression, heeding any parameterization-difference issues.

4. Derive the inverse-CDF formula for generating random variates from a continuous uniform distribution between real numbers a and b $(a < b)$.

5. Derive the inverse-CDF formula for generating random variates from a Weibull distribution. Look up its definition (including its CDF) in one of the references, either book or online, cited in this chapter. Check the Simio definition in its documentation to be sure you have your formula parameterized properly.

6. Derive the inverse-CDF formula for generating triangular variates between a and b with mode m $(a < m < b)$. Take care to break your formula into different parts, as needed. Check the Simio definition in its documentation to be sure you have your formula parameterized properly.

7. Recall the general inverse-CDF method for generating arbitrary discrete random variates, discussed in Section 6.4 and shown in Figure 6.17. Let X be a discrete random variable with possible values (or *support*) 0, 0.5, 1.0, 1.5, 2.0, 3.0, 4.0, 5.0, 7.5, and 10.0, with respective probabilities 0.05, 0.07, 0.09, 0.11, 0.15, 0.25, 0.10, 0.09, 0.06, and 0.03. Compute the exact (to four decimal places) expected value and standard deviation of X. Then, write a program in your favorite programming language, or use Excel, to generate first $n = 100$ and then a separate $n = 1000$ IID variates, and for each value of n, compute the *sample* mean and standard deviation and compare with the exact values by both error and percent error; comment. Use whatever random-number generator is built-in or convenient, which is good enough for this purpose. If you use Excel, you might consider the VLOOKUP function.

8. Re-do the produce-stand spreadsheet simulation (Problem 17 from Chapter 3), but now use more precise probability mass functions for daily demand; Walther has taken good data over recent years on this, and he also now allows customers to buy only in certain pre-packaged weights. For oats, use the distribution in Problem 7 in the present chapter; note that the pre-packaged sizes are (in pounds), 0 (meaning that a customer chooses not to buy any packages of oats), 0.5, 1.0, 1.5, 2.0, 3.0, 4.0, 5.0, 7.5, and 10.0. For peas, the pre-packaged sizes are 0, 0.5, 1.0, 1.5, 2.0, and 3.0 pounds with respective demand probabilities 0.1, 0.2, 0.2, 0.3, 0.1, and 0.1. For beans, the pre-packaged sizes are 0, 1.0, 3.0, and 4.5 pounds with respective demand probabilities 0.2, 0.4, 0.3, and 0.1. For barley, the pre-packaged sizes are 0, 0.5, 1.0, and 3.5 pounds with respective demand probabilities 0.2, 0.4, 0.3, and 0.1. Refer to Problem 7 in the present chapter for the method to generate simulated daily demands on each product.

9. Write out a step-by-step algorithm, using only random numbers as input, to generate a random variate representing the interest paid on a monthly credit-card bill. There's a 60% chance that the interest will be zero, i.e., the cardholder pays off the entire balance by the due date. If that doesn't happen, the interest paid is a uniform random variate between $20 and $200. Then, develop a Simio expression to accomplish this, using relevant built-in Simio random-variate expressions.

10. Show that the method developed in Section 3.2.3 for generating demands for hats distributed uniformly on the integers $1000, 1001, \ldots, 5000$ *is* exactly the same thing as the inverse-CDF algorithm for this distribution, except implemented more efficiently in a search-free way.

Chapter 7

Working With Model Data

There are many different types of data used in a model. So far we've largely entered our data directly into the properties of Standard Library objects. For example, we entered the mean interarrival time directly into a Source object and entered the parameters of the processing-time distribution directly into a Server object. While this is fine for some types of data, there are many cases where other mechanisms are necessary. Specific types of data, such as time-varying arrival patterns, require a unique data representation. In other situations, the volume of data is large enough that it's necessary to represent the data in a more convenient form, and in fact even import the data from an external source. And in situations where the analyst using the model may not be the same as the modeler who builds the model, it may be necessary to consolidate the data rather than having them scattered around the model. In this chapter we'll discuss many different types of data and explore some of the Simio constructs available to represent those data best.

7.1 Data Tables

A Simio *Data Table* is similar to a spreadsheet table. It's a rectangular data matrix consisting of columns of properties and rows of data. Each column is a property you select and can be one of over 30 Simio data types, including Standard Properties (e.g., Integer, Expression), Element References (e.g., Tally Statistic, Station), or Object References (e.g., Entity, Node List). Typically, each row has some significance; for example it could represent the data for a particular entity type, object, or organizational aspect of the data.

Data tables can be imported, exported, and even bound (see Section 7.1.7) to an external file. They can be accessed sequentially, randomly, directly, and even automatically. You can create relations between tables such that an entry in one table can reference the data held by another table. In addition to basic tables, Simio also has Sequence Tables and Arrival Tables, each a specialization of the basic table. All of these topics will be discussed in this section.

While reading and writing disk files interactively during a run can significantly slow execution speed , tables hold their data in memory and so execute very quickly. Within Simio you can define as many tables as you like, and each table can have any number of columns of different types. You'll find tables to be

Table 7.1: Simio Standard Properties for representing table data columns.

Property Type	Description
Boolean	True (1 or non-zero) or False (0)
Color	A property for graphically setting color
Date Time	A specific day and time (7:30:00 November 18, 2010)
Day Pattern	A reference to a Day Pattern for a schedule
Enumeration	A set of values described in a pre-defined enumeration
Event	Event that triggers a token release from a step
Expression	An expression evaluated to a real number (1.5+MyState)
Integer	An integer number (5, or −1)
List	A set of values described in a string list
Rate Table	A reference to a rate table
Real	A decimal number (2.7, or −1.5)
Schedule	A reference to a schedule
Selection Rule	A reference to a selection rule
Sequence Table	A reference to a sequence table
State	A reference to a state
String	Textual information (Red, Blue)
Table	A reference to a data table or sequence table

valuable in organizing, representing, and using your data as well as interfacing with external data.

7.1.1 Basics of Tables

A data table is defined using the *Tables* panel in the Data window. To add a new table you click on `Add Data Table` in the Tables section of the Table ribbon. Once you've added a table you can rename it and give it a description by clicking on the tab for the table and then setting the table properties in the Property window.

Tip: If you add multiple tables, each one has its own tab. Recall our discussion in Section 4.1 about configuring window placement. That can be particularly handy when working with data tables.

To add columns to a table you select a table to make it active and then click on property types under `Standard Property`, `Element Reference`, `Object Reference`, or `Foreign Key`. A table column is typically represented by the Standard Properties illustrated in Table 7.1. Use an Object Reference when you want a table to reference an object instance or list of objects such as an `Entity`, `Node`, `Transporter`, or other model object. Likewise use an Element Reference if you want a table to reference a specific element like a TallyStatistic or Material.

7.1.2 Model 7-1: An ED Using a Data Table

Let's illustrate these table concepts by representing the data for a simple healthcare example. Consider an emergency department (ED) that has some known data concerning how patients of various severities are handled. Specifically, we

Table 7.2: Model 7-1 ED Basic patient data.

Patient Type	Priority	Treatment Time (Minutes)
Routine	1	Random.Triangular(3,5,10)
Moderate	2	Random.Triangular(4,8,25)
Severe	3	Random.Triangular(10,15,30)
Urgent	4	Random.Triangular(15,25,40)

have four patient types, their priority value, and their typical treatment time as given in Table 7.2. The first steps in building our model involve defining the model entities and the data table:

1. Load Simio and start a new model as we've done before. The first thing we'll do is drag four instances of ModelEntity into the model facility view. Click on each one and use the properties or the F2 key to rename them to `Routine`, `Moderate`, `Severe`, and `Urgent`, respectively.

2. Let's take a minute to animate those entities a little better, so that we can tell them apart. Follow the same procedure described in Section 4.7. Click on an entity instance, then click on the symbol library. Scroll down to the `People` category and select `Man6`. Repeat this using `Man6` for all four entity instances. Although all entities will use the same symbol, we can tell them apart by giving each person a different shirt color. Zoom in so you can see your new symbols clearly. Click on the `Routine` entity. On the right side of the Symbols ribbon is a Color button. Clicking the lower half of the color button will display a color pallet. Select a green color and then apply it to `Routine` by clicking on the shirt of the man symbol. Repeat this procedure to apply a light blue to the `Severe` patients and Red to the `Urgent` patients. We'll leave `Moderate` with the default color.

3. Next, let's create the data table. Select the Data tab just under the ribbon, then select the Tables panel on the left if it's not already selected. The Table ribbon will appear, although most of it's unavailable (grayed out) because we don't yet have an active table. Click on the `Add Data Table` button to add a blank table, then click on the Name property in the properties window and change the name to `PatientData` (no spaces are allowed in Simio names).

4. Now we'll add our three columns of data. Our first column will reference an entity type (an object), so click on `Object Reference` on the ribbon and then select `Entity` from that list. You've now created a column that will hold an Entity Instance. Go to the Name property (*not* the Display Name property) in the properties window and change it to `PatientType`. Our second column will be an integer, so click on `Standard Property` in the ribbon and select `Integer`. Go to the Name property and rename it to `Priority`. Finally, our third column will be an expression so click on `Standard Property` in the ribbon and select `Expression`. Go to the Name property and rename it to `TreatmentTime`. Since this represents a time, we need a couple of extra steps here: In the Logic category of

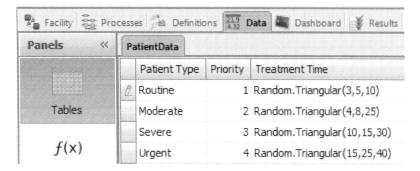

Figure 7.1: Model 7-1 ED basic patient data in Simio table.

the properties window specify a Unit Type of `Time` and specify `Default Units` of `Minutes`.

5. Now that we have the structure of the table, let's add our four rows of data. You can add the data from Table 7.2. You can enter the data a row at a time or a column at a time; here we'll do the former. Click on the upper left cell and you'll see a list containing our four entity types. (If you don't see that list, go back two steps.) Select `Routine` as the `Patient Type` for row 1. Move to the `Priority` column, type 1 and press enter and you'll be moved to the `TreatmentTime` column. Here we'll type in the expression from Table 7.2, `Random.Triangular(3,5,10)`. *Tip*: If the values of your data are partially hidden, you can double click on the right edge of the column name and it will expand to full width. Move to the next row in the `PatientData` table and follow a similar process to enter the remaining data from Table 7.2.

When you're finished your table should look similar to Figure 7.1. We've now defined the patient-related data that our model will use. In the next section, we'll demonstrate how to access the data from the table.

Referencing Data Tables

Table data can be used by explicitly referencing the table name, the row number, and the column name or number using one of the following syntax choices: `TableName[RowNumber].ColumnName` or
`TableName[RowNumber,ColumnNumber]`.
We could continue building our model now using that syntax to reference our table data. For example, we could use `PatientData[3].TreatmentTime` to refer to the treatment time for a `Severe` patient. While using this syntax is useful for referencing directly into a cell of a table, we often find that a particular entity will *always* reference a specific table row. For example in our case, we'll determine the row associated with a patient type, and then that entity will always reference data from that same row. You could add your own property or state to track that, but Simio already builds in this capability. The easiest method to access that capability is by setting the properties in

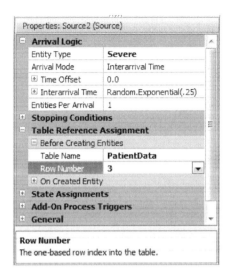

Figure 7.2: Associating an entity with an explicit row in a table.

the Table Reference Assignment category of the Source object. If we had a separate arrival stream for each of our four patient types, we'd probably use this technique. We'd have a source object to create each patient type. Figure 7.2 illustrates how the source object for `Severe` patients could be configured by specifying the table name and an explicit row number. Once you've made this association between an entity and a specific row in a table, you can use a slightly simpler syntax to reference the table data: `TableName.ColumnName` because the row number is already known. For example, we could now use `PatientData.TreatmentTime` to refer to the treatment time for *any* patient type.

Selecting Entity Type

Before we finish our model, we'll explore one more aspect of tables. It's very common to have data in a table where each row corresponds to an entity type, as we have in our model. It's also common to have the entity type be selected randomly. Simio allows you to do both within the same construct. You can add a numeric column to your table that specifies the weighting of each row (or entity type). Then you can specify that you'll randomly select a row based on that column by using the function `TableName.ColumnName.RandomRow`.

Let's follow a few final steps to complete our model.

1. In our ED, historical information suggests that our patient mix is `Routine` (40%), `Moderate` (31%), `Severe` (24%), and `Urgent` (5%). We need to add this new information to our table. Return to the Data tab and the Tables panel. Click `Standard Property` and select `Real`. Go to the Name property and rename it to `PatientMix`. Then add the above data to that new column. When you're finished your table should now look

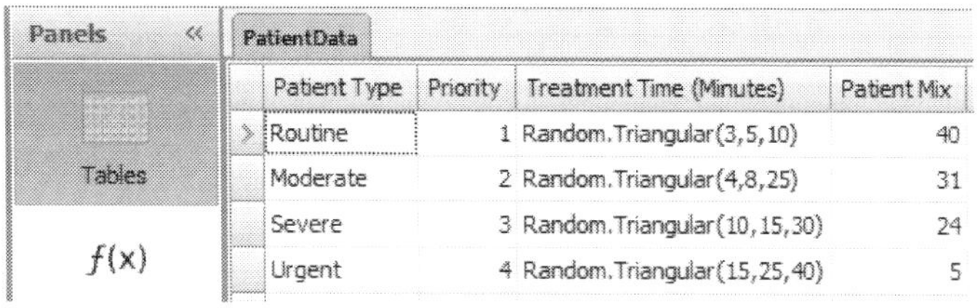

Patient Type	Priority	Treatment Time (Minutes)	Patient Mix
Routine	1	Random.Triangular(3,5,10)	40
Moderate	2	Random.Triangular(4,8,25)	31
Severe	3	Random.Triangular(10,15,30)	24
Urgent	4	Random.Triangular(15,25,40)	5

Figure 7.3: Model 7-1 ED Enhanced patient data in Simio table.

like Figure 7.3.[1]

2. Now we can continue building our model. The last change we made allows us to have a single Source that will produce the specified distribution of our four patient types. Place a Source in your model and specify an Interarrival Time of `Random.Exponential(4)` and units of `Minutes`. Instead of specifying one patient type in the Entity Type property with a specific row number as we did in Figure 7.2, we'll let Simio pick the row number and then we'll select the Entity Type based on the `PatientType` specified in that row. We must select the table row before we create the entity; otherwise the entity would already be created by the time we decide what type it is. So in the Table Reference Assignment, Before Creating Entities category, we'll specify the Table Name of `PatientData` and the Row Number of `PatientData.PatientMix.RandomRow`. After the row's selected, the Source will go to the Entity Type property to determine which Entity Type to create. There we can select `PatientData.PatientType` from the pull-down list. This is illustrated in Figure 7.4.

3. To finish our model is pretty painless. Add a Server, set its Initial Capacity to `3`, and specify a Processing Time of `PatientData.TreatmentTime`. We're using the data in the table for this field, but note that we're using the short reference method. Since no explicit row is specified, we're telling Simio to use the row that's already associated with each specific entity. When an entity of type `Routine` arrives, it will use row one and a treatment time sampled from the distribution `Random.Triangular(3,5,10)`. However, when an entity of type `Severe` arrives, it will use row three and a treatment time sampled from the distribution `Random.Triangular(10,15, 30)`.

   Add a Sink, then connect Source to Server and Server to Sink with Paths. Your model should look something like Figure 7.5.

Before enhancing our model, we'll do a small verification step so that we'll be confident that we've correctly implemented the data table. Using the proportions of patient types and the expected service times for each patient type,

[1] As noted in Section 5.2, the values in the Patient Mix column are interpreted by Simio only in relation to their *relative* proportions. While we entered the values here thinking of the *percent* of patients of each type, they could have equivalently been entered as *probabilities* (0.40, 0.31, 0.24, 0.05), or as any other positive multiple of the values we used (e.g., 4000, 3100, 2400, 500).

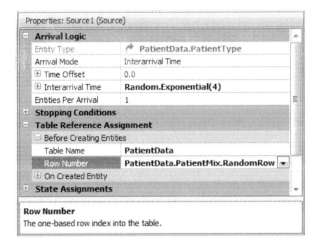

Figure 7.4: Selecting an entity type from a table.

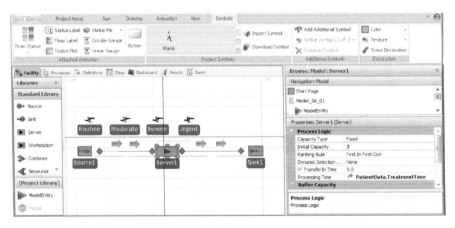

Figure 7.5: Model 7-1 Completed ED model.

we can compute the overall expected service time (11.96 minutes). With the overall arrival rate of 15 patients/hour, we expect a steady-state server utilization of 99.64%. We ran Model 7-1 for 25 replications of length 1100 days with a 100-day warmup period and the resulting average scheduled utilization was 99.59% ± 0.0854 (the 95% confidence interval half-width). This illustrates an important point about model verification — it's often much easier to verify the model as you build it rather than waiting until the model is "finished." Since our sampled utilization matched our expectation quite well, we're now confident that we've property implemented the patient data table and can move on to our model enhancements.

7.1.3 Sequence Tables

A *sequence table* is a special type of data table used to specify a sequence of destinations for an entity. In a manufacturing job shop, this might be the stations or machines that must be visited to complete a part (e.g., Grinding, Polishing, Assembly, Shipping). For a transportation network the sequence might be a series of stops in a bus route (e.g., MainStreet, FifthStreet, NinthStreet, UpTown).

You create a sequence table in the Tables panel of the Data Window just like for normal data tables, but use the `Add Sequence Table` button. This will create the table and automatically add a column named Sequence for specifying a routing sequence for an entity. This required column is the major difference between a normal table and a sequence table. In most other ways everything that applies to a data table also applies to a sequence table. Just as the properties (columns) of a table can be used however you wish, the same is true in sequence tables and you can reference these values the same way you'd reference the values in any other table (e.g., `TableName.PropertyName`).

There are actually two different ways of configuring sequence tables: *simple sequence tables* and *relational sequence tables*. Simple sequence tables are preferred when you have somewhat isolated use of a single sequence, for example if you have only one sequence or entity type in use. Relational tables have the advantage of more easily supporting the more complex use of sequences that you might encounter with multiple entities following different sequences through the same objects. They're both used in the same way, but differ in how the tables themselves are configured. We'll start by explaining how to configure and use simple sequence tables.

Simple Sequence Tables

Each simple sequence table defines one routing plan. If you have multiple routing plans (e.g., several bus routes), then each would be defined in its own sequence table. Each row in the sequence table corresponds to a specific location. The data items in that row are usually used for location-specific properties, for example the processing time, priority, or other properties required at a particular location.

After your sequence table has been created, you must create an association between the entity and the sequence table (in other words, you need to tell the entity which sequence to follow). There are a couple ways to do that, but the easiest is to do so on an entity instance you've placed in a model. It will have a property named Initial Sequence in its Routing Logic category. Specify the sequence table name in this property. The entity will start at row one in this sequence table and work its way through the rows as stations are visited. Although this happens automatically, it's possible to change the current row, and in fact even change the sequence being followed. This can be done at any time using the SetRow step in an add-on process.

At this point the astute reader (that's you, we hope) will be asking "But how does the entity know when to move between sequences?" You must tell the entity when to go to the next step in its sequence. The most common place to do that is in the Routing Logic category of a Transfer Node (recall that the outbound node from every Standard Library object is a Transfer Node). Here you'd specify that the Entity Destination Type is `By Sequence`. This causes three things to happen:

1. The next table row in the entity's sequence table is made current.

2. The entity's destination is set to the destination specified in that row.

3. Any other table properties you've specified (e.g., ProcessingTime) will now look to the new current row for their values.

Because you're explicitly telling the entity when to move to the next sequence, you also have the option to visit other stations in between sequences if you choose — simply specify the Entity Destination Type as anything other than `By Sequence` (e.g., `Specific`). You can do this as many times as you wish. The next time you leave an object and you want to move sequentially, just again use `By Sequence` and it will pick up exactly where you left off.

Relational Sequence Tables

Many of the above concepts are the same for relational sequence tables. Relational tables are used in much the same way, but configured a bit differently. One difference is that you can combine several different sequences (e.g., the set of visitations for a particular entity) into a single sequence table. And instead of setting the sequence to follow on the entity instance (using the Initial Sequence property) you can provide that information in another table. Both of these capabilities are accomplished by linking a main data table with a relational sequence table using special columns identified as Key and Foreign Key. This technique will be demonstrated in Model 7-2.

7.1.4   Model 7-2: Enhanced ED Using Sequence Tables

Let's embellish our previous ED model (Model 7-1) by describing the system in a little more detail. All patients first visit a sign-in station, then all except the `Urgent` go to a registration area. After being registered they'll go to the first available examination room. After the examination is complete the `Routine` patients will leave, while the others will continue for additional treatment. `Urgent` patients visit the sign-in, but then go to a trauma room that's equipped to treat more serious conditions. All `Urgent` patients will remain in the trauma room until they're stabilized, then they're transferred to a treatment room, and then they'll leave.

You may recall that when we started Model 7-1, the first thing we did was place the entities. More generically, we started by placing into the model the objects that we'd be referencing in the table. We'll do that again here. Since Sequence Tables mainly reference locations (more specifically, the input nodes of objects), we'll start by placing the Server objects that represent those locations. Then we'll build our new table, then return to add properties to our model.

1. Start with Model 7-1. Delete the path between the Server and the Sink as we'll no longer be needing it. Likewise, go to the properties of Server1, right click on Processing Time, and select `Reset`.

2. Double click on the Standard Library Server and then click four times in the model to place four additional servers. Hit Escape or right click to exit the placement mode. Name the five servers that you now have `SignIn`, `Registration`, `ExamRooms`, `TraumaRooms`, and `TreatmentRooms`. Add an additional Sink. Name the two sinks `NormalExit` and `TraumaExit`.

3. Move to the Data Window and the Tables panel. We'll start by designating our existing `PatientType` column as a unique Primary Key so that our sequences table can reference this. Select the `PatientType` column and click on `Set Column as Key` in the Ribbon. At this point we can also delete the `TreatmentTime` column — in a few moments we will replace it with a new location-specific ServiceTime column specified in the Treatments table. Click on the `TreatmentTime` column heading, then click on `Remove Column` in the ribbon.

4. Now we can create our Sequence table. Click on `Add Sequence Table`. Click in the properties area of the sequence table and set the name to `Treatments`. You see that one column named `Sequence`, for the required locations, is automatically added for you.

5. We need to add another column to identify the treatment type. Click on `Foreign Key`. This will create a column that uniquely identifies the specific treatment type. In our case the treatment type corresponds exactly to the `PatientType` in the `PatientData` table. Go to the properties of this new column and name it `TreatmentType`. Because it's a foreign reference, that means that it actually gets its values from somewhere else. The Table Key property specifies from where the Treatment Type value comes. Select `PatientData.PatientType` from the pull down list.

6. We have one final column to add to our `Treatments` table — the processing time. Click on `Standard Property` and select `Expression`. This will create a property into which you can enter a number or expression to be used while at that location. Since that value has slightly different meanings at the different locations, we'll go to the properties window and change the Name (not the Display Name) to a more generic `ServiceTime`. Since this represents a time, in the Logic category of the `ServiceTime` properties, we want to specify that Unit Type is `Time` and that Default Units is `Minutes`.

7. Before we enter data into our table, there's one more thing we can change to make data entry easier. If you look at the pull down list in the first cell under `Sequences`, you'll see a list of all of the nodes in your model because Sequences allow you to specify any node as a destination. This is valuable when you have stand-alone nodes as potential destinations, but our model doesn't have any of those. We can take advantage of an option to limit the list to just input nodes at objects, and in fact that option just displays the object name to make it even simpler. Click on the Sequence column, then go to the properties and change Accepts Any Node to `False`. Now if you look at that same pull-down list you'll see a shorter and simpler list.

Sequence	Treatment Type	Service Time
SignIn	Routine	2
Registration	Routine	Random.Uniform(3,7)
ExamRooms	Routine	Random.Triangular(5,10,15)
NormalExit	Routine	0.0
SignIn	Moderate	2
Registration	Moderate	Random.Uniform(3,7)
ExamRooms	Moderate	Random.Triangular(10,15,20)
TreatmentRooms	Moderate	Random.Triangular(5,8,10)
NormalExit	Moderate	0.0
SignIn	Severe	1
Registration	Severe	2
ExamRooms	Severe	Random.Triangular(15,20,25)
TreatmentRooms	Severe	Random.Triangular(15,20,25)
NormalExit	Severe	0.0
SignIn	Urgent	.5
TraumaRooms	Urgent	Random.Triangular(15,25,35)
TreatmentRooms	Urgent	Random.Triangular(15,45,90)
TraumaExit	Urgent	0.0

Figure 7.6: Model 7-2 Relational sequence table to define Treatments.

8. Let's enter our treatment data. In row 1, select SignIn from the pull-down list under Sequence. Then select Routine from the pull-down list under Treatment Type, then enter 2 for the Service Time. Continue entering data for additional rows until your table looks like Figure 7.6.

9. Now that we've entered our data (whew!), we can move back to the Facility window and finish our model.

10. Recall from the general discussion above that we must specify on each outbound node if the departure is to follow a sequence. Click on the blue TransferNode on the output side of the Source. In the routing Logic category change the Entity Destination Type to By Sequence. You could repeat this process for each outbound node, but Simio provides a shortcut. In our case all movements will be by sequence, so we can change them all at once. Click on any blue node, then Ctrl-click one at a time on every other blue node (you should have six total). Now that you've selected all six transfer nodes you can change the Entity Destination Type of all six nodes at once.

Table 7.3: Number of servers in each area of ED.

Service Area	Initial Capacity
SignIn	1
Registration	3
ExamRooms	6
TreatmentRooms	6
TraumaRooms	2

11. Connect the nodes by all the paths that will be used. If you add unnecessary paths, it adds clutter, but does no harm — for example a path from `Registration` to `TraumaRooms` would never be used because `Urgent` patients bypass `Registration`.

12. We've not yet told the Servers how long a patient will stay with them — the Process Time property on the servers. For all servers, the answer will be the same — use the information specified in the sequence associated with that specific patient and the sequence step that's active. The expression for this is in the form `TableName.PropertyName`, or specifically in this case it's `Treatments.ServiceTime`. Enter that value for Process Time on each server. Note that the entity (the patient) carries with it current table and row associations.

13. Some patients are more important than others. No, we're not talking about politicians and sports stars, but rather patient severity. For example we don't want to spend time treating a chest cold when another patient may be suffering from a severe heart attack. We already have patient priority information in our patient table, but we need to change the server-selection priority from its default First In First Out (FIFO). For each server you need to change the Ranking Rule property to `Largest Value First`. That will expose a Ranking Expression property that should point to our table: `PatientData.Priority`. This will guarantee that an `Urgent` patient (priority 4) will be treated ahead of all other patients (priorities 1-3).

14. Each of our servers actually represents an area of the ED, not just a single server. We need to indicate that by specifying the Initial Capacity for each Server. Click on each server and complete the Initial Capacity field as indicated in Table 7.3. The green line above each server displays the entities currently in process. You should extend those lines and make them long enough to match the stated capacity. Likewise, the line to the left of each server displays patients waiting for entry. Those lines may also need to be extended. You can do this while the model is running to make it easier to set the right size.

If you've followed all of the above steps you've completed the model. Although you've probably arranged your objects differently, it should look something like Figure 7.7 (shown with Browse window collapsed).

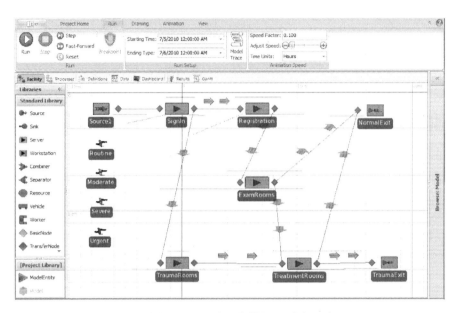

Figure 7.7: Model 7-2 completed ED model with sequences.

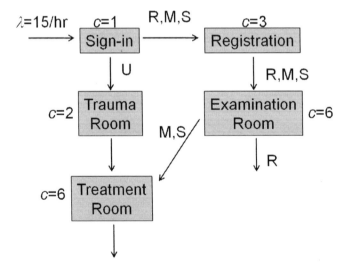

Figure 7.8: Queueing-network model for the ED model with sequences.

Table 7.4: Model 7-2 verification results.

Server	Expected Utilization	Simulation Results
Sign-in	0.4213	0.4214 ± 0.0371
Registration	0.3358	0.3360 ± 0.0307
Trauma Rooms	0.1563	0.1568 ± 0.0443
Examination Rooms	0.5604	0.5633 ± 0.0510
Treatment Rooms	0.4032	0.4033 ± 0.0560

Continuing our strategy of verifying our models as we go, we'll use a queueing-network model (shown in Figure 7.8) to approximate the expected steady-state utilization of the five servers. Table 7.4 gives the expected utilizations from the queueing-network model along with the results from Model 7-2 (based on 25 replications of length 1100 days with 100-day warmup). Clearly our experimental results are in line with our expectations so we're confident about our model and can continue enhancing it.

7.1.5 Arrival Tables and Model 7-3

One more type of table that we've not yet discussed is an *Arrival Table*. This is used with the Arrival Mode property of the Source object to generate a specific set of arrivals as specified in a table. The arrivals can be either stochastic or deterministic.

While *deterministic arrivals* are uncommon in a typical stochastic simulation model, two important applications are Scheduling and Validation. In a scheduling-type application, you have a fixed set of work to get done, and you want to determine the best way to process the items to meet system objectives. Similarly, one validation technique is to run a known set of events through your model and analyze any differences. In either case you could use an arrival table to represent the incoming work or other activities like material deliveries. Each entry in the table corresponds to an entity that will be created.

The same table can be used for *stochastic arrivals* by taking advantage of two extra properties[2] in the Source object. *Arrival Time Deviation* allows you to specify a random deviation around the specified arrival time. *Arrival No-Show Probability* allows you to specify the likelihood that a particular arrival may not occur. These properties make it easy to model anticipated arrivals of items like patients (as we will illustrate below), incoming materials, and trucks for outbound deliveries.

Any table can be used as an Arrival Table as long as it contains a column with a list of arrival times. Arrival tables are not limited just to providing an arrival time; the rest of the table could have a full set of properties and data to determine expected routing and expected performance criteria. Let's explore how we could use an arrival table with our ED model. We'll add a table with an arrival time and patient type onto Model 7-2.

---

[2]Like other model aspects, these stochastic features are disabled when the model is run in deterministic mode by selecting the *Disable Randomness* option in the *Advanced Options* on the Run tab.

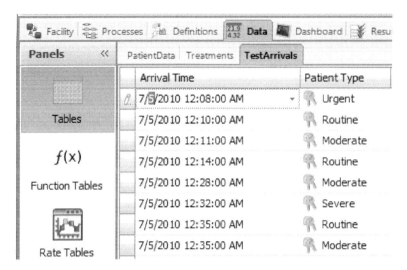

Figure 7.9: Sample arrival table.

1. Go to the Tables Panel of the Data window and click `Add Data Table`. Name it `TestArrivals`.

2. Click on `Standard Property` and select `Date Time`. Also, name this property `ArrivalTime`.

3. Click on `Foreign Key`. This will add a property that will determine the arrival type. Name it `PatientType`, set the Table Key to `PatientData.PatientType`, and the default value to `Routine`. Use of the Foreign Key links this to the other data tables we've already entered — no extra effort is necessary. In fact, we can eliminate the Table Reference Assignments that we previously entered on the Source object.

4. Now we can enter some sample data, perhaps the actual patient arrivals from yesterday. Enter the data as shown in Figure 7.9. If you found it more convenient to enter the time as an elapsed time from the start of the simulation (midnight), that's also possible. Instead of creating a `Date Time` property, create a property of type `Real` and specify that it's a `Time` and your preferred units (e.g., `Minutes`). It's not necessary that the entries be in strict arrival time order. The records will be sorted by increasing time before they're created.

5. To trigger use of this table we must specify it on our Source. For the Arrival Mode property select `Arrival Table`. The Arrival Time Property is where you specify the table and property name that will determine the arrivals — select `TestArrivals.ArrivalTime` as shown in Figure 7.10.

6. Also on the Source we will specify an Arrival Time Deviation of `Random.Triangular(-0.2,0,0.2)` to support random arrival times. This indicates that each individual patient may arrive up to 0.2 hour early or up to 0.2 hour late (i.e., within 12 minutes of the scheduled time).

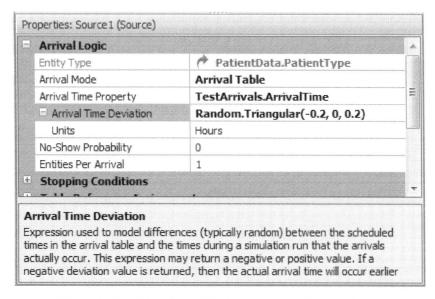

Figure 7.10: Using Arrival Tables with the Source object.

7.1.6   Relational Tables

As the name might imply, *Relational Tables* are tables that have a defined relationship to each other rather than existing independently. These relationships are formed by using the *Table Key* and *Foreign Key* capabilities. The `Set Column As Key` button allows you to indicate that the highlighted column may be referenced by another data table. This button makes that column a Key for this table and is what links the tables together. This column must have exactly one instance of each key value — none of the key values may be repeated. However, you may have multiple columns that are Key in a single table as long as each one contains a set of unique values.

Any other table, potentially multiple tables, may link to the main table by including a column of type Foreign Key that specifies the main table and its Table Key. Once that link is created, you can use any column of any linked table without having to traverse the link explicitly. For example, in Model 7-2 our `PatientData` table specified `PatientType` as a Key and that key contained four unique values. We then created our `Treatments` sequence table, which used that `PatientType` in a Foreign Key column. When we associated an entity with a particular row in `PatientData`, it was automatically associated with a set of related rows in the Treatments table and allowed us to use the values in `Treatments.ServiceTime`.

Relational tables include a Master-Detail view, which allows the relationships between various tables to be seen. For those tables that have a column that's designated as a Key, you can see a detail view that's a collection of rows pointing to a specific keyed row in a table. The small + sign in front of the row indicates that the detail view is available. When the + is pressed, it will show those portions of the related table specific to the entry that was expanded.

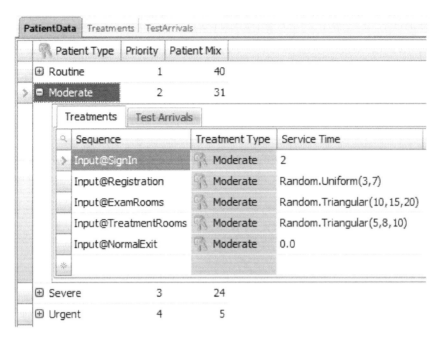

Figure 7.11: Master-Detail view of a relational table.

We can see this if we go back to Model 7-3. If you look at the `PatientType` table, each `PatientType` entry is preceded by a +. If you press the + in front of the `Moderate` patient type you'll see something like Figure 7.11. Under the Treatments tab, it displays all of the associated rows in the `Treatments` table. In this case it's showing you all of the locations and data associated with the treatment sequence for a `Moderate` patient. You'll notice that there is also a `TestArrivals` tab. On that tab you'll find the data for all of the `Moderate` arrivals specified in the `TestArrivals` table. Click on the − to close the detail view and you can open and close the different keys independently.

Relational tables provide a very powerful feature. They allow you to represent your data set efficiently and reference it simply. Moreover this is very useful in linking to external data. For the same efficiency reasons, you may often find that the external data needed to support your simulation is available in a relational database. You can represent the same relationships in your data tables.

7.1.7   Table Import/Export

We trust that by this point you've begun to appreciate the value and utility of tables. For small amounts of data like we've used so far, manually entering the data into tables is reasonable. But as the volume of data increases, you'll probably want to take advantage of the file import and export options.

You can export a table to a comma-separated-value (CSV) file by clicking on *Export* in the ribbon menu. This is useful to create a properly formatted file when you first use it or simply to write an existing table to a file for external editing (e.g., in Excel) or for backup. You can also import the data from a CSV, Excel, or many common database file formats into the table by using the *Import* feature. It's recommended that the initial table be generated by first exporting an existing table from Simio. Before you can Import you must first *Bind To* the external file. This establishes the file name and any other parameters necessary to read the file. Once you have bound the table to a specific file, you may also specify the *Binding Options*. This provides you the option of either importing the data manually (only when you click the Import button) or automatically each time you start a model run. The latter is desirable if the data collection is something that changes frequently like initialization to current system status.

When importing from Excel, it will always assume the first row selected is a column name. Also, the import logic always tries to match up already existing column names in the table with column names from the data. If Simio cannot match a column name from the Excel data with a column name in an existing table, it discards the unmatched data column. If you run into this problem, delete all the columns in the table and do the import again with a blank Simio table. You can also use copy and paste to transfer data in either direction. For example, you can copy the data from Excel, click on the appropriate row and column of the Simio table, and finally click Ctrl-V to paste the data into the Simio table.

7.2 Schedules

As noted in Section 5.1.5, any object can be a resource. Those resources have a capacity (e.g., the number of units available) that may possibly vary over time. As introduced in Section 5.3.3, one way to model situations where the capacity of an object varies over time is to use *Schedules*. Many objects support following a Work Schedule that allows capacity to change automatically over time. The numerical capacity is also used to determine if a resource is in the On-Shift state (capacity greater than zero) or Off-Shift state (capacity equals zero). For some objects (Worker) capacity can be only either 0 (off-shift) or 1 (on-shift). For most objects a schedule can represent a variable capacity (e.g., five for 8 hours, then four for 2 hours, then zero for 14 hours).

7.2.1 Calendar Work Schedules

A calendar work schedule[3] has three main components — Day Patterns, Work Schedules, and Exceptions.

A *Day Pattern* defines the working periods for a single day. You can have as many working periods as desired. Any period not specified is assumed to be non-working. Simio includes a sample Day Pattern (Figure 7.12) called `Standard Day` that has two four-hour work periods separated by a one-hour meal break. You can, of course, revise or replace this sample with your own. For each work period in the day pattern you must specify a *Start Time* and either the *Duration* or the *End Time*. If the number scheduled (e.g., the resource capacity) is not

---

[3]Simio version 4 introduced a more intuitive table-like schedule to replace the graphically defined schedule in version 3, but the concepts are similar in case you're using older software.

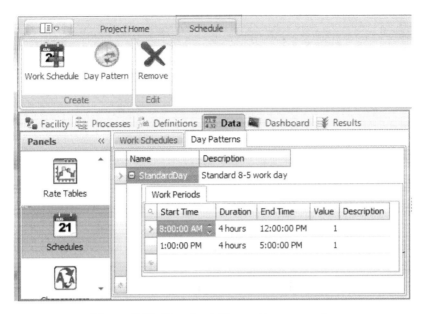

Figure 7.12: Standard Day pattern sample.

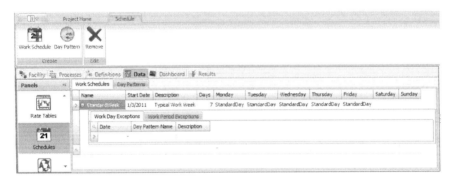

Figure 7.13: Standard week schedule sample.

the default of 1 then you must also specify that in the *Value* column. A *Cost Multiplier* column is also provided for use when one or more periods accrue cost at other than the normal resource busy rate. For example if you added 2 hours of overtime work to a normal schedule, you might specify the cost multiplier for those two hours as 1.5.

A *Work Schedule* includes a combination of Day Patterns to make up a repeating work period. The repeating period can be between 1 and 28 days; the typical situation is a seven-day week. A work period may begin on any calendar day based on the *Start Date*. If all of your work schedules are seven days long and start on the same day of the week (e.g., they may all start on Monday), then the column labels will be the actual days of the week as illustrated in Figure 7.13. If you have schedules that start on different days of the week (e.g., one starts on Monday and another starts on Sunday), or you

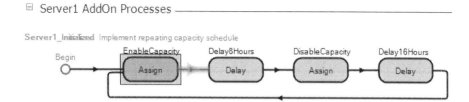

Figure 7.14: Manually created schedule of 8 working hours per day.

have any schedules that are not seven days, the column labels will simply be Day 1, Day 2, ..., Day n. In this case, the days that do not apply to a given schedule will be grayed out to prevent data entry.

The third component of a schedule is an *Exception*. An Exception overrides the repeating base schedule for the duration of the exception. Exceptions may be used to define overtime, planned maintenance, vacation periods, etc. Exceptions may be accessed by clicking the "+" to the left of the work-schedule name. There are two types of exceptions. The first, a *Work Day Exception*, indicates that on a particular day you'll use a work pattern that differs from the normal work pattern. For example, you might define a Day Pattern called AugustFridays indicating an early quit time, then specify for all Fridays in August that you use that work pattern. The other type of exception is a *Work Period Exception*. This indicates that for a specific date and time range, the schedule will operate at the indicated Value, regardless of what's specified in the schedule; this is often used for shut-down periods and extended work periods.

Search for "Schedule" in the Simio Help or Simio Reference Guide for a full explanation of how to configure and use calendar work schedules.

7.2.2 Manual Schedules

While calendar schedules are convenient for some applications, in other applications you might want the additional control, or perhaps simplicity, provided by a manual schedule. At the most basic level, manual schedules can be quite simple. You simply assign the capacity to the desired value, delay by the time it will remain at that value, and repeat as needed.

For example, to create a repeating eight-hour-per-day schedule for Server1:

1. Double click on the Initialized Add-on Process Trigger. This process will be automatically executed once when the Server object is initialized.

2. Add an Assign step, a Delay step, another Assign step, and another Delay step. Use these to set Server1.CurrentCapacity to 1, Delay by 8 hours, set Server1.CurrentCapacity to 0, and Delay by 16 hours.

3. Drag the End connector back to the first Assign step as illustrated in Figure 7.14. This will cause the four steps to keep repeating indefinitely.

Manual schedules also provide great flexibility. You can add as much detail as you want, as well as add transition logic. For example, you might implement

logic that says at lunch, the resource would just immediately stop working or
Suspend any work in progress and then at the end of lunch he would Resume
that work where he left off. But you could also implement logic so that 30
minutes before the end of the shift, he'd stop accepting new work (to allow
time for cleanup), but if work was still in progress at the end of the shift he'd
continue working until it was complete. This was just one example, but with
the full power of processes available, sophisticated schedule behavior can be
modeled. You can make objects as intelligent as they need to be.

7.3   Rate Tables and Model 7-4

A *Rate Table* is very different from the previous tables we've discussed — it's
not general-purpose and doesn't allow adding your own columns. Rather, it's
dedicated to a single purpose: specifying time-varying entity arrival rates by
time period. While our applications so far have involved only constant arrival rates for the duration of a run, many commonplace applications have
arrival rates that vary over time, especially service applications. For example,
customers come into a bank more frequently during some periods than other
periods. Callers call for support more at certain times of the day than at others.

For previous models we've often assumed a *stationary Poisson process*, in
which independent arrivals occurred one at a time according to an exponential
interarrival-time probability distribution with a fixed mean. To implement
arrival rates that vary over time we need a *nonstationary Poisson process*. See
Section 6.2.3 for a general discussion of this kind of arrival process, and the need
to include arrival-rate "peaks" and "valleys" in the model if they're present in
reality, for the sake of model validity.

While you might think that you could stick with an exponential interarrival-
time distribution and simply specify the mean interarrival time as a state and
vary the value of that state at discrete points in time, you'd be incorrect. Following that approach yields incorrect results because it doesn't properly account
for the transition from one period to the next. Use of the Rate Table provides
the correct time-varying arrival-rate capability that we need.

A Rate Table is used by the Source object to generate entities according
to a nonstationary Poisson process with a time-varying arrival rate. You can
specify the number of intervals as well as the interval time duration. The
Rate Table consists of a set of equal-duration time intervals and specification
of the mean arrival rate during each interval, which is assumed to stay constant
during an interval but can change at the start of the next interval. The rate
for each interval is expressed in Arrivals per *Hour*, regardless of the time units
specified for the intervals of the Rate Table (e.g., even if you specify a time
interval of ten minutes, the arrival rate for each of those ten-minute intervals
must still be expressed as a rate per *hour*). The Rate Table will automatically
repeat from the beginning once it reaches the end of the last time interval.
Though it might seem quite limiting to have to model an arrival-rate function
as piecewise-constant in this way, this actually has been shown to be a very
good approximation of the true underlying rate function in reality; see [33].

To use a Rate Table with the standard Source object, set the Arrival Mode
property in the Source to Time Varying Arrival Rate and then select the ap-

Table 7.5: Model 7-4 historical arrival data.

Time Period	Average Number of Patients Arriving in Time Period
0:00 to 4:00	49
4:00 to 8:00	31
8:00 to 12:00	38
12:00 to 16:00	36
16:00 to 20:00	60
20:00 to 24:00	70

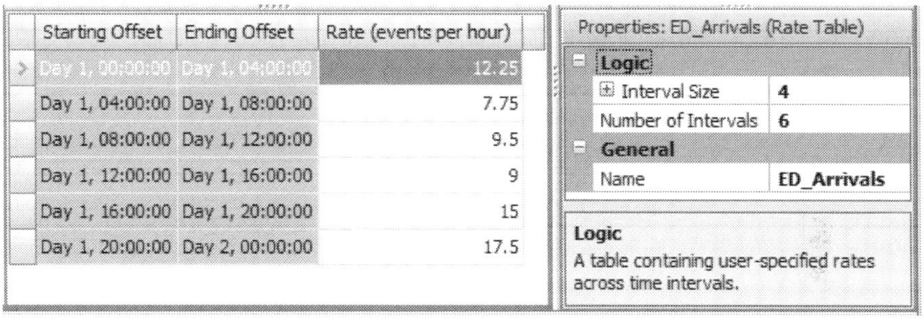

Starting Offset	Ending Offset	Rate (events per hour)
Day 1, 00:00:00	Day 1, 04:00:00	12.25
Day 1, 04:00:00	Day 1, 08:00:00	7.75
Day 1, 08:00:00	Day 1, 12:00:00	9.5
Day 1, 12:00:00	Day 1, 16:00:00	9
Day 1, 16:00:00	Day 1, 20:00:00	15
Day 1, 20:00:00	Day 2, 00:00:00	17.5

Figure 7.15: Rate Table showing time-varying patient arrival rates.

propriate Rate Table for the Rate Table property.

Let's illustrate this concept by enhancing our ED Model 7-2 with more accurate arrivals. While our initial model approximated our arrivals with a fixed mean interarrival time of 4 minutes, we know that the arrival rate varies over the day. More specifically, we have data on the average number of arrivals for each four-hour period of the day as shown in Table 7.5. So let's add that to our model.

1. Start with Model 7-2. Select the Data window and the Rate Tables panel.

2. Click the `Rate Table` ribbon button to add a new Rate Table. Change the name to `ED_Arrivals`.

3. While we could leave the default and specify the arrival rate for 24 one-hour periods, we have information for only six four-hour periods. So we'll set the table property Interval Size to `4 Hours` and set the Number of Intervals to `6`.

4. Our information is specified as the average number of patients arriving in a *four-hour* period, but we need to specify our rates as patients per *hour*. So, divide each rate in Table 7.5 by 4 and then enter it into the Simio Rate Table. When you're done it should look like Figure 7.15.

5. We have one more task. We need to go back to the Facility Window and change our Source to use this Rate Table. For the Arrival Mode property select `Time Varying Arrival Rate`. For the Rate Table property, select `ED_Arrivals` as shown in Figure 7.16.

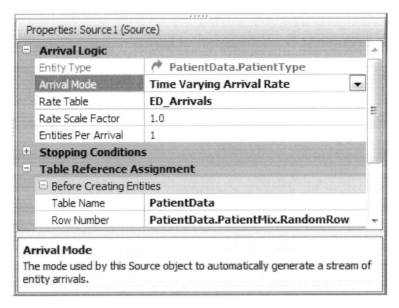

Figure 7.16: Source using Time Varying Arrival Rate.

Now if you run the model you'll see different results, not only because of the overall lower patient load, but also because the model now has to deal with peak arrival periods with a high incoming-patient load.

Although we did not use it in this model, the Source object also includes a Rate Scale Factor property that can easily be used to modify the values within the table by a given factor instead of changing the values separately. For example, to increase the Rate Table values by 50%, simply specify the Rate Scale Factor within the Source to 1.5. This makes it easy to experiment with changing service loads (e.g., how will the system respond if the average patient load increases by 50%?).

7.4   Lookup Tables and Model 7-5

Sometimes you need a value (e.g., processing time) that depends on some other value (e.g., number of completed cycles). Sometimes a simple formula will do (e.g., `Server1.CycleCount * 3.5` or `Math.Sqrt( Server1.CycleCount )`) or you can do a straight table lookup (e.g., `MyTable[ Server1.CycleCount ].ProcessTime`). Other applications can benefit from a non-linear lookup table. A Lookup Table[4] is a special-purpose type of table designed to meet this need. It supplies an $f(x)$ capability where x can be time, count, or any other independent variable or expression. One common application of this is modeling a learning curve where the time to perform a task might depend on the experience level of the person, measured in time or occurrences of the activity. Another application

---

[4]In earlier versions of Simio, lookup tables were known as Function Tables and the reference notation was slightly different (although the concepts are the same).

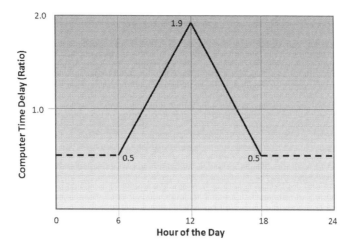

Figure 7.17: Computer responsiveness (speed factor) by time of day.

might be to determine something like processing time or battery discharge rate based on a part size or weight.

You create lookup tables in the Data Window on the Lookup Tables panel. You add a table by clicking `Lookup Table` on the ribbon. The table contains columns for the x (independent) value and the $f(x)$ (dependent) value. You can use a lookup table in an expression by specifying it in the general format `TableName[X_Expression]`, where TableName is the name of the lookup table, and X_Expression (specified as any valid expression) is the independent index value. For example `ProcessingTime[Server1.CycleCount]` returns the value from the lookup table named ProcessingTime based on the current value of the Server1.CycleCount state. The lookup table value returns the defined value or a linear interpolation between the defined values. If the index is out of the defined range then the closest endpoint (first/last) point in the range is returned.

Let's enhance Model 7-4 to add a lookup table for adjusting the Registration Time. Let's assume that the registration process uses a computer system that's faster in the early morning and evening, but slower during mid-day, as shown in Figure 7.17. We'll use a lookup table to apply an adjustment factor to the processing time to account for that computer speed.

1. Start with Model 7-4. Select the Data window and the Lookup Tables panel.

2. Click `Lookup Table` to add a new lookup table. Change the name to `ComputerAdjustment`.

3. Add the data to represent what's in Figure 7.17. You have only three data points: At time 6 the value `0.5`, at time 12 the value `1.9`, and at time 18 the value `0.5`. Any times before 6 or after 18 will return the first or last value (in this case, both 0.5). A time of exactly 12 will produce the exact value of 1.9. Any other times will interpolate linearly between

the specified values. For example at time 8, it's 1/3 of the way between the two x values, so it will return a value 1/3 of the way between the $f(x)$ values: $0.5 + 0.466 = 0.966$.

4. The last thing we need to do is go back to the `Registration` server in the Facility window and amend our Processing Time. Start by right clicking on the Process Time property and selecting `Reset`. We want to multiply the processing time specified in the Treatments table by the computer responsiveness factor we can get from our new lookup table. We need to pass the lookup table a value that represents the elapsed hours into the current day. We can take advantage of Simio's built-in set of DateTime functions to use the one which returns the hour of the day. So the entire expression for the Registration Process Time property is `Treatments.ServiceTime * ComputerAdjustment[DateTime.Hour(TimeNow)]`

Now our revised model will shorten the registration time during the night and lengthen it during the day to account for the computer's mid-day slow response.

7.5   Lists and Changeovers

In Section 5.4 we introduced the concept of Lists. Lists are used to define a collection of strings, objects, nodes, or transporters. Lists are also used to define the possible changeover states for a changeover matrix (e.g., color, size, etc.), or to provide a list from which a selection is to be made (e.g., a resource to seize, transporter to select, etc.). A List is added to a model from the Lists panel within the Definitions window.

The members of a list have a numeric index value that begins with 0. A List value may be referenced in an expression using the format *List.ListName.Value*. For example, if we have a list named `Color` with members `Red`, `Green`, `Blue`, and `Yellow`, then `List.Color.Yellow` returns a value of 3. If we have a ListProperty (i.e. a property whose possible values are list members) named `BikeColor` we can test conditions like `BikeColor == List.Color.Yellow`.

We will discuss the application of other types of Lists in Section 8.2.7, but for now we will just consider a String List. As you can easily infer from the name, a *String List* is simply a list of Strings. These are used primarily when you want to identify an item with a human readable string rather than a numeric equivalent.

String lists are easily constructed. In the Lists panel within the Definitions window, simply click on the String button to create a new String List. In the lower section of the window type in the strings for each item in the list like the example above as shown in Figure 7.18.

A *changeover* is the general term used to describe the transition required between different entity types. In manufacturing this could be from one part size to another (e.g., Large to Small) or some other characteristic like color. Changeovers are not as commonly used in service applications, but are still occasionally encountered. For example the cleanup time or transition between two Routine patients might be significantly less than that of a Serious or Urgent patient. Changeovers are built into the WorkStation object, but can also be added to other objects using add-on processes.

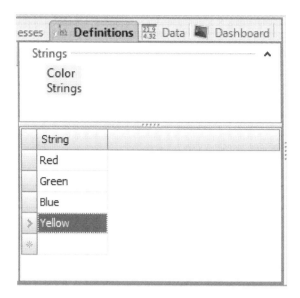

Figure 7.18: Example of a string list for colors.

There are three types of changeovers: `Specific`, `Change-Dependent`, and `Sequence-Dependent`. The simplest changeover is *Specific*, where every entity uses the same expression. It could be a simple constant-value expression (e.g., `5.2 minutes`) or a more complex expression perhaps involving an entity state or table lookup (e.g., `Entity.MySetuptime` or `MyDataTable.ThisSetupTime`). The other types of changeovers require that you keep track of some information about the last entity that was processed and compare that to the entity about to be processed. In *Change-Dependent* we don't care what the exact values are, just if it has changed. For example, if the color as changed we need one changeover time, and if it has not changed we need another (possibly 0.0) changeover time. In *Sequence-Dependent*, we're usually tracking change between a discrete set of values. That set of values is typically defined in a Simio List (e.g., Small, Medium, and Large). Each unique combination of these from and to values might have a unique changeover time. This requires another Data Window feature called a Changeover Matrix.

A *Changeover Matrix* has a matrix based on a list. In many cases the list will be a list of Strings that identify the characteristic by name (e.g., a List named Colors that contains items `Red`, `Green`, `Blue`, and `Yellow`). The entity will typically have a property or state (e.g., `Color`) that contains a value from the list. The Changeover Matrix starts with the values in the specified list and displays it as a From/To matrix. In each cell you'd place the corresponding changeover time required to change from an entity with the From value to an entity with the To value. When the changeover Matrix is applied, the row is selected based on this value for the previous entity, and the column is selected based on this value for the current entity.

You can define a changeover matrix on the Changeovers panel of the Data window. Click the Changeover Matrix button on the ribbon to add a new

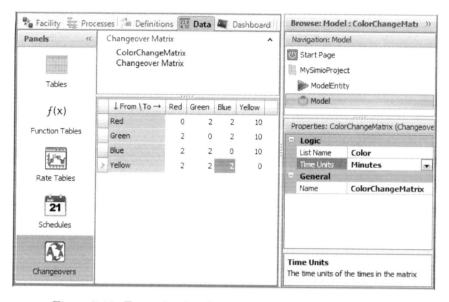

Figure 7.19: Example of a changeover matrix based on color.

changeover. Then specify the name of your previously created list in the List Name property. At this point the lower window will expand to show a matrix of all the members of your list. You can then change the value at each intersection to indicate the transition time between pairs. In Figure 7.19 it shows a 10 minute delay to change to Yellow, but only a 2 minute delay to change to any other color, and no delay if the color remains the same.

7.6   State Arrays

In Section 5.1.2 we introduced states, but deferred two important topics — arrays and initialization. A *state array* is simply a set of related states. By default a state is scalar — it has a Dimension (or set size) of 0.

A state can be made into an array by changing the value of the Dimension property. If the Dimension property is set to `Vector` (or a `1` is typed into the Dimension property), the State is one-dimensional and the Rows property will determine the number of rows in the array. If the Dimension property is set to `Matrix` (or a `2` is typed into the Dimension property), the Rows and Columns properties will determine the number of rows and columns in the matrix. If you'd like an array larger than 2 dimensions, Simio supports up to 10 dimensions — just type an integer up to 10 into the Dimension property.

Referencing state arrays uses the general form `StateName[row]`, `StateName[row, column]`, or `StateName[row, column, dimension 3, ...,` `dimension 10]`. For example, to reference row 7, column 5 of a state array named Weight, you'd use `Weight[7,5]`. Note that all state-array indexing is based at 1 (e.g., a vector with 5 array elements is addressed using 1-5, not 0-4).

Each state has an Initial Value property that can be used to initialize all items in the array to the same value. If you want to initialize the items to unique values you can do so using the Assign step in a process — possibly the OnInitialized process for an object. Alternatively, you could initialize a state or state array using a Read step to read the initial values from an external data source.

An even-easier approach takes care of dimensioning and initialization all at once — set the Dimension property of the state to `[Table]`. First, this will automatically dimension the state array based on the table contents. Each row in the table will generate a row in the array, and each *numeric* column in the table (e.g., an integer, expression, date-time, ..., but *not* an Element or Object Reference) will generate a column in the state array. Second, during initialization it will evaluate each numeric field and copy the values into the appropriate state-array item. If you combine this with the capability to Bind a table to an external file, this provides the capability easily to create and populate an array that exactly matches an external data source.

7.7 Macro Modeling

We'll close this chapter with one final data concept — writing generic models. *Macro modeling* is the concept of creating generic models with much of the key data "passed in" rather than explicitly specified in the objects. Each visiting entity would supply its own data to the object to determine how it is processed. The object would be configured to process the entity by following instructions from the entity. This then allows the objects to be relatively simple and of relatively few types, since the customization is primarily the responsibility of the entity more than that of the object. The data is abstracted from the objects – instead of the data being spread across many objects, it can be consolidated in one place. This is often referred to as a *data driven model.*

Simio provides several features to support this and we have already covered some of them. Late in Section 5.1.2 we discussed using Referenced Properties as Controls to make key parameters easy to find and change. Using a control is a simple way of making a model somewhat data driven because key data can be supplied via the control values in model properties or an experiment.

In Model 7-2 we implemented a slightly more comprehensive form of data driven modeling by supplying patient type, patient mix, processing location, and processing times in a table to avoid embedding that data directly into the model objects. So instead of specific actions, our model objects essentially said "Create the patient types specified in the table, in the percentage specified in the table, move to the location specified in the table, then process for the time specified in the table." So now someone with little model knowledge can experiment with our model by simply changing the data tables.

The object reference and element reference states and properties let us take this concept even further. Recall that, as the name implies, object references specify an object or a list of objects. This would allow us to extend our previous model for example, to specify the list of employees or equipment (resources) that are needed to treat or escort each type of patient. Likewise element references specify an element such as a material or statistic. This would allow you to specify any extra materials (like a surgery pack) needed by a patient or perhaps the specific statistic where a patient's data should be recorded.

The primary purpose of macro modeling is to create a somewhat abstract object that can handle many different things – like an object that represents a number of similar servers. The primary purpose of data driven modeling is to abstract the data so it can be maintained in a single place to make experimentation and model maintenance easier. While macro modeling and data driven modeling have different goals, they share the same techniques of creating somewhat generic objects that get their data from outside the object rather than directly in object properties. And these techniques often allow your model to be simpler to understand, use, and maintain.

7.8 Summary

Importing, storing, and accessing data is a very important part of most models. We've discussed several different types of data and several ways to store data in Simio. The table constructs, and particularly the relational tables, provide extremely flexible ways of handling data and a way to abstract the data from the model objects for easier model use and maintenance.

7.9 Problems

1. Adjust the Server capacity of example Model 7-1 to 4. Create an experiment containing responses for Utilization (Server1.Capacity. ScheduledUtilization) and Level of Service (Sink1.TimeInSystem.Average). Run it for 25 replications of 100 days with a 10-day warmup period. Compare patient waiting times if the Patient Mix of the modified example Model 7-1 changed seasonally so that Urgent patients increased to 10%, Severe patients to 30%, and Moderate and Routine patients decreased to 24% and 36% respectively.

2. In addition to the patient categories of modified example Model 7-1, a small-town emergency hospital also caters to a general category of presenting patients who are called "Returns." These are previous patients returning for dressing changes, adjustment of braces, removal of casts and stitches, and the like. Include this category into the Patient Mix of Model 7-1 such that proportions now become Returns 8%, Routine 30%, Moderate 26%, Severe 26%, and Urgent 10%. The treatment time for this category can vary anywhere from three minutes to thirty minutes. Compare the performance with that of Problem 1.

3. More realistically, about two thirds of the patients of Problem 2 arrive during the daytime (8:00 a.m. – 8:00 p.m.). Without changing the overall daily arrival rate, modify the model of Problem 2 to accommodate this updated arrival pattern. Compare the performance with that of Problem 2. Collect Level of Service (LOS) for daytime separate from night time. What change would you make to minimize staffing, while keeping average LOS for both periods under 0.5 hour?

4. On further analysis it is found that only 10% of returning patients of Problem 2 require treatment times between twenty and thirty minutes; the balance require between 3 and 20 minutes. What is the new LOS? Offer suggestions to management as to how to improve staff efficiency while keeping the LOS under the target of 0.5 hour.

5. Routine and Return patients have a somewhat low tolerance (30 to 90 minutes) for waiting for treatment and will sometimes leave without being seen (LWBS). The other categories of "really sick" patients will always wait until they are treated. Modify the model of Problem 2 so that any patient who waits more than his or her tolerance will exit the system. Hint: Use Search and Transfer steps. Report on the waiting times and the percentage of total patients who left without treatment.

6. A small free clinic has a single doctor who sees patients between 8:00 a.m. and noon. She spends 6 to 14 minutes (an average of 10) with each patient and so can theoretically see 6 patients per hour or a total of 24 each day. Staff is currently scheduling patient arrivals every 10 minutes, but have found that patients arrive as much as 15 minutes early or 30 minutes late, thus causing disruption to other patients. Worse, about 10 percent of patients never show up at all causing the doctor's time to be wasted and wasting an appointment that could have been used by other sick patients. Staff would like to evaluate alternative scheduling strategies such as scheduling 2-3 arrivals every 20 minutes with the main objective of maximizing the doctors utilization (e.g., she would like to really see 24 patients each day if possible). Assume that the doctor will stay until all scheduled patients are seen, but she is very unhappy if she needs to stay past 12:30. Measure system performance mainly on the number of patients actually seen, but also consider the average patient waiting time, and how late the doctor must typically stay to see all scheduled patients.

7. Of great concern to emergency-department staff are those patients who on arrival are classified Severe but whose condition before receiving treatment deteriorates to Urgent, requiring immediate attention and stabilization in the Trauma Rooms. If such deterioration is noticed during the examination stage in 10% of the Severe patients, adjust Model 7-2 to reflect this and compare the impact on waiting times and throughput. What assumptions have you made about "immediate attention?"

8. Emergency-department staff work 8.5 hour shifts according to an "ideal" schedule that includes a one-hour meal break each shift — preferably around mid-shift. During meal breaks, services in each section are maintained. One of "Registration" staff members will cover for the "Sign In" officer during her meal break. The minimum staffing capacity of each area is given in Table 7.6. When a new shift arrives there is a 30 minute "briefing period" during which the two shifts team together to share information and handover. Develop a defensible staffing schedule and update example Model 7-2. Re-estimate staff utilization. What assumptions do you make concerning "handover time?"

Table 7.6: Minimum staffing of service areas during breaks for Problem 9.

Service Area	Minimum Capacity
Sign-In	1
Registration	2
Examination	4
Treatment Rooms	4
Trauma Rooms	1

9. Clearly, during emergencies staff in the Trauma Rooms continue working with patients and forgo breaks — "grabbing a bite" when they can. Update Problem 8 to reflect this. Estimate the "real shift working time."

Chapter 8

Animation and Entity Movement

The goal of this chapter is to expand our knowledge of animation and entity movement by adding more context and additional concepts to what you've seen in earlier chapters. We'll start by exploring the value of 2D and 3D animation to a typical project. Then we'll discuss some additional animation tools that are at our disposal.

At that point we'll "move" on to discuss the various types of entity movement, providing a brief introduction to free space (off-network) movement, and a somewhat more detailed exploration into the Standard Library objects that support entity movement.

8.1 Animation

In Section 4.7 we introduced some basic animation concepts so that we could animate our early models. In this section we'll revisit and expand on those topics. We'll start by discussing animation, why it's important, and how much is enough. Later in this section we'll introduce some additional tools and techniques available to produce the animations you need.

8.1.1 Why Animate?

Every model we've built so far includes some amount of animation. While a few commercial simulation products still exist that are text-based with no animation, in general at least some level of animation has been included in most simulation products for decades. There is a reason for this — animation makes it much easier to build and understand models.

Understanding the System

Simulation models are often large and complex. From a modeler's perspective, it's hard to manage this complexity and understand both the detail and the "big picture" of the project. Visualizing with even a simple animation aids understanding. Ideally the simulation software you're using will make simple animation very easy to do. Most modern products do just that, and in fact the default 2D animation provided in Simio goes well beyond the minimum requirements and takes little or no extra effort on the part of the modeler.

While there are many components to effective model verification, careful review of an animation is usually a key. You can study the details by watching model progression step by step. You can quickly analyze common or unusual situations by studying model reaction around those events. And you can often discover problems and opportunities just by studying an animation in fast-forward mode, stopping from time to time to study the details.

Caution: While animation is important in verification it should *never* be the only, or even primary, mechanism for verification. An effective verification requires the use of many different techniques.

Communication

Two common high-level simulation goals are to understand the system and to provide benefit to stakeholders. Animation helps with both. If you provide a page of numbers or a diagram of model logic to a stakeholder, the typical response will involve eyes glazing over, often quickly followed by entry into a near-catatonic state. But show them an animation that looks familiar and they'll instantly perk up. They'll immediately start comparing the components, movements, and situations in the animation to their own knowledge of the facility. They may start discussing design aspects amongst themselves just as though they were looking through a window at the real system. At this point, you're well along the way to having delivered on both high-level objectives. At the top end, animation quality may be good enough for the stakeholders to use it for outside communication — not only promoting the project, but often even using it to promote the organization.

Of course getting to this point is rarely free. Depending on the modeling software you use, building an animation suitable for use with stakeholders may be quite involved, especially with products for which animation is built independent of the modeling. Sometimes animations are created in a totally separate animation product. And sometimes the animation can only be run in a *post-process mode* after the model run has completed. Such a post-process analysis severely limits the level of model experimentation possible.

Fortunately, the animation that Simio creates by default is often fairly close to "stakeholder ready." Simio's animation is built-in, in fact created automatically as you build the model. And it runs *concurrently* so you have full interactive capabilities while watching the animation.

The Importance of 3D

So far in this discussion we've not even mentioned 3D animation. You might be wondering if it's necessary or why anyone would use it. Ten years ago we might have told you that 3D animation was unnecessary in most projects and not worth the effort. But a great deal has changed in that time. First, high-quality graphics have become routine and even expected in our daily lives. Most people now easily relate to 3D animation and can immediately appreciate its use. Because of this new frame of reference, 2D animations tend to look primitive and don't engage the stakeholder or inspire their understanding and confidence

Figure 8.1: Simio View ribbon.

to the same degree as 3D. Second, the effort involved in creating 3D animations has decreased tremendously. A modeler no longer requires extensive drawing skills or access to a large custom symbol library. And they no longer need skills in a complex drawing package. Modern software allows a modeler with minimal artistic skills to generate compelling animations without any special skills, tools, or custom libraries.

Animation is often important and many people find it enjoyable and rewarding to work on. But you can have too much of a good thing. It's easy to get so involved in making your animation look "just right" that you spend inadequate time on model-building, validation, analysis and other important parts of the project. Keep in mind that, just as in the rest of your modeling, the animation is an approximation of the real system. We're trying to develop animation that's sufficient to meet the project objective, and no more. Determine with your stakeholders the appropriate level of animation required and then stick to that plan. Save animation "play time" for after all other important aspects of the project have been completed.

8.1.2 Navigation and Viewing Options

If you've not yet disabled it, you may still be seeing a gray area at the top of your Facility Window. If you've already disabled it or want to bring it back as a reminder, recall that pressing h while in the Facility window will toggle it on and off. Let's explore these navigation controls in a bit more detail.

- To toggle between 2D and 3D you can press the 2 and 3 keys respectively, or you can click on the 2D and 3D buttons in the View Ribbon (refer to Figure 8.1).

- You can pan (move) the facility window contents by left clicking in an empty space and dragging the window. *Caution*: If you happen to click on an object (for example a large background object) you'll move the object instead of pan the view. To avoid this you can either click and drag using the middle mouse button[1], (which will not select an object) or you can right click on the background object and enable the Lock Edits option to prevent that object from moving.

- If you have a mouse with a scroll wheel (highly recommended) you can zoom the view in and out by scrolling the mouse wheel. If you don't, then you can accomplish the same action with a Control-Right Click and move the mouse up and down.

---

[1]Often the same thing as clicking down on (but not spinning) the mouse's scroll wheel.

Figure 8.2: Visibility ribbon.

- You can resize or rotate an object using its handles, the green dots that appear when you select an item. If you click and drag one of the handles, it will resize the object. If you Control-Left Click on a handle you can rotate the object.

- Clicking the `View All` button on the View ribbon will resize the Facility window so that your entire model is in view. This is particularly useful to return to a familiar view if you've been zooming and panning and "lost" your model objects.

- The `Background Color` button on the View ribbon changes the background color of both 2D and 3D views. Although some older models may have a black background in the 3D view, it is no longer possible to create different background colors for 2D and 3D.

- The `Skybox` and `Day/Night Cycle` buttons control how the background above your animation looks and changes. This is particularly useful for outdoor models like ports and transportation networks.

There are a few actions applicable only when you're in the 3D view:

- Right-click and drag left and right to rotate the view in 3D.

- Right-click and drag up and down to zoom in and out.

- Hold down the Shift key while moving an object and you'll move it up and down in relation to the floor.

- Hold down the Control key while moving an object and you can stack it on top of other objects rather than overlapping it.

- Pressing the R key in the 3D view or clicking the `Auto Rotate` button on the View ribbon will start the 3D view rotating. Pressing Escape will stop the rotation.

The Visibility ribbon illustrated in Figure 8.2 helps you fine tune the look of the animation: :

- The Visibility group contains buttons that toggle on and off specific display components. In many cases you may want to disable display of most or all of those items to make your animation look more realistic. Note that you may have to enable some of those buttons later if you want to interact with the model. For example, you cannot select a node if `Nodes` animation is disabled.

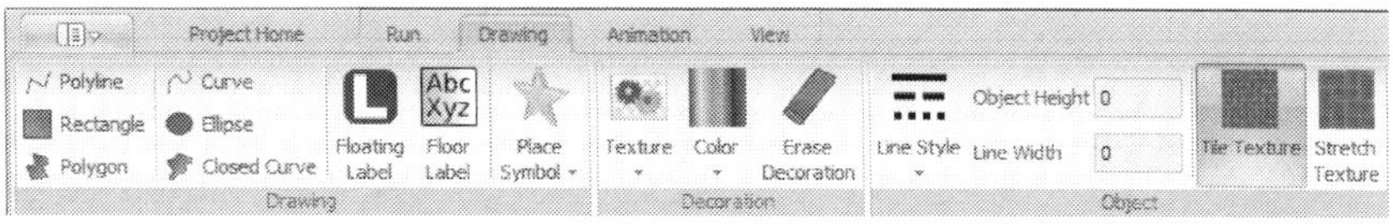

Figure 8.3: Facility window Drawing ribbon.

- The `Direct Shadows` and `Diffuse Shadows` buttons in the Visibility group provide finer control over how shadows appear around your objects.

- The buttons in the Networks grouping provide a flexible way to see individual or sets of networks of your travel paths.

When you have your model ready to show, there are a few more helpful features on the View ribbon:

- A *View* is a particular way of looking at your model including 2D or 3D, its zoom factor and rotation. The Named Views grouping allows you to add views with the `Add View` button — this allows you to provide a name for the view you currently have on your screen. `Manage Views` allows you to edit or remove an existing view. The `Change View` button is a two-part button. The top part cycles you through sequentially all named views. The bottom section provides a list so you can select which view to display.

- The Camera Tracking group contains items to determine the vantage point of the camera as well as the ability to focus the camera on a particular object or even have the camera ride just in front of or behind a moving object like an entity or vehicle.

- The Camera Sequences group allows you to define the timing and sequence of multiple camera moves that can help prepare a presentation to "tell a story".

- And finally the Video group allows you to record video (avi) files of your animation and related activities.

8.1.3 Background Animation With the Drawing Ribbon

Let's talk next about the things that appear in your model but don't move, or static animation. While Simio doesn't intend to be a drawing tool (there are very powerful tools like Google Sketchup available for free), it does provide basic drawing and labeling tools as well as the ability to import symbols and the capability to change how those drawn and imported symbols look. Taking advantage of these tools can be a quick and easy way to improve the realism and credibility of your model. Let's look at the components on the Facility window Drawing Ribbon illustrated in Figure 8.3.

The Drawing Group on the left side of Figure 8.3 starts with six drawing tools to create some basic objects: `Polyline`, `Rectangle`, `Polygon`, `Curve`,

Ellipse, and Closed Curve. If you're unfamiliar with any of these shapes, we suggest that you experiment with them to see how they work. When you're drawing most shapes, after you click the last point, right-click to end. Hitting the Escape key will abort the attempt and erase your drawn points. You may notice that many of them appear to have an extra point. That's the rotation point. You may move that rotation point to any position. If you Ctrl-click and drag any of the other points, the entire object will rotate around that rotation point.

If you skip over to the Decoration and Object groups in the drawing ribbon, you'll see tools to change the looks of these basic objects including the color, texture (pattern), options to tile or stretch that texture, line style and width, and object height. We'll leave it to you to explore the full capabilities of these tools, but here are a few applications and tricks that might stimulate your imagination:

- Making a wall: Add a line the length of the wall. Set the Line Width to perhaps 0.1 meter. Set the Object Height to perhaps 2 meters. Go to 3D view. Click on the Texture button and select a texture (perhaps a brick pattern), then click on the wall object to apply the texture.

- Creating a company logo: Find a .jpg file with the logo you want. Save that file in your Public Documents (the exact name of this varies depending on your operating system) .. Simio .. Skins folder. This will then automatically appear as a choice under the Textures button. Draw a rectangle similar in shape to the logo. Apply your logo texture to the top or sides of your rectangle.

- Creating a simple building: Create a rectangle. Give it height. Apply a texture to the sides. Extra credit — take a photo of the face of your favorite building and apply it to your building. (Of course you can get much better buildings, perhaps even your own, from Google Warehouse or Google Maps.)

To finish our discussion of the Drawing ribbon, there are three buttons we've skipped. Floating Label creates a simple label that "floats" in the air and will always face you no matter your viewing direction. It also always displays at the same size regardless of zoom level. A Floor Label, as the name might imply, appears as though it has been painted on the floor. It can have multiple lines and you can choose color, size, and formatting options. Don't miss the fact that you can embed expressions inside the text so you can create dynamic and informative labels.

We saved the best button for last. Place Symbol provides many different options to place a symbol as part of your model background. The top part of the button provides easy access to place another of the last symbol you selected (if any). The bottom part of the Place Symbol button brings up a large dialog that provides three main choices:

- You can scroll around in the built-in Simio symbol library to select a symbol to place, similarly to what we did in Section 4.7.

- You can `Import Symbol` from a local file already available to you. Perhaps this is a file you've created using Sketchup, or maybe a DXF file exported from some CAD program. Or you can import an image file like a JPG, BMP, or PNG file. A caution here is that DXF files can be quite large. They are often created at a detail inappropriate for simulation (e.g., they contain things like the threads on the nuts and bolts that hold the building structure together). If you plan to use a DXF file, it's best to delete the unnecessary detail before you export to the DXF file. *Tip*: Often, a 2D building layout as found in a JPG or even .pdf file is a good basis for an animation. It's a very efficient way to provide scale and background and then you can add 3D objects (possibly including walls) to provide the depth.

- You can `Download Symbol` from Google 3D Warehouse as we also discussed in Section 4.7. *Tip*: When possible, search for "low poly" symbols. This refers to symbols with a low polygon count (a measure of display complexity). Even some Google 3D Warehouse symbols have the same problem as DXF files — their complexity and detail is inappropriate for simulation animation. So you might select a very nice picture of the "perfect" fork lift truck only to find that it inflates your model size by 10 MB because it's detailed down to the bolts that hold the muffler bracket in place. In some cases you can load such a symbol into Sketchup and remove the unnecessary detail.

8.1.4 Status Animation With the Animation Ribbon

In addition to seeing objects moving on your screen, often you want other types of visual feedback and interaction to help assess model performance. Not only does Simio provide a set of tools for this, it also provides a choice on how to display them: in the Dashboard window or in the Facility window.

Dashboard window

Every model has a *Dashboard window* designed as a place to display selected graphical status indicators and interactive buttons. The Dashboard window for any model is defined using the Dashboard tab. Under that tab you'll see something that looks similar to the Facility window, except that you can't place library objects here; only the items described in this section can be placed in a Dashboard window.

There are two advantages to putting status items in the Dashboard window:

- It allows you to keep business graphics separate from your facility animation. You can then hide the business graphics or dedicate a certain portion of your screen to viewing them.

- If your model is later used as an object, this dashboard is available from that object's right click menu. The user of your object will see the status display that you designed. The dashboard from an embedded object has the additional advantage that it can be displayed outside of the Simio window space, even on a second monitor if one is available. *TIP*: The

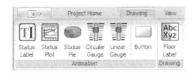

Figure 8.4: Dashboard window Drawing ribbon.

Figure 8.5: Facility window Animation ribbon.

dashboard can be a great place to store documentation of your object. Just add it in a floor label.

Facility Window

You can also place your status animations in the Facility window. The advantage of placing them here is that they can be arranged adjacent to the objects of interest, or you can place them elsewhere in your model and associate one or more named views with them. The main disadvantage is that these are primarily 2D business graphics and they don't always look good in the 3D Facility window.

Status Animation Tools

Regardless of the window you choose to display your status objects, you have a similar set from which to select. When you display the Dashboard window, the default ribbon will be a Drawing ribbon (Figure 8.4) that's quite different from the Facility window Drawing ribbon we discussed in Section 8.1.3. If you want to add status objects to your Facility window you'll have to click on the Animation ribbon (Figure 8.5). You'll note that the tool set in each of these two ribbons is very similar:

- `Status Label`: Displays static text or the value of any expression.

- `Status Plot`: Displays one or more values as they change over time.

- `Status Pie`: Compares two or more values as a proportion of the total.

- `Circular Gauge` and `Linear Gauge`: Provide a more visually compelling display of a model value.

- `Button`: Provides a way for a user to interact with the model. Each time the button is clicked an Event is triggered, which could be linked to a process.

Figure 8.6: Symbols ribbon for symbol editing.

- `Floor Label` (Dashboard only): Similar to the Floor Label we discussed in Section 8.1.3

- `Detached Queue` (Facility only): Adds animation to show entities that are waiting in a queue.

8.1.5 Editing Symbols with the Symbols Ribbon

When you click on most objects in the Facility window, the active ribbon automatically changes to the Symbols ribbon (Figure 8.6) . This ribbon allows you to customize the animation of that object by modifying the symbol(s) and adding animation features. Although at first glance it looks like a lot of new options, if you take a second look, you'll discover many familiar features. In fact, way back in Section 4.7 we used the Project Symbols category of this ribbon to select a new ATM Customer. And the `Import Symbol` and `Download Symbol` ribbon buttons are similar to options on the `Place Symbol` button we discussed on the Drawing ribbon. Likewise, the items in the Decoration group are identical to those in the Drawing ribbon Decoration group.

Attached Animation

The Attached Animation group looks very similar to the Animation ribbon in Figure 8.5 but there is one very important difference summed up by the word "Attached." Since you got here by selecting an object, if you place any items from the Attached Animation group, those items will be *attached* to that selected object. In the case of fixed objects like a Server, this just means that if you later move the object on the screen, the attached items will move along with the object. But an interesting situation occurs when you attach an item to a dynamic object like an Entity, Vehicle or Worker. In this case the attached items will travel with each dynamic object as it travels through the model. Let's examine a few interesting applications of this:

- The green horizontal line across the default Vehicle symbol is an attached queue that animates `RideStation.Contents` to show the entities that are being carried by (i.e., riding on) the vehicle.

- In a similar fashion, if you're grouping your entities by using the Combiner object, you can display the group members by adding an attached queue to your parent entity object. If you animate the `BatchMembers` queue you'll see the members currently in the batch of the parent. You can find examples of this in SimBits `CombineThenSeparate` and `RegeneratingCombiner`.

- You can also display text or numeric information with an entity. In Sim-Bit `OverflowWIP`, we wanted to be able to verify the creation time of each entity, so we added an attached status label using the expression `TimeCreated` on the entity. We also rotated the label so we could still read it when it's in a queue.

Adding attached information can be both visually compelling and valuable for verifying your model. Since attached animation becomes part of (or even "inside") the object, the scope for attached animation is the object itself. The advantage of this is that you can reference any expression from the perspective of the object. So in the first bullet above, for example, we referenced `RideStation.Contents`, not `Vehicle1.RideStation.Contents`. The latter wouldn't work because it would be at the scope of the containing model, rather than the object itself.

Additional Symbols

The last item group on the Symbols ribbon is the Additional Symbols group. These features allow you to add, edit, and remove symbols from a set of symbols associated with an object. By default each object has only one symbol[2], but in many cases you'd like to have a set of symbols for that object. Some useful examples include:

1. A entity that's a single entity type, but you want to illustrate some variety (such as a Person entity, but you want to show 5 or 10 different person symbols).

2. An entity that changes as it progresses through the model (e.g., a part that changes its picture as it gets processed or when it fails an inspection).

3. A Server (or any object) where you want to show different pictures for each state it may enter (e.g., Busy, Idle, OffShift).

In the Properties window of most objects you'll find a category named Animation (Figure 8.7) that contains properties Current Symbol Index and Random Symbol. Note that these properties are disabled unless you've added at least one additional symbol to your object. Using these properties allows you to select from between the specified symbols.

The *Random Symbol* property is looked at when an object is first created. If this is set to `True`, then it will automatically select one symbol from the set of defined symbols. For example, in case 1 above, if you had defined 3 person symbols: a man, a woman, and a child, then setting Random Symbol to `True` would result in each entity having a 1/3 chance of using each symbol.

The *Current Symbol Index* property is used to tell Simio where to look to determine that object's symbol number. Symbols are numbered starting at 0, so if you have 5 symbols they'll be numbered 0 through 4. To animate case two above, set Current Symbol Index to `ModelEntity.Picture` and then assign

---

[2]Multiple symbols per object were being added as this book went to press and the expression for `Current Symbol Index` is changed accordingly.

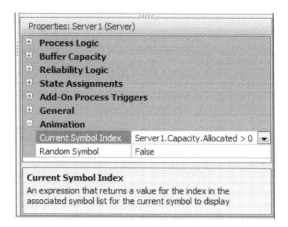

Figure 8.7: Default animation properties for a server.

that entity state to the index of the picture you want to display as the entity progresses through the model.

The Current Symbol Index property doesn't have to be an object state — it can be any expression. For example in SimBit OverflowWIP, we wanted an obvious visual cue to show in which hour a part was created, so we used the Current Symbol Index expression of Math.Floor(ModelEntity.TimeCreated) on the entity. Since time is in hours, it uses a new symbol for each hour of the day.

The default Current Symbol Index property for a Server is *ServerName*. Capacity.Allocated > 0. This assumes two symbols — it uses symbol 0 when it's not busy and symbol 1 when it's busy. You could use other expressions to cover other states or conditions (e.g., *ServerName*.Capacity would use symbol 0 when off-shift and symbol 1 when on-shift). Using the expression *ServerName*.ResourceState would use symbols 0 through 4 for Starved, Processing, Blocked, Failed, and Offshift, respectively. *Tip*: View the "List States" topic in help to see the state definitions for other objects.

8.1.6 Model 8-1: Animating the PCB Assembly

Let's use some of this new-found knowledge to continue work on our PCB model. But before we start, recall our earlier admonition to *develop animation that's sufficient to meet the project objective, and no more*. Sometimes you may not be able to find exactly the right symbol or display things in exactly the precise way, but that's okay — a reasonable approximation is often good enough. You can always improve the animation later as time and objectives dictate.

Let's start by loading our PCB Model 5-3 discussed in Section 5.4 and save it as Model 8-1. Since this is a fictional system, we can take some liberties in animating it. First let's find a better symbol to use for our PCB object.

- Click on the PCB object and then click on the Symbols library. Search the library to see if there's a symbol there that could be used as a PCB (recall that this is a small board with computer chips on it). Unfortunately nothing appears to be close, so let's keep looking.

- Again click on the PCB object and then click on the `Download Symbol` button. If you're on-line, this will bring up the Trimble 3D Warehouse search screen. Do a search on PCB. In our search we found 315 symbols involving that term (most having nothing to do with *our* topic) but there were several good candidates. Select the symbol named `Gumstix Basix R1161` and then select `Download Model`. TIP: You might think that `Nat's PCB` on page one is an appropriate symbol, and it could be. But after you download it you would find that your model size has increased by several megabytes. This is because that symbol has a very *high polygon count* - a measure of the *complexity* of the symbol. If you happen to download such a large symbol and you do not want it bloating your model, you can replace the symbol with a smaller one and then delete the large symbol from the Project > Symbols window.

- After the symbol downloads it will be displayed in the Simio Import window. Here you have a chance to rotate, resize, and document the symbol. Let's do all those. Use the `Rotate` button to rotate it so the short edge is to the right. We want the longest dimension to be 0.3 meters, so change the Width to be 0.3 (you'll note that the other dimensions change proportionally). And finally we'll change the Name to `PCB` and the Description to `Gumstix Basix`.

- The symbol will appear quite small, but no worries — we'll deal with that later.

Let's follow a similar process to change our chip-placement machine.

- Since there are no placement machines in the Simio symbol library, we'll skip that step. Click on the Placement object and then click on the `Download Symbol` button to bring up the Trimble 3DWarehouse search screen. Do a search on `placement machine`.

- We selected the `Fuji QP 351` and downloaded it. It downloaded in the correct orientation and size, so we can just click OK to accept it as is and apply it to the `Placement` object.

The three parallel stations are also placement machines, so we could simply use the same symbol for them. But they actually represent higher-tech machines that do more accurate placement, so let's pick a different symbol for those machines. Repeat the above process starting on the upper machine. Select a different machine (we chose the `HSP 4796L` because it looks high-tech) and download it. Adjust the rotation and size of your selection appropriately — we had to rotate ours 90 degrees and change the width to 1.5 meters.

You may have noticed that each time you download a symbol, it gets added to your project library. So now you can click on each of the other two parallel machines and apply the same placement symbol you just downloaded. Both the inspection and the rework are manual operations, so we'll animate both of them using a suitable table symbol. You can download another symbol if you like,

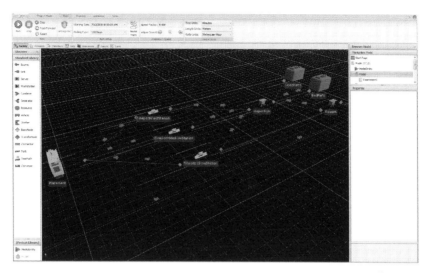

Figure 8.8: Good symbol selection but drawing not yet to scale.

but we think the TableSaw in the symbol library under the Equipment category looks close enough to a work table to meet our needs, so we just applied that to both manual stations.

If you've followed along as instructed, you may be feeling unhappy now. Our symbols don't look very nice at all against the white background in 2D and in fact they look pretty small. The first problem is easy to fix by changing to 3D and zooming in a bit (Figure 8.8). The second problem is one of model detail. In our earlier models we were not concerned about the details of part and machine sizes and their proximity because we were making the simplifying assumption of instantaneous movement between stations. While this is often a very good assumption at the beginning of a modeling project, you'll usually come to a point where it does matter — that's where we are now.

In many cases you'll have a JPG or other image file of the system layout that you could use as a background image. That can then be used for sizing and placing your equipment appropriately. In this case we don't have that. As we were selecting symbols, we've been providing some selected dimensions. But in our original model we placed the machines arbitrarily about 10 meters apart and in the real system they are only about 1-2 meters apart. Before we start moving machines around, let's clean up the associated objects for each of our servers:

- Zoom into the three parallel stations so they fill the screen.

- Start at the bottom of the screen (the easiest to see) and click on the input node (the one named `Input@FinepitchSlowStation`) and drag it so that it's adjacent to the input side of the placement machine symbol. Repeat that with the output node. It may look good in the 3D view, but we often flip back and forth between 2D and 3D to fine tune the adjustments to

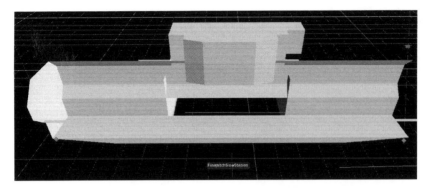

Figure 8.9: Editing animation on FinepitchSlowStation.

look good in both views. The nodes and paths in your animation should now look similar to those in Figure 8.9.

- The queue named `Processing.Contents` (the green line above the machine in the 2D view) shows what's currently processing on the machine. It would look better if this were on (or actually inside) the machine. In the 2D view, drag the queue so it looks like it's in the bed of the placement machine. At this point, it's still on the floor. Move to the 3D view and use the Shift-Drag to lift the queue up so it's visible above the bed. Then move both ends of the line so the queue is centered on the bed as illustrated in Figure 8.9. *Tip*: It's easy to "lose" a small symbol inside of a large one, so it's generally best to leave the queue longer than the object on which you're displaying it until you're sure it's in the right place. Only then should you adjust the length.

- Shorten the length of the Output Buffer queue and move it adjacent to the output side. Since we're using a zero capacity input buffer, the queue on the input side will never be used — delete it to avoid confusion.

- Now repeat that whole process (the above three steps) with the other two placement machines.

- Follow the same process for Placement, Inspection, and Rework except that you should not delete their input-buffer-queue animations because they may still be in use.

After completing all of the above, your model still looks odd. If you run it now, the PCB symbols that are moving around will look like little dots. We've fixed the scale of the symbols and their associated objects, but we've not yet adjusted the scale of the layout to reality. Let's take care of that now.

- In the 2D view, select FinepitchSlowStation and drag it upwards so it's parallel to and about 1 meter away from FinepitchMediumStation. (You can take advantage of the grid lines to help.) Then select FinepitchFastStation and drag it downwards so it's parallel to and about 1 meter away

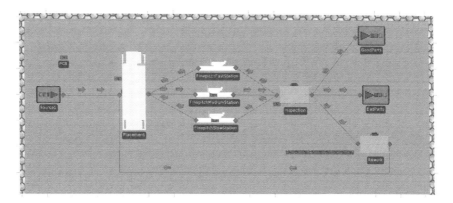

Figure 8.10: 2D Animation of PCB to scale.

from FinepitchMediumStation. Note that your queues and nodes will move along with each object. Consider why that is.

- Move the rest of the objects so that there is approximately 2 meters of horizontal distance between the objects.

- The objects that are still using the standard library symbols now appear quite large. You may replace them with other symbols if you wish. We just dragged a corner of the symbol to reduce the sizes an appropriate amount.

As a last quick enhancement, let's add a floor and some walls.

- In the 2D Facility view, use the `Rectangle` button on the Drawing ribbon to draw a rectangle that completely covers the floor area of the equipment. Use the `Color` button to make it gray. If you'd like the floor grid to show through, go to the 3D view and use Shift-drag to move the floor slightly down. Now right-click on the floor and enable Lock Edits to prevent accidentally moving it.

- Back in the 2D view, draw a polyline around the left, back, and right edges of the floor. Still in the Drawing panel, give the wall a `Width` of 0.2 and a `Height` of 1 meter (so we can still easily see over it). In the 3D view, select an interesting texture from the `Texture` button and apply it to the wall. Again, right click to Lock Edits and prevent accidental movement of the wall.

In this section we've used some of the simple techniques we learned in earlier sections to improve our animation a bit and make it roughly to scale (Figure 8.10). Sometimes it's convenient to do this "after the fact" as we just did, but generally it's easiest to build the model at least roughly to scale from the start (preferably starting with some sort of scale drawing). But either way, you can see that, with just a little effort, we've converted a rather plain engineering tool into one that's a bit more realistic as in Figure 8.11. We could continue improving our animation with more attention to the detail (perhaps upgrade

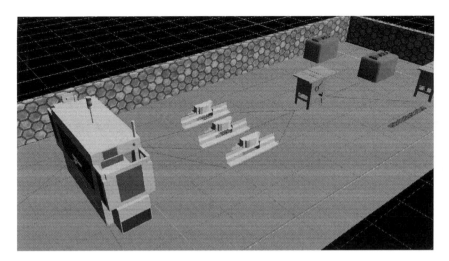

Figure 8.11: 3D Animation of PCB to scale.

our "table saw"), but we'll leave that as an exercise for you. But we'll come back for some additional enhancements later.

8.2   Entity Movement

We've incorporated the movement of entities from one location to another in most of our previous models. In some models you've seen the entities moving; in others it's been an instantaneous movement. In fact there are many ways that an entity can move in Simio. Figure 8.12 illustrates some of the movement alternatives. The easiest and most intuitive alternatives are those using the network capabilities built into the Standard Library. We'll spend considerable time discussing each of those in the following sections. But we'll start with a very brief introduction into movements not relying on those networks.

8.2.1   Entity Movement Through Free Space

While in most cases you'll want to take advantage of the movement support built into the Standard library, there are a few cases where you might want more flexibility. *Free Space* is the term Simio uses to describe the area of a model that's not on the network, e.g., the "spaces" between the objects. When an entity isn't located in a "physical location" (a station, a node, or a link, as illustrated in Figure 8.13,) it's said to be in *free space*. Entities can exist in free space, can move through free space, and can be animated during that move. The two options under Free Space Movement in Figure 8.12 differ primarily by whether physical locations are involved.

If you want an off-network instantaneous move, you can use a Transfer step in a process. The Transfer step initiates a (possibly instantaneous) move *from* a station, a link, a node, or free space *to* a station, a link, a node, or free space. One somewhat common use for this is in combination with a Create

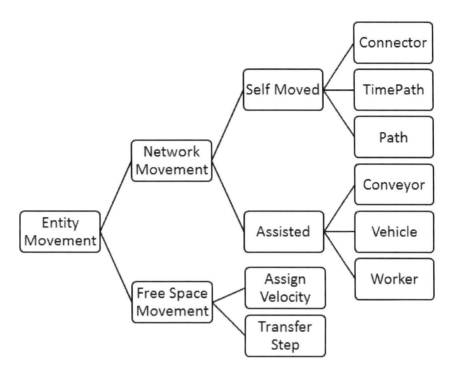

Figure 8.12: Ways entities can move in Simio.

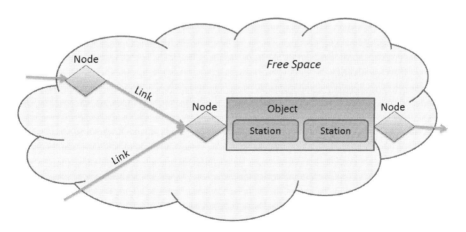

Figure 8.13: Entities can be in a node, a link, a station, or free space.

step. When entities are created, they're created in free space. If you don't immediately make a change, they may start moving through free space at their default speed. So it's common to follow a Create step with a Transfer step that initiates transfer into the specified location.

Of course, you sometimes want an entity to move through free space without a physical location as its immediate destination. This is particularly common in *Agent-Based Modeling* (ABM). You can set an entity's movement state to specify its location, heading, speed and acceleration. Specifically, you may set:

- Location parameters: Movement.X, Movement.Y, Movement.Z

- Heading parameters: Movement.Pitch, Movement.Heading,
  Movement.Roll

- Motion parameters: Movement.Rate, Movement.Acceleration, and
  Movement.AccelerationDuration

Note that most of these parameters work only in free space and don't work on links (e.g., don't try to use these to set acceleration while on a link). But to make this a bit easier, Simio also provides a Travel Step which provides the above functionality and more in a somewhat easier to use fashion. One of the additional features provided is to move directly through free space to any absolute or relative coordinate or to any specified object.

To make things easier still, you can simply change the *Initial Network* property of any entity (which includes Vehicles and Workers) from the default of `Global` to the value `No Network (Free Space)`. As that name might imply, it releases that entity from the constraints of traveling on a network, to instead travel directly from object to object through free space. This can be particularly useful when you have a large number of locations (like a country-wide distribution network) and you don't need your entities to follow a specific path between locations.

Most of the concepts we learned in Section 5.4 about selecting entity destinations still apply with free space moves. Unless you change the defaults, the entity will always move to the closest destination. But you have the complete set of options in *Entity Destination Type* (i.e., `By Sequence`, `Continue`, `Select From List`, and `Specific`) available for your use.

For some example models refer to the SimBit `FreeSpaceMovement` which contains three models illustrating free space movement concepts.

8.2.2   Using Connectors, TimePaths, and Paths

As we discussed above, in most cases you'll be moving your entities across a network using features provided by the Standard Library. The first choice you need to make is whether your entity will be moving by itself (e.g., a person or autonomous object moving on its own), or if it will require assistance to move (e.g., a person, vehicle, conveyor, or other device). The options under Network Movement in Figure 8.12 will be explored in the next sections. But first let's discuss what links are and what they have in common.

General Link Properties

Simio defines a *link* as a fixed object representing a pathway between nodes that may be traveled by an entity. There are a lot of concepts buried in that sentence:

- A link is an object. It's a fixed object, meaning that it can't move during a run.

- A link connects nodes. It can't exist by itself or with a single node.

- Entities are the only objects that may travel on link. This sounds more constraining that it is. Vehicles and Workers are also entities (with extra behavior), since they are derived from entities.

Here are a few more concepts common to all links:

- A link must have a start node and an end node.

- A link is a member of one or more networks, which is simply a collection of links. A link is always a member of a special network called the "Global" network.

- A link object has a length that may be separated into equally spaced locations (cells).

- Link control and position are determined by the leading edge of an entity. When an entity's leading edge moves onto a link, the entity movement is controlled by the logic built into that link.

- Simio provides tracking of the leading and trailing edges of each entity, and detects collision and passing events. This makes it relatively easy to define and customize your own link behavior in processes.

Standard Library link objects include the Connector, Path, TimePath, and Conveyor. Let's discuss those first three objects next.

Connectors

Connectors are the simplest type of link. They connect two objects directly with no internal time delay. Since the movement is instantaneous there is no interference between entities; in fact, only one entity at a time is traversing the link. It works as though the entity were transferring directly from the start node to the end node, or directly to the object associated with the end node.

Connectors have a *Selection Weight* property that can be used to select between outbound links when more that one outbound link is connected to a node. We used this feature in Model 5-1, discussed in Section 5.2.

TimePaths

TimePaths have all of the basic link capabilities including the Selection Weight discussed above. But instead of being a zero-time transfer like a Connector, it has a *Travel Time* property to designate the time the entity should take to travel the length of the path. It also has a *Traveler Capacity* so you can limit the number of entities concurrently traveling on the path. Because of that capacity limitation, it also has an *Entry Ranking Rule* to use in selecting which of the potential waiting entities should be granted access next.

TimePaths also have a *Type*, which determines the travel direction. This can be either `Unidirectional` or `Bidirectional`. Unidirectional allows travel only from the start node to the end node. Bidirectional allows travel in either direction, *but only in one direction at a time*, much like a one-lane bridge on a two lane road. If you want to model a fully bidirectional path (e.g., a two-lane road) you'd need to model each lane as a separate unidirectional path. For a bidirectional path you can control the direction in your logic by assigning the path state DesiredDirection to:

`Enum.TrafficDirection.Forward`,

`Enum.TrafficDirection.Reverse`,

`Enum.TrafficDirection.Either`, or

`Enum.TrafficDirection.None`.

Note that TimePaths support *State Assignments* for On Entering and Before Exiting the path. They also support a set of *Add-on Process Triggers* to allow you to execute logic at key situations. These are very powerful features for customizing links' behavior and their interaction with surrounding objects.

Paths

The Standard Library *Path* object has all the features of the TimePath except for the TravelTime property. Instead of that, the Path calculates the travel time based on each individual entity's DesiredSpeed and the length of the Path (for example, an entity with Desired speed of 2 meters/minute would require *at least* 3 minutes to traverse a path of 6 meters). There are other parameters that can also affect the travel time. If the *Allow Passing* parameter is set to `False`, then if a faster entity catches up to a slower entity, the fast entity will automatically slow to follow (see SimBit `VehiclesPassingOnRoadway` for an example of this). Also, the path itself may impose a *Speed Limit* that may prevent the entity from traveling at its Desired Speed.

Path also has a *Drawn To Scale* property. If you set Drawn To Scale to `False`, you can specify a logical length that's different from the drawn length. This is useful in situations where you need an exact length, or in situations where you want to compress the drawn length for animation purposes (e.g., if you're modeling movement within and between departments, but the departments are a long distance away from each other).

8.2.3 Using Conveyors

General Conveyor Concepts

Conveyors are fixed-location devices that move entities across or along the device. Although conveyors are links (like paths), we're discussing them separately because, unlike the other Standard Library links, a conveyor represents a device. There are many different types of conveyors varying from powered or gravity-fed *roller conveyors* that you might see handling boxes, to *belt conveyors* or *bucket conveyors* that move everything from food to coal, up to overhead *power-and-free conveyors* like you might see moving appliances or auto bodies. While heavy manufacturing has relied on conveyors for decades, they have also become prominent in light industry and a surprising variety of other applications. The food and consumer-goods industries use sophisticated high-speed conveyors for bottling, canning, and packaging operations. Mail-order and institutional pharmacies use light-duty conveyors to help automate the process of filling prescriptions. Many larger grocery stores even incorporate belt conveyors into their checkout counters.

The Standard Library Conveyor object shares many features with the Path object discussed above, but adds significant new features as well. Current Simio conveyors are not reversible, so there is no Type property — all conveyors move from start node to end node. Conveyors also don't have a Speed Limit property, having instead a *Desired Speed* property that indicates the (initial) conveyor speed when it's moving. And finally, there is no Allow Passing property — passing never occurs on a conveyor because all entities that are moving are traveling the same speed (the speed of the conveyor link).

An entity is either moving at the speed of the conveyor or the entity is stopped. An entity is said to be *engaged* if its speed (whether moving or stopped) exactly matches the speed of the conveyor. An entity is *disengaged* if it's possible for the conveyor to be moving while the entity isn't moving (e.g., the conveyor is "slipping" under the entity). We'll discuss more on how that could happen in a moment.

Entity Alignment is a property that determines what is a valid location for an entity to ride on the conveyor. If Entity Alignment is `Any Location`, then an entity can engage at any location along the link. If Entity Alignment is `Cell Location`, then you must specify the *Number of Cells* that the conveyor has — an entity can engage only when its front edge is aligned with a cell boundary. This property is used when there are discrete components to a conveyor, such as buckets on a bucket conveyor or carriers on a power-and-free system.

The *Accumulating* property determines whether an entity can disengage from the conveyor. If Accumulating is set to `False`, then every entity on the conveyor is always engaged. If any entity must stop (e.g., it reaches the end) then the conveyor and all entities on it must also stop. If Accumulating is set to `True`, then when an entity stops at the end, that entity will disengage from the conveyor — while the conveyor and other entities conveying on it may keep moving. As each entity reaches (or "collides with") the stopped entity in front of it, that entity will also disengage and stop. These entities are said to have *accumulated* at the end of the conveyor.

The reverse process occurs when the entity at the end moves off the conveyor.

The blockage is removed and the second entity will attempt to re-engage. If the Entity Alignment is `Any Location`, then that attempt will always be immediately successful. If the Entity Alignment is `Cell Location`, then the second entity must wait until its leading edge is aligned with a cell boundary, then it can re-engage and start moving.

A final way in which a conveyor differs from a Path is that, because a conveyor represents a device, it has *Reliability* properties (similar to those of a Server, discussed in Section 5.3.4) and it has additional Add-on Process Triggers associated with reliability.

8.2.4   Model 8-2: PCB Assembly with Conveyors

Let's use some of this new-found knowledge to continue work on our PCB model. All of the placement machines are fed by and deliver to conveyors. And they have no buffer space on inbound or outbound sides — they feed directly from and to the conveyors and *Block* when the part cannot move on. Select the three stations and change the Buffer Capacity of the Output Buffer to 0 (the Input Buffer should have already been 0).

Use Click and Ctrl-click to select all six connectors so we can convert them and configure their properties as a group, then:

- Right click on any object in the group, then select `Convert to Type >` `Conveyor` to convert all six connectors to conveyors.

- Change the Desired Speed to `1 Meters per Minute`.

- We'd like to make it look like a conveyor. We could apply the Simio built-in conveyor path decorators, but they are too large for our use. So we'll just make the line that represents the conveyor path look more realistic.

  - Go to the General Category > Size and Location > Size > Width and set it to `0.2` meters.

  - Click on the bottom half of the `Texture` button in the Draw panel and select the `Corrugated Metal` pattern — it looks similar to a roller conveyor. Apply that to one of the conveyors.

  - Then double click on the top half of the `Texture` button and apply the texture to the other five conveyors.

- Since we're enhancing the looks of our conveyors, now is a good time to adjust the height of the end points so the parts are delivered to the correct point on the connected machines. Use Shift-Drag in the 3D window to move the height of each of the nodes so they attach at the correct height to the machine.

The Inspector can store only three parts on his table. All other parts must remain on the incoming conveyors. Set the Inspector InputBuffer to 3. Adjust the Inspection input-buffer queue (InputBuffer.Contents) so that it's a short line just large enough to display three parts and arrange it on top of the back edge of the inspector's table as in Figure 8.14.

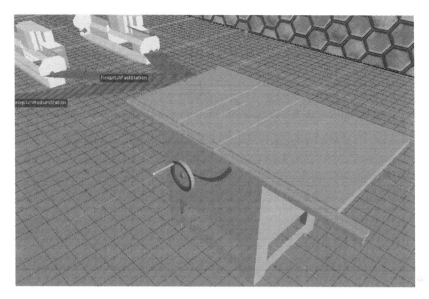

Figure 8.14: 3D view illustrating node and queue positioning.

Run the model and study the behavior closely. Notice that, on accumulating conveyors, parts can be stopped at the delivery end, while other parts are still moving along the conveyor. Note that you'll start to see backups on the Inspector's incoming conveyors, especially when it goes Off-Shift. And because the FinepitchFastStation has a faster production rate, it tends to back up faster. But that's exactly the machine that we don't want to block. Let's give entities on conveyors servicing faster machines a better priority by changing the entity priority as it enters that conveyor:

- Select the top outbound conveyor and go to the OnEntering property under State Assignments. Clicking on the ellipses on the far right will bring up a Repeating Property Editor for making assignments. Set the State Variable property to `ModelEntity.Priority` and set the New Value property to 3.

- Repeat this procedure with the middle outbound conveyor, only this time set the New Value property to 2.

- Select the entry node for Inspection (Input@inspection) and set its Initial Capacity property to 3, one for each conveyor. This allows all three entities (one from each conveyor) to enter the node and be considered for selection by Inspection. Note that they have not yet left the conveyor, but their front edges are in the node.

- Select Inspection and change the Ranking Rule to be `Largest Value First` and the Ranking Expression to be `Entity.Priority`. This will pull the highest-priority entity from the conveyors first, rather than just selecting the one that's been waiting the longest.

Actually, when the inspector is Off-Shift, no parts get processed regardless of priority, so this change won't make much difference in model behavior. But it will help the top conveyor to remain empty and help it get back to empty as quickly as possible, to minimize any blocking of the fastest machine.

We've barely scratched the surface of the available conveyor features. The various properties can be combined to model many different types of conveyors. You can also supplement with add-on processes on the conveyor or surrounding objects to model complex systems, in particular complex merging and diverging situations. And since the Conveyor is implemented entirely in Process logic that's open, you can create your own Conveyor objects with custom behavior.

8.2.5   Using Workers

In the preceding sections we've discussed various ways an entity can move from place to place. Assisted entity movement is a rather broad category, including any time an entity needs some constrained resource like a fixture, cart, person, or bus to assist in getting to its destination. In the simplest case, a fixed resource will suffice. If the movement time from the current location of a resource isn't important and the animation of the movement isn't important, then just seizing a fixed resource when you need it and releasing it when you're done might suffice. This same approach is an alternative when all you need is essentially permission or authorization to move (e.g., a traffic-control officer who points to you and gives you permission to drive past).

The more common case that we'll discuss next is where a transporter is necessary, and the current location and other properties/states of that device are important. The Simio Standard Library provides two objects for just that purpose: *Worker* and *Vehicle*. Both Workers and Vehicles are derived from Entity so they have all the capabilities and behaviors of entities, and more. They can move across networks similarly to entities, except that their behavior when reaching a node may be a bit different.

Workers

A *Worker* is an entity with additional capabilities to carry other entities. It's dynamic, in that the runtime object (the RunSpace) is created during the run. By specifying the Initial Number in System, you can determine how many Worker copies will be available. Each copy (a separate object) would represent an individual, so Worker doesn't have an Initial Capacity property; nor will it allow assigning a capacity greater than one. Since Worker is intended primarily to represent a person, it doesn't have device-specific properties like Reliability, but it does support Work Schedules as would normally be followed by a person. Please consult Figure 8.15 as we review some of the key properties of Worker.

Workers must travel on a designated network, although the default Initial Network is the Global network that includes all links. Under the Routing Logic Category, Workers must be provided a value for Default Node (Home) to specify where on the network the Worker starts (also its "Home" location). There are two actions that must be specified. *Idle Action* specifies the action to take when the worker runs out of work and work requests. *OffShift Action* specifies the

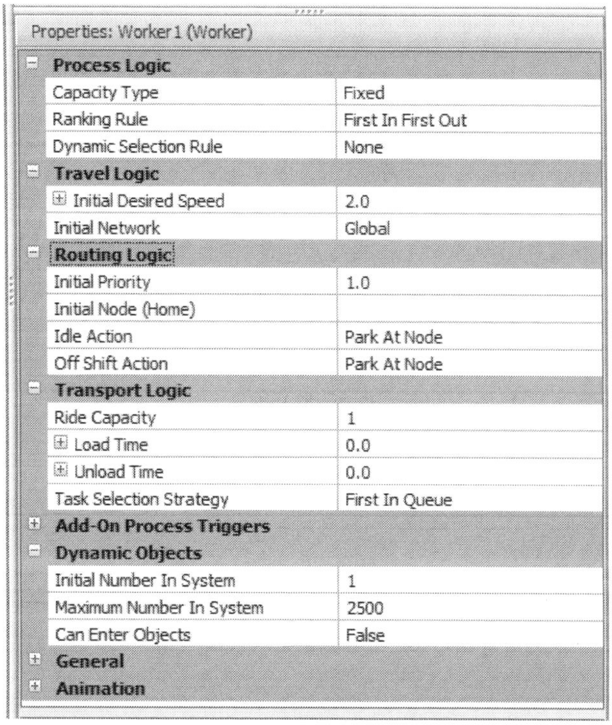

Properties: Worker 1 (Worker)	
Process Logic	
Capacity Type	Fixed
Ranking Rule	First In First Out
Dynamic Selection Rule	None
Travel Logic	
⊞ Initial Desired Speed	2.0
Initial Network	Global
Routing Logic	
Initial Priority	1.0
Initial Node (Home)	
Idle Action	Park At Node
Off Shift Action	Park At Node
Transport Logic	
Ride Capacity	1
⊞ Load Time	0.0
⊞ Unload Time	0.0
Task Selection Strategy	First In Queue
⊞ **Add-On Process Triggers**	
Dynamic Objects	
Initial Number In System	1
Maximum Number In System	2500
Can Enter Objects	False
⊞ **General**	
⊞ **Animation**	

Figure 8.15: Default properties of the Worker object.

action to take when the worker becomes Off-Shift (capacity is zero). For both of these actions, you have two choices. The first choice is whether to remain at your present node location, or return to the designated home-node location. The second choice is whether to *Park* (move off the network into a parking area that may or may not be animated), or to stay on the network, potentially blocking any traffic that may want to pass.

Under the Transport Logic Category, a Worker has a *Ride Capacity* that specifies how many entities it can concurrently carry. It also has a *Load Time* and *Unload Time*, which determine how long is required per entity for loading and unloading.

Just like an entity, under Dynamic Objects a Worker has a *Maximum Number in System*, which is provided as a validation tool. You can set this to the maximum number of Workers that you might expect, and if you dynamically create Workers beyond this limit, Simio will generate an error. There is also a *Can Enter Objects* property that controls whether a Worker can enter an object through its external nodes (e.g., whether it's permitted to enter a server). This defaults to `False` because in most cases Workers may provide a service to an entity or other object, but they usually don't execute the same processing logic at a server that an entity would.

The Worker has dual modes: as a *moveable resource* that's seized and released and travels between model locations, and as a *transporter* that can pick up, carry, and drop off entities at different locations.

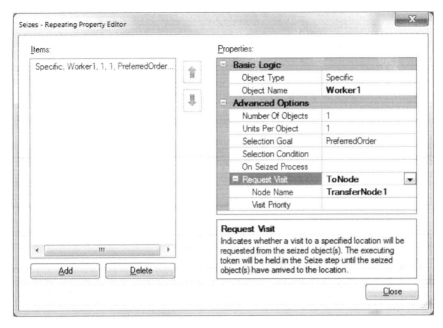

Figure 8.16: Properties used to seize a moveable object.

Using Worker as a Resource

Worker can be can be seized and released just like any other resource object. The Workstation object has provisions for doing that directly, using the Secondary Resources property in the Other Requirements category. Other standard-library objects can also seize and release resources but they must use the Seize and Release steps in an add-on process.

In either case, you'll ultimately be presented with a Repeating Property Editor similar to Figure 8.16. The simplest use is to leave Object Type set to Specific, meaning that you already know which resource you want. Then enter the name of the object you want (e.g., Worker1) in the Object Name property. Note that, even if you specify the Specific resource, if it's a transporter (e.g., a Worker or Vehicle), there could be multiple units of that name. Alternately, if you want to select from a set of resources, choose Object Type FromList, which will then allow you to specify the name of a list of resources from which you want to choose.

If you've provided more than one possible resource from which to choose, the Advanced Options category provides the rules (Selection Goal and Selection Condition) to select from the choices. The Advanced Options category also allows you to specify the quantity needed, and other options.

If moving the resource to a location is important, you can specify ToNode for the Request Visit property, and then the location the resource should visit in the Node Name property, as illustrated in the bottom section of Figure 8.16. While you may use any node as a destination, it's often simpler to add a node adjacent to the fixed object, and request that the resource use this new node as a destination for that moveable resource.

Requesting a visit works only for moveable resources — either an Entity or an object derived from an entity like a Worker or Vehicle. When you specify a Request Visit, the token will not leave the Seize step until the resource has been seized and moved to the specified node. When the resource is eventually released, it will either get automatically reallocated to another task, or start its Idle or OffShift action if appropriate.

Using Worker as a Transporter

A Worker can be thought of as a light-duty Vehicle. It has basic transportation capabilities that permit a Worker to carry or escort one or more entities between locations. Specifying that you want to use a Worker is identical to specifying that you want to use a Vehicle or any type of Transporter:

- Select a TransferNode (including the transfer nodes on the output side of most Standard Library objects).

- In the Transport Logic category, set the property Ride On Transporter to True. You'll see several additional properties displayed. Until you specify which transporter to use, your model will be in an error condition and unable to run.

- The Transporter Type property works very similarly to the Object Type property for resources described above. You can use it to request a specific transporter by name or choose from a list using an appropriate Selection Goal and Selection Condition.

- The Reservation Method specifies how you wish to select and reserve a transporter. There are three possible values:

  - Reserve Closest — Select the transporter that's currently the closest and use the Selection Goal to break any ties. Then Reserve that transporter. A *Reservation* is a two-way commitment: the transporter commits that it will go there (although possibly not immediately), and the entity commits that it will wait for the reserved transporter.

  - Reserve Best — Select the best transporter using the Selection Goal and use the closest to break any ties. Then Reserve that transporter.

  - First Available at Location — Don't make a reservation, but instead just wait until the first appropriate transporter arrives for a pickup. Note that it's possible that no transporter will ever arrive unless the transporter is following a sequence (like a bus schedule) or directed via custom logic.

8.2.6   Using Vehicles

A Vehicle is in many ways similar to a Worker. It has the same capability to be seized and released, travel between locations, and to act as a transporter that can pick up, carry, and drop off entities. In fact it has many of the same features. The primary differences are due to its intended use to model a device rather than a person. It doesn't follow a schedule, but it does have a few extra features.

Vehicles have a *Routing Type* that specifies the routing behavior of the vehicle. If it's specified as On Demand, then the vehicle will intelligently respond to visit requests, scheduling pickups and dropoffs appropriately. If it's specified as Fixed Route, then the vehicle will follow a sequence of node visitations (Section 7.1.3). In urban-transportation terminology, On Demand performs like a taxi service, while Fixed Route performs more like a bus or subway.

Vehicles have *Reliability Logic*, which recognizes that devices often fail. Vehicles can have Calendar Time Based or Event Count Based failures. The latter is extremely flexible. It can respond not only to external events, but you can also trigger internal events based on tracking things like distance traveled, battery-charge state, or other performance criteria.

Beyond the above differences, Vehicles and Workers behave and are used in a similar fashion.

8.2.7   Model 8-3: ED Enhanced with Hospital Staff

Let's use what we've just learned to enhance our Emergency Department model. We'll start with our ED Model 7-5 (save it as Model 8-3), and enhance it with staffing using Workers. But before we enhance the ED capabilities, let's make a minor animation improvement. We discussed earlier that the Trauma, Treatment, and Exam-room objects did not represent individual beds, but rather represented areas that could handle multiple patients. While we could certainly enhance the model to represent each bed individually, we don't yet need that level of detail. Instead let's just create a slightly better animation symbol to represent a suite of beds.

Creating a New Symbol

Your Project name is by default the same as your file name (e.g., Model_08_03). Select the Project name in the navigation window and it will display a window that contains all of your project components. This window allows you to add, remove, and edit your project components. Click on the Symbols icon on the left to view your active symbols. This is where you'd come to edit symbols or to remove unused symbols from your project. In our case we want to add a symbol:

- Click on Create New Symbol. This provides a drawing window similar to the Facility window, including 2D and 3D viewing and several familiar ribbons.

- We want a room that's about 10 meters by 3 meters. Use the Polyline tool to create one four-segment line for the four walls. (Feel free to leave some breaks for doorways if you like, but we didn't do so.) On the Drawing ribbon set the Object Height to 1, the Line Width to 0.05, and optionally apply a Texture (we used Banco Nafin).

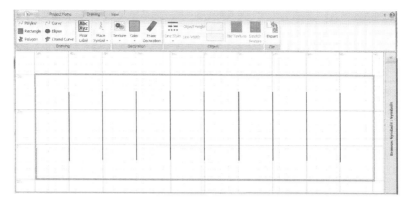

Figure 8.17: Drawing a simple suite of treatment rooms.

- We want to put dividers between each patient. Create the first divider, again using the Polyline. On the Drawing ribbon, set the Object Height to 1, the Line Width to 0.04, and optionally apply a Texture (we used Dark Green Marble).

- Copy and paste your first divider for a total of nine interior dividers, placed one meter apart.

- In the Properties window set the Name property to HospitalSuite.

- With the properties panel collapsed, your symbol should look something like Figure 8.17.

- When you're done, use the Navigation window to switch back to the Project window or the Model window and your symbol will be automatically saved.

The symbol you just created is now part of the project library. Use it to apply new symbols to ExamRooms, TreatmentRooms, and TraumaRooms. Because these are a very different size from the original symbols, we should probably pause and do some related model cleanup (refer to Figure 8.18. for an example):

- Adjust the locations of the objects so the layout looks natural.

- Adjust the node locations so they're placed outside the objects — perhaps adjacent to any doors if you created them.

- Delete the original paths connecting those objects to others. Replace them with multi-segment paths that are appropriate.

- Adjust the locations of animation queues for each of your rooms: move the InputBuffer.Contents to somewhere near your input node, move the OutputBuffer.Contents near your output node, and move the Processing.Contents so that it stretches across the length of the room. This is a quick way to display the "occupants" in their respective rooms.

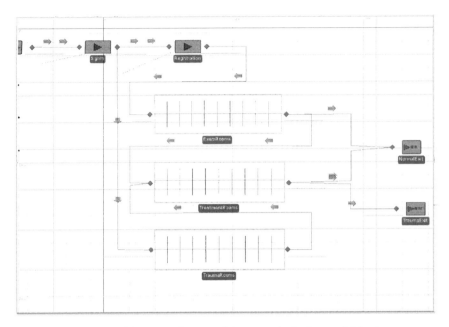

Figure 8.18: Model 8-3 with new object symbols for multi-bed rooms.

Table 8.1: Type of escort required to destination.

Patient Type	Exam	Treatment	Trauma	Exit
Routine	Aide	–	–	AidePreferred
Moderate	AidePreferred	AidePreferred	–	AidePreferred
Severe	AidePreferred	AidePreferred	–	AidePreferred
Urgent	–	NursePreferred	NursePreferred	AidePreferred

Updating the Model

Now that we've updated the ED layout, we're in a better position to add the
staffing. We'll model two types of staff; Nurse and Aide. We want all patients
to be accompanied to the exam, trauma, and treatment rooms as specified in
Table 8.1. Routine patients will wait until an Aide is available. Moderate and
Severe patients will be escorted by an aide if available; otherwise they'll take
the first Aide or Nurse who becomes available. Urgent patients will be escorted
by a Nurse if available, otherwise by an Aide. All patients will be escorted to
the exit by an Aide if available, otherwise a Nurse.

Let's start by creating a place for the Workers to call home:

- Draw a 3 × 5-meter Nurses' station using polylines, as we did earlier. Set
  the wall height to 1 meter. Add a basic node at the entrance; name it
  `NurseStation`.

- We want to turn off the automatic parking queue for that node and replace
  it with one we'll draw inside the nurse-station room. When `NurseStation`
  is selected, select the Appearance ribbon button named Parking Queue to

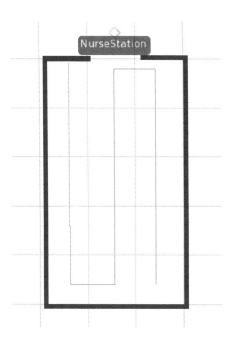

Figure 8.19: Nurse station with node and attached parking queue.

deselect the automatic parking queue. Use the **Draw Queue** button and select the **ParkingStation.Contents** queue, then draw the queue inside the room. You can wrap the queue line around the room so you can fit in a "crowd" of nurses if we need them.

- Your nurse station should look something like Figure 8.19.

- Repeat the above three steps to create an **AideStation** that will be the home location for the Aides.

Now let's create and configure the Workers:

- Place a Worker in your model and name it **Nurse**. Set Initial Desired Speed to **0.5** (Meters per Second). In the Routing Logic category: set Initial Node (Home) to **NurseStation**; set the Idle Action and Off Shift Action to **Park At Home**. In the Population category set the Initial Number in System to **4** to indicate that we will have four nurses.

- Place another Worker in your model and name it **Aide**. Set Initial Desired Speed to **0.5** (Meters per Second). In the Routing Logic category: set Initial Node (Home) to **AideStation**; set the Idle Action and Off Shift Action to **Park At Home**. In the Population category set the Initial Number in System to **6** to indicate that we will have six aides.

Defining and Using Lists

As we discussed in Section 7.5, a Simio *List* is a collection of objects that's referenced when a choice or selection must be made. For example, you may often need to choose between destinations (nodes) or between resources (objects). Selections from a list often support several different mechanisms like `Random` or `Smallest Value`, but a common default is *Preferred Order*, which means select the first one available starting at the top of the list.

Looking at Table 8.1 again, you'll see that sometimes we need a particular type of worker for an escort (e.g., an Aide), and at other times we want to choose from a preferred order list (e.g., we prefer an Aide if available, otherwise we'll use a Nurse). Since we're already using sequence tables to move around, it's easiest to add this information as a new property (column) in the table. Recall that Workers are a type of transporter, so in this case the property type can be either a transporter or a transporter list. Each property must have just one specific data type, so we'll make them all transporter lists and just make the list contain exactly one member when there's no choice. We'll get back to the table shortly, but let's first create our lists:

- A list is defined using the *Lists* panel in the Definitions window. To add a new list you click on the type of list you want to create in the Create section of the List Data ribbon. Referring again to Table 8.1, there are three lists from which we'll need to choose for an escort: AideOnly, AidePreferred, and NursePreferred. Click on the Transporter button three times to create those three lists, then name them `AideOnly`, `AidePreferred`, and `NursePreferred`.

- When you select the AideOnly list, the lower part of the screen will display the (currently empty) list of transporters in that list. Click on the right side of the first cell (it may require several clicks) and select `Aide` from the pull-down list of transporters. Since there's only a single choice in the `AideOnly` list, we're done defining this list.

- Repeat the above process for the `AidePreferred` list, putting `Aide` in the first row and `Nurse` in the second row. This list order means that when you select from this list using the Preferred Order rule, you'll select an Aide if available, otherwise select a Nurse if available. (If neither is available, you'll select the first one that becomes available.)

- Repeat the above process one final time for the `NursePreferred` list, putting `Nurse` in the first row and `Aide` in the second row.

Now that we've created our lists, we can add data to our `Treatments` table. In the Tables panel of the Data window, select the interior tab labeled `Treatments`. Recall that this table is referenced by the patients to determine which location they should visit next, and what properties to use while traveling to and at that location. This covers all of the locations visited, not just the exam-room and treatment locations. We'll add a property for the transportation:

- Click on the `Object Reference` button and select the `Transporter List` type. Name that property `EscortType`. We'll use this to specify which worker group from which we want to select for escort to each location.

	Sequence	Treatment Type	Service Time (Minutes)	Escort Type
>	Input@SignIn	Routine	2	null
	Input@Registration	Routine	Random.Uniform(3,7)	null
	Input@ExamRooms	Routine	Random.Triangular(5,10,15)	AideOnly
	Input@NormalExit	Routine	0.0	AidePreferred
	Input@SignIn	Moderate	2	null
	Input@Registration	Moderate	Random.Uniform(3,7)	null
	Input@ExamRooms	Moderate	Random.Triangular(10,15,20)	AidePreferred
	Input@TreatmentRooms	Moderate	Random.Triangular(5,8,10)	AidePreferred
	Input@NormalExit	Moderate	0.0	AidePreferred
	Input@SignIn	Severe	1	null
	Input@Registration	Severe	2	null
	Input@ExamRooms	Severe	Random.Triangular(15,20,25)	AidePreferred
	Input@TreatmentRooms	Severe	Random.Triangular(15,20,25)	AidePreferred
	Input@NormalExit	Severe	0.0	AidePreferred
	Input@SignIn	Urgent	.5	null
	Input@TraumaRooms	Urgent	Random.Triangular(15,25,35)	NursePreferred
	Input@TreatmentRooms	Urgent	Random.Triangular(15,45,90)	NursePreferred
	Input@TraumaExit	Urgent	0.0	AidePreferred

Figure 8.20: Treatment table with escort staff added.

- Use the data from Table 8.1 to complete the data in the new column. When completed it should look like Figure 8.20.

Updating the Network and Logic

We'll finish our enhancements by updating the networks and the logic to take advantage of the staff. Before we start adding to the network, though, we need to make one minor change. Most patients leaving SignIn will go directly to Registration without an escort. For those patients, we need the Ride On Transporter property to be False. But Urgent patients will need an escort to TraumaRooms and need the Ride on Transporter property set to True. One easy way to solve this dilemma is simply to have Urgent patients pass through an extra TransferNode adjacent to SignIn. Hence, all patients will leave SignIn without an escort, but Urgent patients will almost immediately pass through a node where they'll wait for an escort. To make this change, you must delete the existing path from SignIn to TraumaRooms, add a new node adjacent to SignIn (let's name it WaitUrgentEscort), then add a path from SignIn to WaitUrgentEscort and another from WaitUrgentEscort to TraumaRooms.

Now we're ready to enhance our network. Although our staff will be escorting patients on the existing network, they'll also be traveling on their own

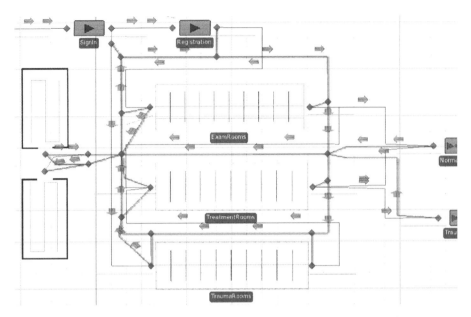

Figure 8.21: Highlighted paths identifying the staff network.

paths to move between locations. And they need to be able to travel freely from any input or output node to any other input or output node. So we'll add an additional set of paths to facilitate staff movement.

There are several ways to approach the creation of this network. One extreme approach is to draw two paths (one in each direction) between every pair of nodes. This would obviously be quite tedious, but complete. Another extreme would be to have a unidirectional cycle of paths. This would likely represent the fewest number of paths, but would also result in inefficient and unrealistic movements. We'll adopt a middle approach — we'll have a cycle of unidirectional paths, but add extra nodes to support paths to connect the destination nodes as well as support some shortcuts to make travel more efficient.

We've added that set of nodes and paths for the "middle approach" as shown in the highlighted paths in Figure 8.21. Add paths and nodes to your model in the same fashion. In case you're wondering, we achieved the highlighting by selecting all the new paths and then right clicking to add them to a network called StaffNetwork. While a network can be used to limit access to only certain entities, in this case we don't need that capability. But when your path networks start to get complex, assigning them to a named network can be handy for visualization and debugging. To highlight those paths as we've done here, go to the View ribbon and select the View Networks button.

So far we've created our staff, created lists of them to support selection, added data to our tables so we know what to select, and we now have our paths set up to support escorting. With all that groundwork in place, we just need to tell our entities when to wait for an escort. That's done on the transfer nodes.

Lets start with our new node, WaitUrgentEscort. Select it and set Ride On Transporter to True. Set Transporter Type to From List. We want to get the list for this patient and location from our table, so for Transporter List select the value `Treatments.EscortType`. And finally set the Reservation Method to `Reserve Best`, leaving the Selection Goal (e.g., how we define "Best") set to `Preferred Order`.

Select the three other output nodes where we need escort (`Registration`, `ExamRooms`, and `TraumaRooms`) and make the identical changes to those properties as well. Note that you can do them individually, or do a group-select and then do them all at once.

If you've not already done so, now would be a good time to save and run the model. If you've done everything right, you should see your patients moving just as they did in earlier versions of this model, but they'll now be accompanied by staff in some of their movements. You can adjust the location of the RideStation queues on your workers so that the patients travel beside the workers rather than in front of them.

At this point, you're probably thinking something like — "So what about the actual *treatment* – are the patients treating themselves? The staff escorts patients, but isn't involved in patient treatment!" We will add this functionality to the model in Chapter 9. Also, there are many ways that we could enhance the animation and patient movement (the things we focused on in this chapter) but, as authors like to say, we'll leave such enhancements as an exercise for the reader.

8.3 Summary

In this chapter we have illustrated the value of 2D and 3D animation and some of its uses in verification and communication. We also discussed some pitfalls in placing too much emphasis on animation to the exclusion of higher-priority project tasks and other analysis techniques.

We briefly explored many of the basic animation and viewing options to give you a sound basis for animating your own models. Then we discussed how to animate dynamic objects (e.g., entities) when they are moving through free space, on links, on conveyors, or being assisted by other dynamic objects like the Standard Library Worker or Vehicle.

In addition to the above topics there are many other modeling situations and approaches that are outside the scope of this book. Many Simio extensions can be found in the *Shared Items* user forum that can be found via the Support Ribbon.

For example modeling multiple *overhead bridge cranes* sharing a common runway can be demanding because of the potential interference and the need to yield to or even back up from another load. This is even more demanding when acceleration is a factor. Simio has a crane library which not only handles this very well, but also fully models the new *Underhung Cranes* which allow crane cabs to move between crane bridges and essentially form transportation networks on the ceiling.

Efficiently modeling fluid-like flow is another modeling challenge. There are many aspects to this problem depending on the materials being moved, the types of devices or containers used for movement, the volume being moved, and the type of emptying/filling operations. Again the Shared Items user forum and the built-in Flow Library offer many interesting alternatives including continuous flow, mass flow, containerized flow, and pipes.

Other movement-related topics discussed in the Shared Items user forum include transportation using multi-car trains, local movement of parts using rotating or articulated robots, and transfer devices for use with conveyors. There are also many useful topics unrelated to movement that can be found in this forum.

8.4 Problems

The goal for the first four problems is to better understand animation issues, not necessarily to generate a useful analysis. To this end, unless otherwise specified you can make up reasonable model parameters to help you illustrate the assigned task.

1. Create a new model similar to Model 4-3. Animation is of primary importance in this model, so use what you have learned in this chapter to make it as realistic as possible. For the customers, use multiple animated (walking) people including at least 2 men, 2 women, and 1 child. Use symbols from Trimble 3D Warehouse to add realism.

2. Enhance Model 7-2 by adding a floor label attached to each entity instance to display its Level of Service (elapsed time in system in minutes) and use the background color to indicate patient type. Add one or more floor labels to the model to display in tabular form 3 pieces of information for each patient type: Number In-Process, Number Processed (those who have left the system), and Average Level of Service for those who have left.

3. Create a model with three Source–Path–Sink sets that each require a vehicle to move entities from the Source to its associated Sink. Use a custom vehicle symbol from the included library. Try each of the three Network Turnaround Methods and compare and contrast each approach.

4. Use a wide selection of animated people/characters and a wide selection of their possible movements to recreate your own version of the "Harlem Shake". Google for a YouTube video if you haven't seen it.

5. Speedee Car Rentals is open 24 hours and has two classes of customers – 25% are premium customers who are promised a waiting time of less than 5 minutes from when they enter or they will receive a $15 discount on their rental. The others are regular customers who will wait as long as they have to wait. Customers arrive about 2 minutes apart (exponential) and move about 4 meters at 1 mile per hour to either the premium counter (1 agent) or the regular counter (2 agents). All customers have a uniformly distributed service time of from 2 to 7 minutes. Each agent costs $55 per hour (including overhead) and each customer completed contributes $8 to income (minus any discount they might have received). Model this

system without using any links. Run for 10 days per replication. For each scenario, report the utilization of each type of agent and the waiting time of each type of customer. Both on-screen and in the output, report the number of premium customers who received a discount, the total discounts paid, and the net profit (after discounts and costs) of the system.

Scenario a) Assume that each agent only handles customers of his designated type.

Scenario b) Relax the above assumption. Consider alternate work rules (who services which customer under which conditions) for this same staff. Evaluate each proposed scenario to see the impact on the metrics.

Scenario c) Model other possible solutions to improve the system. You may use links if that helps.

Chapter 9

Advanced Modeling With Simio

The goal of this chapter is to discuss several advanced Simio modeling concepts not covered in the previous four chapters. The chapter is comprised of three separate models, each of which demonstrates one or more advanced concept. One common thread throughout the chapter is the comprehensive experimentation and use of simulation-based optimization with the OptQuest® add-in. In Section 9.1 we revisit the Emergency Department model originally introduced in Chapter 8 and focus on the decision-making aspects of the problem. As part of the decision-making function, we develop a single "total cost" metric and use the metric to evaluate alternative system configurations. In this section, we also introduce the use of the OptQuest simulation-based optimization add-in and illustrate how to use this add-in. In Section 9.2 we introduce a Pizza Take-out restaurant model and incorporate a Worker object to demonstrate a pooled resource system. We then use the model to evaluate several alternative resource allocations, first using a manually-defined experiment, and then using OptQuest. Finally, in Section 9.3, we develop a model of an assembly line with finite capacity buffers between stations and demonstrate how to set up an experiment that estimates line throughput for different buffer allocations. Of course there are many, many advanced features/concepts that we are not able to discuss here so this chapter should be considered as a starting point for your advanced studies. Many other simulation and Simio-specific concepts are demonstrated in Simio's SimBits (as described in the Preface) and in other venues such as the proceedings of the annual Winter Simulation Conference (www.wintersim.org) and other simulation-related conferences.

9.1 Model 9-1: ED Model Revisited

In Section 8.2.7 we added hospital staff to our Emergency Department (ED) model and incorporated entity transportation and animation (Model 8-3). However, we didn't account for the actual patient treatment and we didn't perform any experimentation with the model at that time. Our goal for this section is to demonstrate how to modify Model 8-3 primarily to support *decision making*. Specifically, we are interested in using the model to help make *resource allocation* and *staffing* decisions. While we could have extended Model 8-3 by adding the patient treatment and performance metric components to that

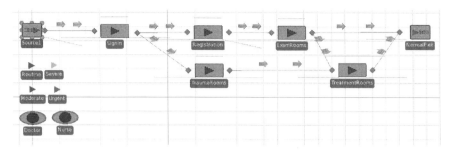

Figure 9.1: Facility view of Model 9-1.

model, we chose to create an abstract version of the model by removing the worker objects and transportation paths and using Connector objects to facilitate entity flow. Our motivation for this was two fold: First, the abstract model is smaller and easier to present in the book; and second, to reinforce the notion that the "best" simulation model for a project is the simplest model that meets the needs of the project. In Model 8-3, our goal was to demonstrate the transportation and animation aspects of Simio incorporating Standard Library Worker objects. For Model 9-1, our goal is to demonstrate a model that can be used for decision-making and to demonstrate *simulation-based optimization* using Simio.

Figure 9.1 shows the facility view of Model 9-1. Note that the model retains the four patient entity objects (Routine, Moderate, Severe, and Urgent) along with SignIn, Registration, ExamRooms, TraumaRooms, and TreatmentRooms server objects. The Nurse and Aide Worker objects have been replaced with Doctor and Nurse Resource objects. Also, the transportation networks and Path objects have been replaced with a simplified network using Connector objects and the animation symbols for the patient entities and various rooms have been removed, as we are not primarily interested in modeling the physical patient movement (and because we wanted to simplify the model).

Model 9-1 maintains the same basic problem structure where patients of four basic types arrive and follow different paths through the system using a Sequence Table to specify their routes. The only difference in the patient routing is that Urgent patients now exit through the NormalExit object whereas in Model 8-3 they exited through the TraumaExit object. Figure 9.2 shows the updated PatientData data table and Treatment Sequences table for Model 9-1. Note that the Escort Type column has been removed from the Treatments table and a Waiting Time column has been added to the PatientData table. As part of our resource allocation model, we track patient waiting time — defined as any time that the patient is waiting to see a doctor or nurse (including the interval from when the patient first arrives to when they first see a doctor or nurse). We will use the waiting time to determine patient service level and to balance the desire to minimize the resource and staffing costs. The Waiting Time entries in the PatientData table allow us to track waiting time by patient type. We removed the Escort Type column as we are no longer modeling the physical movement of patients and staff.

In addition to the ExamRooms, TreatmentRooms, and TraumaRooms resources,

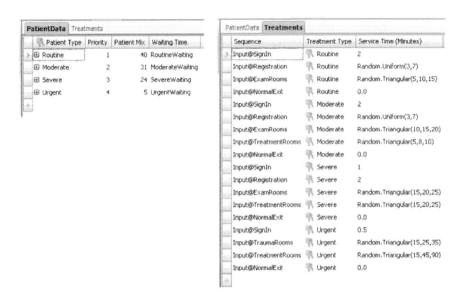

Figure 9.2: PatientData and Treatment data tables for Model 9-1.

Table 9.1: Patient examination/treatment resource requirements.

Room	Exam/Treatment Resource(s) Required
Exam Room	Doctor or Nurse, doctor preferred
Treatment Room	Doctor and Nurse
Trauma Room	Doctor and Nurse or Doctor, nurse preferred

Model 9-1 also incorporates the resource requirements for the patient examination and treatment (Model 8-3 used `Nurse` and `Aide` Worker objects to escort patients, but didn't explicitly model the *treatment* of patients once they arrived at the respective rooms). The exam/treatment requirements are given in Table 9.1. When a patient arrives to a particular room (as specified in the Sequence table), the patient will seize the `Doctor` and/or `Nurse` resource(s) required for treatment. If the resources aren't available, the patient will wait in the respective room until the resources are available (much as it works in a real emergency room!). As such, we separate the physical room resources (modeled using Server objects) from the exam/treatment resources (modeled as Resource objects) so that we can evaluate system configurations with different numbers of each – think of a typical doctor's office that has more exam/treatment rooms that it has doctors.

As one would expect in an emergency room, priority for the `Doctor` and `Nurse` resources should be based on the priority level specified in the `PatientData` table (as we did in Model 8-3). So, for example if there are `Routine` and `Moderate` patients waiting for a doctor and an `Urgent` patient arrives, the `Urgent` patient should move to the front of the queue waiting for the doctor. Similarly, if there are only `Routine` patients waiting and a `Severe` or `Moderate` patient arrives, that patient should also move to the front of the

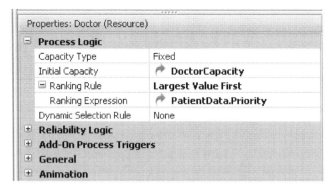

Figure 9.3: Properties for the `Doctor` resource.

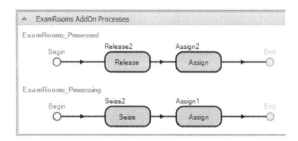

Figure 9.4: Add-on processes for the `ExamRooms` object.

doctor queue. We can implement this using the RankingRule property for the `Doctor` and `Nurse` resources. Figure 9.3 shows the properties for the Doctor resource. Setting the Ranking Rule property to `Largest Value First` and the Ranking Expression property to `PatientData.Priority` tells Simio to order (or *rank*) the queue of entities waiting for the resource based on the value stored in the `PatientData.Priority` field[1]. One might also consider allowing Urgent patients to *preempt* a doctor or nurse from a Routine patient, but we will leave this as an exercise for the reader.

In order to model the allocation of exam/treatment resources to patients, we will use *Add-on Processes* (see Section 5.1.4) associated with the respective room resources. Recall that Add-on processes provide a mechanism to supplement the logic in existing Simio objects. In the current model, we need to add some logic so that when a patient arrives in a room (exam, treatment, or trauma), the doctor and/or nurse who will provide the treatment are "called" and the patient will wait in the room until the "treatment resources" arrive. The current Server object logic that we have seen handles the allocation of room capacity, but not the treatment resources. As such, we will add-on a process that calls the treatment resource(s) once the room is available.

Figure 9.4 shows the two add-on processes for the `ExamRooms` object. The

---

[1]Note that when a resource is seized from different places in a model, Simio has a single internal queue for the waiting entities. This simplifies implementing the type of global priority that we need for this case.

Figure 9.5: List and Seize details for the ExamRooms add-on process.

process ExamRooms_Processing is executed when an entity has been allocated server capacity (the ExamRooms Server) and is about to start processing and the ExamRooms_Processed add-on process is executed when an entity has completed processing and is about to release the server capacity. When a patient enters the ExamRooms server, we do two things — seize the exam/treatment resource and record the waiting time once the resource is allocated to the patient. As shown in Table 9.1, patients in the exam room need either a doctor or a nurse, with the preference being a doctor if both are available. In Section 8.2.7 we demonstrated how to use Simio *Lists* and the PreferredOrder option for an entity to select from multiple alternatives with a preferred order. We use that same mechanism here — we defined a list (DoctorNurse) that includes the Doctor and Nurse objects (with the Doctor objects being first in the list) and seize a resource from this list in the Seize step in the add-on process. Figure 9.5 shows the DoctorNurse list and the details of the Seize step. Note that we specified the Preferred Order for the Selection Goal property so that the Doctor will have higher priority to be selected. Note that we could also have used the Server object's *Secondary Resources* properties to achieve this same functionality, but wanted to explicitly discuss how these secondary resource allocations work "under the hood."

The Assign step, Assign1 is used to record the initial component of the patient waiting time and the Assign step, Assign2 is used to *mark* the time that the patient leaves the Exam room (more on this later). The Assign1 step will execute when the doctor or nurse resource has been allocated to the entity. Recall that we would like to track all of the time that a given patient waits for a doctor or a nurse. Since patients can potentially wait in multiple locations, we will use an entity state (WaitingTime) to accumulate the total waiting time for each patient. Figure 9.6 shows the properties for the the two Assign steps. In Assign1 we are assigning the value TimeNow-ModelEntity.TimeCreated to the state ModelEntity.WaitingTime. This step subtracts the time that the entity was created (the time the patient arrived) from the current simulation time (stored in the model state TimeNow) and assigns the value to the entity state ModelEntity.WaitingTime. If the entity represents a Routine patient, this will be the total waiting time for the patient. Otherwise, we will also track the time the patient waits after arriving to the TreatmentRooms. In order to track

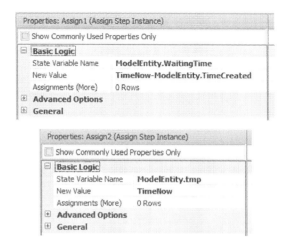

Figure 9.6: Assign step properties.

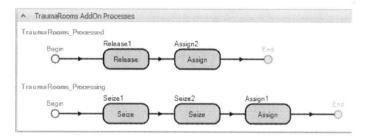

Figure 9.7: Add-on processes for the `TraumaRooms` object.

this additional waiting time `Assign2` marks the entity — stores the current simulation time to the entity state `ModelEntity.tmp`. We will use this time when the entity arrives to the Treatment room and is allocated the necessary resources. The Release step in the `ExamRooms_Processed` step releases the doctor or nurse resource seized in the previous process.

Figure 9.7 shows the add-on processes for the `TraumaRooms` object. The processes are similar those for the `ExamRooms` object. The only difference is that the `TraumaRooms_Processing` process includes two seize steps — one to seize a `Doctor` object and one to seize a `Nurse` object or a `Doctor` object. Since our preference is to have a nurse in addition to the doctor rather than two doctors, we seize from a list named `NurseDoctor` that has the `Nurse` objects listed ahead of the `Doctor` objects (the opposite order from the `DoctorNurse` list shown in Figure 9.5 and used in the `ExamRooms` object).

Figure 9.8 shows the add-on processes for the `TreatmentRooms` object. Patients come to the treatment from either the trauma room (`Urgent` patients) or from the exam room (`Moderate` and `Severe` patients) so we cannot use the entity creation time to track the patient waiting time (as we did before). Instead, we use the previously stored time (`ModelEntity.tmp`) and use a second *Assign* immediately after the Seize in order to add the waiting time in

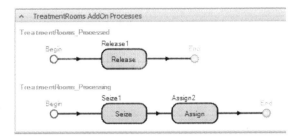

Figure 9.8: Add-on processes for the `TreatmentRooms` object.

Figure 9.9: Assign step properties for adding the second entity wait component.

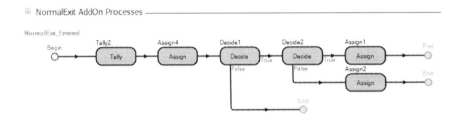

Figure 9.10: Add-on processes for the `NormalExit` object.

the treatment room to the patient's total waiting time (recall that an entity[2] remains in the Seize step until the resource is available — the queue time). Figure 9.9 shows the properties for the first and second *Assign* steps in the `TreatmentRooms_Processing` add-on process. The basic concept of marking an entity with the simulation time at one point in the model, then recording the time interval between the marked time and the later time in a tally statistic, is quite common in situations where we want to record a time interval as a performance metric. While Simio and other simulation packages can easily track some intervals (e.g., the interval between when an entity is created and when it is destroyed), the package has no way to know which intervals will be important for a specific problem. As such, you definitely want to know how to tell the model to track these types of user-defined statistics yourself.

Figure 9.10 shows the add-on processes for the `NormalExit` object. When a patient leaves the system, we will update our waiting time statistics that we will use in our resource allocation analysis. The Tally step (see Figure 9.11 for the

---

[2]Technically, it's a token that waits at the Seize step, but the token represents the entity in this case.

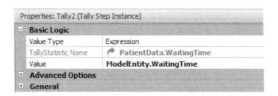

Figure 9.11: *Tally* step properties.

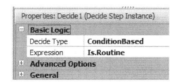

Figure 9.12: Model Elements for Model 9-1.

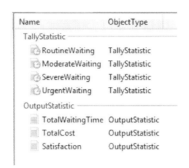

Figure 9.13: Decide step properties for identifying `Routine` patient objects.

properties) records the waiting time for the patient (stored in the `WaitingTime` entity state). The `PatientData.WaitingTime` value for the TallyStatisticName property indicates that the observation should be recorded in the tally statistic object specified in the `WaitingTime` column of the `PatientData` table (Figure 9.2 shows the data tables and Figure 9.12 shows the tally and output statistics for the model). As such, the waiting times will be tallied by patient type. This is another example of the power of Simio data tables — Simio will look up the tally statistic in the table for us and, more importantly, if we add patient types in the future, the reference does not need to be updated. The Assign step adds the waiting time for the entity (stored in the `WaitingTime` entity state) to model state `TotalWait` so that we can easily track the total waiting time for all entities.

The final set of steps in the process are used to track the patient *service level*. For our ED example, we define *service level* as the proportion of `Routine` patients whose total waiting time is less than one-half hour (note that we arbitrarily chose one half hour for the default value). So the first thing we need to do is identify the `Routine` patients — the first Decide step does this (see Figure 9.13 for the step properties). Next, we need to determine if the current

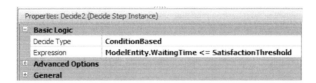

Figure 9.14: Decide step properties for determining whether the patient is satisfied based on waiting time.

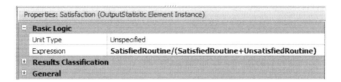

Figure 9.15: Properties for the `Satisfaction` output statistic.

Name	ObjectType	DisplayName
Properties		
DoctorCapacity	Expression Property	DoctorCapacity
ExamCapacity	Expression Property	ExamCapacity
TreatmentCapacity	Expression Property	TreatmentCapacity
TraumaCapacity	Expression Property	TraumaCapacity
NurseCapacity	Expression Property	NurseCapacity
SatisfactionThreshold	Numeric Property	SatisfactionThreshold
Properties (Inherited)		

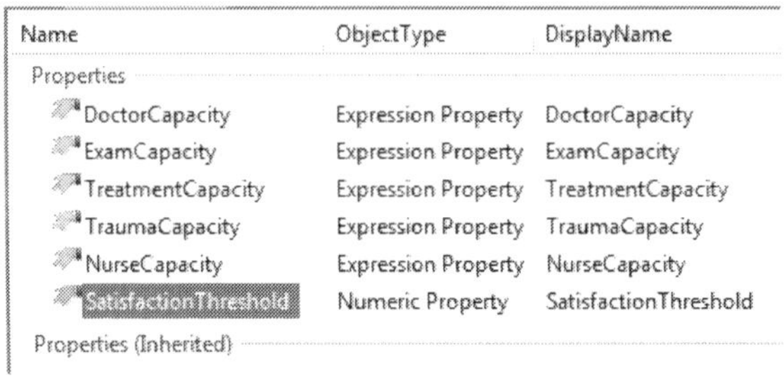

Figure 9.16: Model properties for Model 9-1.

patient is "satisfied" based on that patient's waiting time. The second Decide step does this (see Figure 9.14 for the step properties). Note that we used a referenced property (`SatisfactionThreshold`) rather than a fixed value (0.5 for one-half hour) to simplify the experimentation with different threshold values. If the patient is satisfied, we increment a model state for counting satisfied customers (`SatisfiedRoutine`). Otherwise, we increment a model state for counting unsatisfied customers (`UnsatisfiedRoutine`). The two Assign steps do this. We will use an output statistic to compute the overall service level for a run. Figure 9.12 shows the model output statistics and Figure 9.15 shows the properties for the `Satisfaction` output statistic where we compute the proportion of satisfied patients.

Finally, we will use several referenced properties to support our resource allocation analysis. Since we are interested in determining the numbers of exam, treatment, and trauma rooms and the numbers of doctors and nurses required in the Emergency Department, we defined referenced properties for each of these items. Figure 9.16 shows the model properties and Figure 9.3 shows how the `DoctorCapacity` property is used to specify the initial capacity for the `Doctor` resource. We specified the `Nurse` object capacity and `ExamRooms`, `TreatmentRooms`, and `TraumaRooms` server object capacities sim-

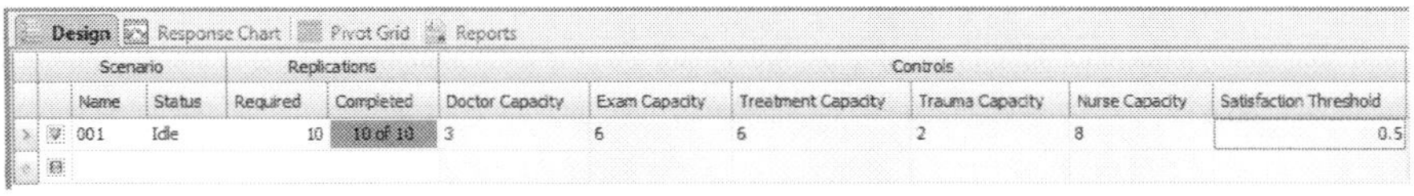

Figure 9.17: Controls for an experiment for Model 9-1.

Table 9.2: Weekly costs for each unit of the resources for Model 9-1.

Resource	Weekly cost
Exam Room	$2,000
Treatment Room	$2,500
Trauma Room	$4,000
Doctor	$12,000
Nurse	$4,000

ilarly. Figure 9.17 shows a portion of an example experiment for Model 9-1. Note that the referenced properties show up as *Controls* in the experiment. As such we can easily compare alternative allocation configurations using different experiment scenarios without having to change the model since these values can be changed for individual scenarios in an experiment.

In order to simulate a resource allocation problem, we need to assign costs for each unit of the various resources to be allocated. In Model 9-1, we are interested in the numbers of exam rooms, treatment rooms, trauma rooms, doctors and nurses. Table 9.2 gives the weekly resource costs that we'll use. To offset the resource and staffing costs, we will also assign a "waiting time cost" of $250 for each hour of patient waiting time. This will allow us to explicitly model the competing objectives of minimizing the resource costs and maximizing the patient satisfaction (no one likes to wait after all). Since our model tabulates the total waiting time already, we can simply set our replication length to one week and can then calculate the "total cost" of a configuration as an output statistic. The expression that we use for the `TotalCost` output statistic is:

$2000 * ExamCapacity$
$+ 2500 * TreatmentCapacity$
$+ 4000 * TraumaCapacity$
$+ 4000 * NurseCapacity$
$+ 12000 * DoctorCapacity$
$+ 250 * TotalWaitingTime.Value$

Now we have two output statistics that we can use to evaluate system configurations — total cost and service level – total cost we want to be as low as possible and service level we want to meet a minimum value set by the stakeholders. Figure 9.18 shows a sample experiment with four scenarios and the two output statistics defined as experiment Responses (`TC` for Total Cost and `Satisfaction` for the service level). Note that we just made up all of the costs in our model so that the model would be interesting. In the real world, these are often difficult values to determine/identify. In fact, one of the more difficult

Scenario			Replications		Controls						Responses	
✓	Name	Status	Required	Completed	DoctorCap...	ExamCapacity	TreatmentC...	TraumaC...	NurseC...	SatisfactionThre...	TC	Satisfaction
✓	001	Completed	10	10 of 10	3	6	6	2	8	0.5	59.4278	0.741979
✓	002	Completed	10	10 of 10	5	6	6	2	10	0.5	21.8805	0.999114
✓	003	Completed	10	10 of 10	4	7	7	3	9	0.5	25.0415	0.966985
✓	004	Completed	10	10 of 10	2	5	1	1	6	0.5	497.105	0.993656

Figure 9.18: Sample experiment for Model 9-1.

things to expose students to in an educational environment is the difficulty in finding accurate model data.

9.1.1 Resource-Level Optimization With Model 9-1

The hospital's management would like to select just the right values for the five capacity levels (one each for `Doctor`, `Exam`, `Treatment`, `Trauma`, and `Nurse`) that will minimize Total Cost (`TC`), while at the same time providing an "adequate" service level (a high proportion of `Routine` patients whose total waiting time is less than or equal to a half hour). Management feels that "adequate" service level means that this proportion is at least 0.8; in other words, no more than 20% of `Routine` patients should have a total wait of more than a half hour (of course this is totally subjective and the "right" values to use for the threshold time and satisfaction percentage will be determined by the stakeholders). Each of the five capacity levels must be an integer, and must be at least one or two (depending on the specific item); management also wants to limit the capacities of the Doctor to at most ten and the Nurse to at most fifteen, and the capacities of Exam and Treatment to at most ten each, and Trauma to at most five.

Let's think a moment about what all this means. We're trying to *minimize* total cost, an output response from the Simio model, so at first blush it might seem that the best thing to do from that standpoint alone would be to set each of the five capacity levels to the minimum allowed, which is one or two for each. However, remember that total cost includes a penalty of $250 for each hour of patient waiting time, driving up total cost with such rock-bottom capacity levels (so maybe that's not such a good idea after all, from the total-cost viewpoint). Also, we need to meet the service-level Satisfaction requirement that it be at least 0.8, which will also demand staffing capacity at some levels that are very possibly above the bare minimum of one each. It helps to write all this out as a formal optimization problem, even though standard mathematical-programming methods (like linear programming, nonlinear programming, or integer programming) can't come close to actually solving it, since the objective function is an output Response from the simulation, so is stochastic (subject to sampling error so we can't even measure it exactly — more on this later). But, we can still express formally what we'd *like* to do:

$$\min \quad TotalCost$$

Subject to:

$$2 \leq DoctorCapacity \leq 10$$
$$2 \leq ExamCapacity \leq 10$$
$$2 \leq TreatmentCapacity \leq 10$$
$$1 \leq TraumaCapacity \leq 5$$
$$2 \leq NurseCapacity \leq 15$$

All five of the above capacities are integers

$$Satisfaction \geq 0.8,$$

where the minimization is taken over *DoctorCapacity*, *ExamCapacity*, *TreatmentCapacity*, *TraumaCapacity*, and *NurseCapacity* jointly. The *objective function* that we'd like to minimize is Total Cost, and the lines after "Subject to" are *constraints* (the first five on input controls, and the last one on another, non-objective-function output). Note that the first five constraints, and the integrality constraint listed sixth, are on the input Controls to the simulation, so we know when we set up the Experiment whether we're obeying all the range and integrality constraints on them. However, Satisfaction is another output Response (different from the total-cost objective) from the Simio model so we won't know until we run it whether we're obeying the requirement that it turns out to be at least 0.8 (so this is sometimes called a *requirement*, rather than a constraint, since it's on an output rather than on an input).

Well, how are you going to go about solving the optimization-formulation problem stated above? You might first think of expanding the sample Experiment in Figure 9.18 for more Scenario rows to explore more (all, if possible) of the options in terms of the capacity-level controls, then sort on the TC (total cost) response from smallest to largest, then go down that column until you hit the first "legal" scenario, i.e., the first scenario that satisfies the service-level Satisfaction requirement (the Satisfaction response is at least 0.8). Looking at the range and integrality constraints on the staffing Capacity levels, the number of possible (legal) scenarios is $8 \times 8 \times 8 \times 5 \times 14 = 35,840$ and that's going to be a lot of Scenario rows for you to type into your Experiment. And, we need to replicate each scenario some number of times to get decent estimates (sufficiently precise) of Total Cost and Satisfaction. Some initial experimentation indicated that to get an estimate of Total Cost that's reasonably precise (say, the 95% confidence interval half width is less than 10% of the average) would require about 120 replications for some combinations of staffing capacities, each of duration ten simulated days — for each scenario. On a quad-core 3.0GHz laptop, each of these replications takes pretty close to 1.7 seconds, so to run this experiment (even if you could type in 35,840 Scenario rows in the Experiment Design) would take $35,840 \times 204$ seconds, which is just under three months. Further, since we are sampling the objective function values using our stochastic simulation (i.e., performing a *statistical sampling experiment*), we have to ensure that the selection of "the best" is statistically valid by controlling the numbers of replications of each scenario. And, bear in mind that this is just a small made-up illustrative example — in practice you'll typically encounter much larger problems for which this kind of exhaustive, complete-enumeration strategy is ridiculously unworkable in terms of run time.

So we need to be smarter than that, especially since nearly all of the three months would be spent simulating scenarios that are either vastly inferior in terms of (high) Total Cost, or unacceptable due to a low Satisfaction proportion. If you think of our optimization formulation above in terms of just the input capacity-level controls, you're searching through the integer-lattice points in a five-dimensional box, and maybe if you start somewhere in there, and then simulate at "nearby" points in the box, you could get an idea of which way to move, both to decrease total cost, and to stay feasible on the Satisfaction-level requirement. It turns out that there's been a lot of research done on such *heuristic-search* methods, mostly in the world of optimizing difficult deterministic functions rather than finding an optimal simulation-model configuration, and software packages have been developed. One of them, OptQuest (from Opt-Tek Systems, Inc., `www.opttek.com`), has been adapted to work with Simio and other simulation-modeling packages, to search for an optimal feasible solution to problems like our hospital-capacities problem. You won't need to be concerned with how OptQuest decides which way to move from a particular point in the five-dimensional box to try to improve things, but basically it uses a combination of heuristic searches (including techniques known as scatter search, tabu search, neural networks, etc.) to find a much more efficient route through the box toward an optimal point — or, at least a good point, as these methods cannot absolutely guarantee that they'll find *the* optimal solution. There's a large research literature on all this, and a good place to start is the handbook [15] edited by Fred Glover and Gary Kochenberger. The following section discusses the use of OptQuest with Simio for Model 9-1. Following that, we will return to the question of how to ensure statistical validity of the results with the hopes of identifying the "best" configuration.

Using OptQuest to Identify Good Configurations

The way OptQuest works with Simio is that you describe the optimization formulation in a Simio Experiment, including identifying what you want to minimize or maximize (we want to minimize total cost), the input constraints (the capacity ranges for us, and that they must be integers), and any other requirements on other, non-objective-function outputs (the Satisfaction-proportion lower bound of 0.8). Then, you turn over execution of your Experiment and model to OptQuest, which goes to work and decides which scenarios to run, and in what order, to get you to a good (even if not provably optimal) solution in a lot less than a month's worth of nonstop computing. One of the time-saving strategies that OptQuest uses is to stop replicating on a particular scenario if it appears pretty certain that this scenario is inferior to one that it's already simulated — there's no need to get a high-precision estimate of a loser.

Here's how you set things up for OptQuest in a Simio experiment. First, we'll create a new Simio Experiment (choose the New Experiment button from the Project Home ribbon) so that we can experiment both with and without OptQuest. Next, we need to tell Simio to attach the OptQuest for Simio *add-*

Figure 9.19: Selecting the OptQuest Add-in.

in. An experimentation add-in is an extension to Simio (written by the user or someone else — see the Simio documentation for details about creating add-ins) that adds additional design-time or run-time capability to experiments. Such an add-in can be used for setting up new scenarios (possibly from external data), running scenarios, or interpreting scenario results. Combining all three technologies can provide powerful tools, as illustrated by the OptQuest for Simio add-in. To attach the add-in, select the Add-In button on the experiment Design ribbon and choose the OptQuest for Simio option from the list (see Figure 9.19). Note that when you select OptQuest, you may get a warning similar to "Setting an add-in on an experiment cannot be undone. If you set the add-in, the current undo history will be cleared. Do you want to continue?" If you need to keep the undo history (and we can't think of a reason why this would ever be the case, but you never know!), you should answer No, make a copy of the model, and open the copy to set up for OptQuest. Otherwise, just answer Yes and Simio will load the add-in (and, of course, clear the undo history). Once you've the added OptQuest add-in, your experiment will have some additional properties that control how OptQuest works (see Figure 9.20). The OptQuest-related properties are:

- Min Replications: The minimum number of replications Simio should run for each OptQuest-defined scenario.

- Max Replications: The maximum number of replications Simio should run for each OptQuest-defined scenario.

- Max Scenarios: The maximum number of scenarios Simio should run. Basically, OptQuest will systematically search for different values of the *decision variables* — the number of Doctors, Exam Rooms, Treatment Rooms, Trauma Rooms, and Nurses, in our case — looking for good values of the objective (small values of Total Cost, in our case), and also meeting the Satisfaction-proportion requirement that it be at least 0.8. This property will limit how many scenarios OptQuest will test. Note that OptQuest will not always need this many scenarios, but the limit will provide a forced stopping point, which can be useful for problems with very large solution spaces. But, you should realize that if you specify this property to be too small (though it's not clear how you'd know ahead of

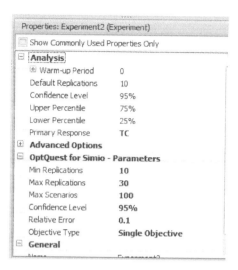

Figure 9.20: Experiment properties after attaching the OptQuest Add-in.

time what "too small" might mean), you could be limiting the quality of the solution that OptQuest will be able to find.

- Confidence: The confidence level you want OptQuest to use when making statistical comparisons between two scenarios' response values.

- Error Percent: Percentage of the sample average below which the half-width of the confidence interval must be. For example, if the sample average is 50 and you set the Error Percent to 0.1 (which says that you want the confidence-interval half-width to be at most 10% of the average, so roughly speaking 10% error), the confidence-interval will be at least as precise as 50 ±5 (5 = 0.1 × 50). After running the default number of replications, OptQuest will determine if this condition has been satisfied. If not, Simio will run additional replications in an attempt to reduce the variability to achieve this goal (until it reaches the number of replications specified by the Max Replications property value).

Attaching the OptQuest add-in also adds properties to the experiment controls. With the OptQuest add-in, we can specify the optimization constraints using either the experiment controls or Constraints — we'll use the experiment controls for the current model.[3] Figure 9.21 shows the properties for the Doctor Capacity control after attaching the OptQuest add-in (note that prior to attaching the add-in, there were no user-editable properties for experiment controls). The "new" control properties tell OptQuest whether it should manipulate the control to search the feasible region (*Include in Optimization*), the "legal" values for the control (between and including the *Minimum Value*

---

[3]In Simio 3.44, simulation output statistics could not be used in experiment Constraints. Rather than mixing Constraints and Responses we elected simply to use the already-defined experiment Controls combined with the experiment Responses to define the constraints, requirements, and objective.

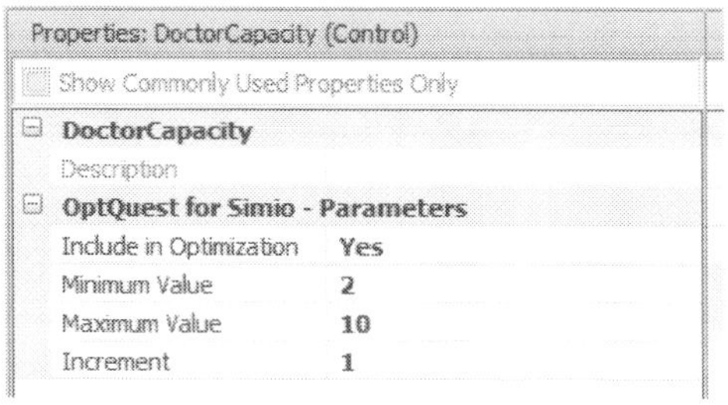

Figure 9.21: Properties for the Doctor Capacity control after attaching Opt-Quest.

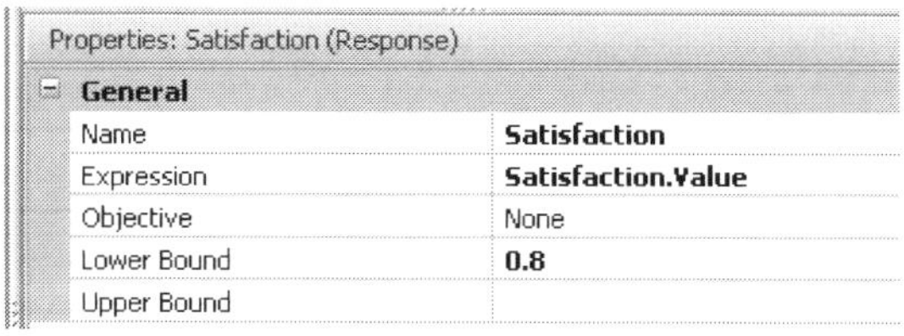

Figure 9.22: Properties for the Satisfaction Response.

and *Maximum Value* property values), and the increment (*Increment* property value) to use when searching the feasible region from one point to the next. The control properties shown in Figure 9.21 specify the constraint on the Doctor Capacity from the optimization model above (i.e., integers between 1 and 10). The corresponding properties for the `Exam Capacity`, `Treatment Capacity`, `Trauma Capacity`, and `Nurse Capacity` controls are set similarly. Since we're not using the Satisfaction Threshold control in our optimization (recall that we used a referenced property to specify the threshold), we set the Include in Optimization property to `No`. For a control that's not used in the optimization, Simio will use the default value (0.5 in this case) for the property in all scenarios.

The only remaining "constraint" from our optimization above is the requirement on the Satisfaction level (recall that we called this a requirement rather than a constraint since Satisfaction is an output rather than an input). Figure 9.22 shows the properties for the `Satisfaction` response (note that, with the exception of setting the Lower Bound property to 0.8, this response is the same as the corresponding `Satisfaction` response in the previous experiment).

The final step for setting up OptQuest is to define the *objective function* — i.e., tell OptQuest the criterion for optimization. Since we'd like to minimize the total cost, we'll define the TC response (as we did before) and specify the Objective property as `Minimize` (see Figure 9.23). Note that we divided the `TotalCost` model state value by 10,000 in the definition of the TC response so that it would be easier to compare the results visually in the experiment table (clearly, the division merely changes the measurement units and does not alter the comparative relationship between scenarios in any way). Finally, we'll set the experiment property Primary Response to TC to tell OptQuest that the TC

Properties: TC (Response)	
General	
Name	**TC**
Expression	**TotalCost.Value/10000**
Objective	**Minimize**
Lower Bound	
Upper Bound	

Figure 9.23: Properties for the TC response.

Name	Status	Required	Completed	Doctor...	Exam...	Treatme...	Trauma...	Nurse...	SatisfactionThreshold	TC ▲	Satisfaction
045	Completed	10	10 of 10	5	6	6	1	9	0.5	21.2601	0.998499
065	Completed	10	10 of 10	5	6	6	1	8	0.5	21.4977	0.988347
020	Completed	10	10 of 10	5	6	6	2	9	0.5	21.6427	0.996414
084	Completed	10	10 of 10	6	6	6	1	7	0.5	21.6724	0.991782
041	Completed	10	10 of 10	5	6	6	1	10	0.5	21.7216	0.999389
085	Completed	10	10 of 10	5	6	7	1	9	0.5	21.7312	0.993893
083	Completed	10	10 of 10	6	6	6	1	8	0.5	21.7453	0.996487
087	Completed	10	10 of 10	5	5	6	1	9	0.5	21.7772	0.992088
042	Completed	10	10 of 10	5	6	6	2	10	0.5	21.8805	0.999114
086	Completed	10	10 of 10	6	6	7	1	8	0.5	21.9482	0.997182
039	Completed	10	10 of 10	5	6	6	3	9	0.5	22.0458	0.996015
035	Completed	10	10 of 10	4	6	6	1	10	0.5	22.2869	0.998759
072	Completed	10	10 of 10	6	6	7	1	9	0.5	22.321	0.997612
019	Completed	10	10 of 10	6	6	6	3	8	0.5	22.3287	0.997071
066	Completed	10	10 of 10	4	5	6	1	10	0.5	22.3291	0.995934
079	Completed	10	10 of 10	6	6	7	3	8	0.5	22.4997	0.996613
001	Completed	10	10 of 10	6	6	6	3	9	0.5	22.575	0.999024

Figure 9.24: Results for the initial OptQuest run.

response (minimizing TC in this case) is the objective of the optimization.

Now you can click the Run button and sit back and watch OptQuest perform the experiment. Figure 9.24 shows the "top" several scenarios from the OptQuest run (we sorted in increasing order of Total Cost, so the top scenario has the minimum total cost of the scenarios tested)[4]. The top scenario in the list is scenario 045, which has 5 Doctors, 6 Exam Rooms, 6 Treatment Rooms, 1 Trauma Room and 9 Nurses, a Total Cost of 21.2601 (in thousands) and a Satisfaction of 0.998. Note that all of the Satisfaction values are quite high – you have to go far down the list of scenarios before you find an infeasible solution (one where the 80% satisfaction requirement is not met). Naturally, the next question should be "Is this really the truly optimal solution?" Unfortunately, the answer is that we're not sure (actually, we do know that this is not the optimal solution for this problem, but we cheated a bit — described below — and we will not know the answer in general). As mentioned earlier, OptQuest uses a combination of heuristic-search procedures, so there is no way (in general) to tell for sure whether the solution that it finds is truly optimal. Remember that we identified 35, 840 possible configurations for our system and

---

[4]Depending on the specific version of Simio that you are running, you may see different scenarios in the list, but specific configurations should have similar responses values. These differences can be caused by differences in the number of processors/cores and processor speed.

that we limited OptQuest to testing 100 of these. It turns out that, in this case, it seems that OptQuest did fairly well — we ran an additional experiment allowing OptQuest to test 1000 scenarios and found a configuration with Total Cost 20.7515 (a little better) and Satisfaction 0.999 — the configuration was 5 Doctors, 7 Exam Rooms, 4 Treatment Rooms, 1 Trauma Room, and 8 Nurses, in case you're curious. Unfortunately, even though we ran ten times more scenarios, we *still* don't really know how good our new "best" solution actually is with respect to the true optimum. Such is the nature of heuristic solutions to combinatorial optimization problems. We're pretty confident that we have good solutions, but we just don't know with absolute certainty that we have "the" optimum, upon which no improvement is possible.

While we can't say with any certainty that OptQuest finds the best solution, it's quite likely that OptQuest can do better than *you* (or we) could do by just arbitrarily searching through the feasible region (especially after you've reached your 37th scenario and are tired of looking at the computer). Where OptQuest can be really useful is in cases where you can limit the size of the feasible region. For example, we constrained the number of Trauma Rooms in our problem to be between 1 and 5, but the expected arrival rate of Urgent patients (the only patients that require the Trauma room) is quite low (about 0.6 per hour). As such, it is *extremely* unlikely that we would ever need more than 1 or 2 trauma rooms. If we change the upper limit on the Trauma Room control from 5 to 2, we reduce the size of the feasible region from 35,840 to 14,336 and can almost certainly improve the performance of OptQuest for a small number of allowed scenarios. There are clearly other similar adjustments we could make to the experiment to reduce the solution space. This highlights the importance of your being involved in the process — OptQuest doesn't know anything about the specific model that it's running or the characteristics of the real system you're modeling, but *you* should, and making use of your knowledge can substantially reduce the computational burden or improve the final "solution" on which you settle. Now we'll turn our attention to evaluating the scenarios that OptQuest has identified, hopefully to identify the "best" scenario.

Ranking and Selection

Since the response values shown in Figure 9.24 are sample means from independent replications of our simulation model, they are subject to *sampling error*. As such, we are not sure that the ordering based on the OptQuest experiment identifies the true optimal of our problem (i.e., the best configuration of our resources). Figure 9.25 shows the SMORE plots for the top five scenarios from our OptQuest run. While the sample mean of total cost (TC) for scenario 045 is clearly the lowest compared to the other scenarios, when we take the confidence intervals into consideration through a visual comparison of the SMORE plots, things are no longer so clear. In fact, in looking a Figure 9.25 it's difficult to tell if there is any *statistical difference* between the five sample means (while visual inspection of the SMORE plots cannot accurately identify true statistical differences, it should not be discounted as a means to guide further analysis). Our uncertainty here is an artifact of sampling error and this is exactly is where the *Subset Selection Analysis* comes into play, as discussed earlier in Section 5.5

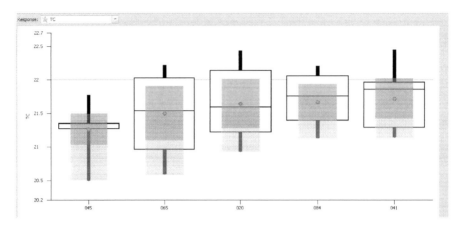

Figure 9.25: SMORE plots for the top five scenarios from the OptQuest run of Model 9-1.

Name	Status	Required	Completed	Doctor...	Exam...	Treatme...	Trauma...	Nurse...	SatisfactionThreshold	TC ▲	Satisfaction
045	Completed	10	10 of 10	5	6	6	1	9	0.5	21.2601	0.998499
065	Completed	10	10 of 10	5	6	6	1	8	0.5	21.4977	0.988347
020	Completed	10	10 of 10	5	6	6	2	9	0.5	21.6427	0.996414
084	Completed	10	10 of 10	6	6	6	1	7	0.5	21.6724	0.991782
041	Completed	10	10 of 10	5	6	6	1	10	0.5	21.7216	0.999389
085	Completed	10	10 of 10	5	6	7	1	9	0.5	21.7312	0.993893
083	Completed	10	10 of 10	6	6	6	1	8	0.5	21.7453	0.996487
087	Completed	10	10 of 10	5	5	6	1	9	0.5	21.7772	0.992088
042	Completed	10	10 of 10	5	6	6	2	10	0.5	21.8805	0.999114
086	Completed	10	10 of 10	6	6	7	1	8	0.5	21.9482	0.997182
039	Completed	10	10 of 10	5	6	6	3	9	0.5	22.0458	0.996015
035	Completed	10	10 of 10	4	6	6	1	10	0.5	22.2869	0.998759
072	Completed	10	10 of 10	6	6	7	1	9	0.5	22.321	0.997612
019	Completed	10	10 of 10	6	6	6	3	8	0.5	22.3287	0.997071
066	Completed	10	10 of 10	4	5	6	1	10	0.5	22.3291	0.995934
079	Completed	10	10 of 10	6	6	7	3	8	0.5	22.4997	0.996613
001	Completed	10	10 of 10	6	6	6	3	9	0.5	22.575	0.999024

Figure 9.26: Results for the initial OptQuest run after running Subset Selection Analysis.

in a smaller example than we consider here.

Figure 9.26 shows the same scenarios shown in Figure 9.24 after running the Subset Selection Analysis function (by clicking on the Subset Selection Analysis icon on the Design ribbon). This function considers the replication results for all scenarios and divides the scenarios into two groups: the "possible best" group, consisting of scenarios whose means are statistically indistinguishable from the sample best mean; and the "rejects" group. Based on the initial experimentation, the scenarios in the "possible best" group warrant further consideration, while those in the "rejects" group are statistically "worse" and do not need further consideration (with respect to the given response). The "rejects" group is identified by the lighter brown shading in the corresponding response cell (see Figure 9.26). Note that if no scenarios in your experiment are identified as "rejects," it is likely that your experiment does not include enough replications for the Subset Selection procedure to identify statistically different groups. In this case, running additional replications should fix this problem (unless there actually *is no statistical difference* between the scenarios).

In our example, the "possible best" group includes eleven scenarios (see Figure 9.26). Clearly the question at this point is "What now?" We have a choice between two basic options: Declare that we have eleven "good" configurations, or do additional experimentation. Assuming that we want to continue to try to find the best configuration, we could run additional replications of just the eleven scenarios in the "possible best" group and re-run the Subset Selection Analysis to see if we could identify a smaller "possible best" group (or, with some luck, the single "best" scenario). We deleted the scenarios in the "rejects" group from the experiment and ran the remaining eleven scenarios for an additional 30 replications each (for a total of 50 replications each), and the resulting new "possible best" group (not shown) contained only six scenarios; note that you must "Clear" the OptQuest add-in prior to running the additional replications, by clicking on the Clear icon in the Add-ins section of the Ribbon. As before, we can either declare that we have six "good" configurations, or do additional experimentation. If we want to continue searching for the "best" configuration, we could repeat the previous process — delete the scenarios in the most recent "rejects" group and run additional replications with the remaining scenarios. Instead, we will use another Simio Add-in to assist in the process.

The add-in "Select Best Scenario using KN," based on [26], is designed to identify the "best" scenario from a given set of scenarios, as mentioned in Section 5.5. The add-in continues to run additional replications of the candidate scenarios until it either determines the "best" or exhausts the user-specified maximum number of scenarios. To run the add-in, choose "Select Best Scenario using KN" using the Select Add-In icon in the Design ribbon, just as we did when we used OptQuest earlier (see Figure 9.19). This will expose some additional experiment properties related to the add-in (see Figure 9.27). In addition to the "Confidence Level" property (that specifies the level to use for confidence intervals), you also specify the *Indifference Zone* and *Replication Limit* for the add-in to use. The Indifference Zone is the smallest "important" difference that should be detected. For example, 21.65 and 21.68 are numerically different, but the difference of 0.03 may be unimportant for the particular problem (clearly the appropriate choice of an indifference-zone value depends on what the response is measuring, and on the context of the study). An indifference zone of 0.10 would cause the add-in not to differentiate between these values during the experimentation. The Replication Limit tells the add-in the maximum number of replications to run before "giving up" on finding the best scenario (and declaring the remaining scenarios "not statistically different"). Figure 9.28 shows the experiment after running the add-in. The add-in identified scenario 045 as the "best" configuration, per the checked box in the leftmost column, after running it and scenario 065 for 70 replications. If the add-in terminates with multiple scenarios checked, it cannot differentiate between the checked scenarios based on the maximum number of replications and the Indifference Zone.

Analysis	
Warm-up Period	3
Units	**Days**
Default Replications	10
Confidence Level	95%
Upper Percentile	75%
Lower Percentile	25%
Primary Response	TC
Advanced Options	
Select Best Scenario using KN - Parameters	
Confidence Level	**0.95**
Indifference Zone	**.1**
Replication Limit	**100**
General	
Name	Experiment2

Figure 9.27: Experiment properties after including the Select Best Scenario using KN add-in.

Scenario			Replications		Controls						Responses	
	Name	Status	Required	Completed	Do...	Exa...	Treatm...	**Traum...**	NurseCap...	SatisfactionThr...	TC ▲	Satisfaction
✓	045	Completed	70	70 of 70	5	6	6	1	9	0.5	21.6437	0.996406
	065	Completed	70	70 of 70	5	6	6	1	8	0.5	21.7484	0.987779
	041	Completed	50	50 of 50	5	6	6	1	10	0.5	21.8574	0.998916
	083	Completed	50	50 of 50	6	6	6	1	8	0.5	21.9011	0.995354
	084	Completed	50	50 of 50	6	6	6	1	7	0.5	21.9926	0.9886
	020	Completed	50	50 of 50	5	6	6	2	9	0.5	22.0285	0.994133
	087	Completed	50	50 of 50	5	5	6	1	9	0.5	22.0896	0.990724
	085	Completed	50	50 of 50	5	6	7	1	9	0.5	22.1086	0.991678
	042	Completed	50	50 of 50	5	6	6	2	10	0.5	22.1324	0.998325
	039	Completed	50	50 of 50	5	6	6	3	9	0.5	22.3761	0.993955
	035	Completed	50	50 of 50	4	6	6	1	10	0.5	22.6629	0.998108

Figure 9.28: Experiment results after running the Select Best Scenario using KN add-in.

So, to summarize the procedures we followed in this example, we started by using the OptQuest add-in to identify potentially promising scenarios (configurations with specific values for the referenced properties) based on our optimization problem. After running OptQuest, we used the Subset Selection Analysis function to partition the set of scenarios into the "possible best" and "rejects" groups. We then did further experimentation with the scenarios in the "possible best" subset. We first ran additional replications to reduce the size of this subset, and finally used the "Select Best Scenario using KN" add-in to identify the "best" configuration. Since we are using heuristic optimization and generally will not be able to evaluate all configurations in the design space, we will generally not be able to guarantee that our process has found the best (or even very good) configurations, but experience has shown that the process usually works quite well.

9.2   Model 9-2: Pizza Take-out Model

In Model 9-2 we will also consider a related resource allocation problem. The environment that we are modeling here is a take-out pizza restaurant that takes phone-in orders and makes the pizzas for pickup by the customer. Customer calls arrive at the rate of 12 per hour and you can assume that the call interarrival times are exponentially distributed with mean 5 minutes. The pizza shop currently has four phone lines and customers that call when there are already 4 customers on the phone will receive a busy signal. Customer orders consist of either 1 (50%), 2 (30%), 3 (15%), or 4 (5%) pizzas and you can assume that the times for an employee to take a customer call are uniformly distributed between 1.5 and 2.5 minutes. When an employee takes an order, the shop's computer system prints a "customer ticket" that is placed near the cash register to wait for the pizzas comprising the order. In our simplified system, individual pizzas are made, cooked, and boxed for the customer. Employees do the making and boxing tasks and a semi-automated oven cooks the pizzas.

The make station has room for employees to work on three pizzas at the same time although each employee can only work on one pizza at a time. The times to make a pizza are triangularly distributed with parameters 2, 3, and 4 minutes. When an employee finishes making a pizza, the pizza is placed in the oven for cooking. The cooking times are triangularly distributed with parameters 6, 8 and 10 minutes. The oven has room for eight pizzas to be cooking simultaneously. When a pizza is finished cooking, the pizza is automatically moved out of the oven into a queue where it awaits boxing (this is the "semi-automated" part of the oven). The boxing station has room for up to three employees to work on pizzas at the same time. As with the make process, each employee can only be working on one pizza at a time. The boxing times are triangularly distributed with parameters 0.5, 0.8, and 1 minutes (the employee also cuts the pizza before boxing it). Once all of the pizzas for an order are ready, the order is considered to be complete. Note that we could also model the customer pick-up process or model a home delivery process, but we will leave this to the interested reader.

The pizza shop currently uses a *pooled* employee strategy for the order-taking, pizza making, and pizza boxing processes. This means that all of the employees are cross-trained and can perform any of the tasks. For the make and box operations, employees need to be physically located at the respective stations to perform the task. The shop has cordless phones and hand-held computers, so employees can take orders from anywhere in the shop (i.e., there is no need for a physical order-taking station).

Figure 9.29 shows the Facility view of the completed version of Model 9-2. Note that we have defined Customer and Pizza entity types and have labeled the entities using *Floor Labels*. The Floor Labels allow us to display model and entity states on the labels. Floor Labels that are attached to entities will travel with the entities. To create a Floor label for an entity, simply select the entity, in the Facility view, click on the Floor Label icon on the Symbols ribbon, and place the label by clicking, dragging, and clicking on the Facility view where you would like the label. This will create a label with default text including instructions for formatting the label. Clicking on the Edit icon in

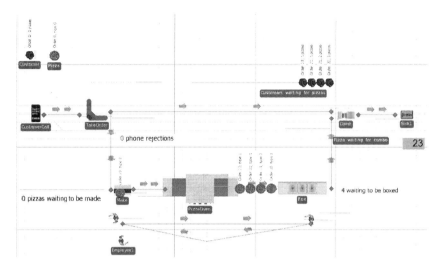

Figure 9.29: Facility view for Model 9-2.

Figure 9.30: Edit dialog box for the Pizza entity Floor Label.

the Appearance ribbon will bring up an edit dialog box (see Figure 9.30) for
the Pizza entity Floor label. Note the use of the OrderNumber entity state
set off by braces. As the model executes, this place-holder will be replaced
with the entity state value (the order number for the specific pizza). The
model also has a Source object (`CustomerCall`) and Server objects for the order
taking (`TakeOrder`), making (`Make`), and boxing (`Box`) processes and the oven
(`PizzaOven`). The model includes a Combiner object (*Register*) for matching
the Customer objects with the Pizza objects representing the pizzas in the
order and a single Sink object where entities depart. Finally, in addition to the
Connector objects that facilitate the Customer and Pizza entity flow, we add a
Path loop between two Node objects (*BasicNode1* and *BasicNode2*) that will
facilitate employee movement between the make station and the box station. If
you are following along, you should now place all of the objects in your model

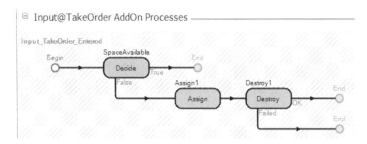

Figure 9.31: Add-on process for the input node of the *TakeOrder* object.

so that the Facility view looks approximately like that in Figure 9.29.

The `CustomerCall` Source object creates the customer calls based on the specific customer arrival process described above (the *Interarrival Time* property is set to `Random.Exponential(5)` minutes). Note that the model includes `Customer` and `Pizza` entity types so that Entity Type property for the Source object is set to `Customer`. Customer calls are then sent to the `TakeOrder` Server object for potential processing. Recall that if there are already four callers in the system, the potential customer receives a busy signal and hangs up. We will model this capacity limitation using an add-on process. Since we need to make a yes/no decision before processing the customer order, we will use an add-on process associated with the Input node for the `TakeOrder` object and will reject the call if the call system is at capacity. We will use the *Entered* trigger so that the process is executed when an entity enters the node. Figure 9.31 shows the `Input_TakeOrder_Entered` add-on process. The Decide step implements a condition-based decision to determine whether there is a phone line available. We model the number of phone lines using the `TakeOrder` object capacity and the associated Input Buffer capacity. Figure 9.32 shows the properties for the `TakeOrder` object. Note that we have set the Initial Capacity property to 1 and the Input Buffer property to a referenced property named `NumLines`. Our Decide step therefore uses the following expression to determine whether or not to let a call in: `TakeOrder.InputBuffer.Capacity.Remaining > 0`. If the condition is true, the call is allowed into the `TakeOrder` object. Otherwise, we increment the `PhoneRejections` user-defined model state to keep track of the number of rejected calls and destroy the entity. So based on the `TakeOrder` object properties and the `Input_TakeOrder_Entered` add-on process a maximum of one employee can be taking an order at any time and the number of customers waiting "on hold" is determined by the value of the `NumLines` referenced property. Note that our use of a referenced property to specify the input buffer capacity will simplify experimenting with different numbers of phone lines (as we will see soon). For the base configuration described above, we would set the referenced property value to 3 to limit the number of customers on the phone to a total of 4.

Once a customer enters the system, an employee will take the call and process the order. We use a *Worker* object (`Worker1`) to model our pooled employees. Figure 9.33 shows the properties for the `Worker1` object. Recall from Chapter 8 that the Worker object will allow us to model workers that

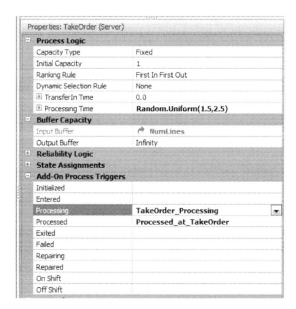

Figure 9.32: Properties for the *TakeOrder* object.

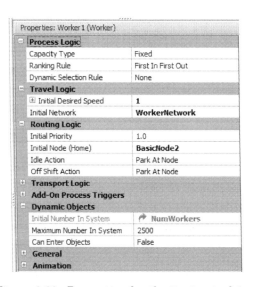

Figure 9.33: Properties for the `Worker1` object.

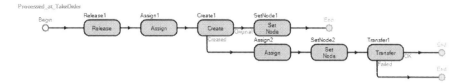

Figure 9.34: *Processed* add-on process for the *TakeOrder* object.

move and have an associated physical location within the model. Below, we will define locations for the make and box stations and will track the location of each of the employees. As such, we set the Initial Desired Speed property to 1 meter per second, the Initial Network to the `WorkerNetwork`, and the *Initial Node (Home)* to `BasicNode2` (we defined the WorkerNetwork using *BasicNode1* below the box station and *BasicNode2* below the make station along with the Path objects comprising the loop that the Worker objects use to move between the two stations). Note that we also used a referenced property (`NumWorkers`) for the Initial Number In System property so that we can easily manipulate the number of workers during experimentation. As we did in Model 9-1, we will use a *Seize* step to seize a *Worker1* resource in the *Processing* add-on process for the *TakeOrder* object (see Figure 9.32). If a *Worker1* resource is not available, the entity will wait at the *Seize* step. Since the employee can take the order from anywhere in the shop, we have no requirement to request the Worker1 object to actually move anywhere, we simply need the resource to be allocated to the order taking task. In the real pizza shop, this would be equivalent to the employee answering the cordless phone and entering the order via the hand-held computer from wherever she/he is in the shop. This will not be the case when we get to the Make and Box stations below.

Once the Worker1 resource is allocated, the entity is delayed for the processing time associated with actually taking the customer order. At this point, we need to create the associated Pizza entities and send them to the make station and send the Customer entity to wait for the completed Pizza entities to complete the order. This is done in the `Processed_at_TakeOrder` add-on process (see Figure 9.34). The first step in the process releases the Worker1 resource (since the call has been processed by the time the process executes). The first Assign step makes the following state assignments [5]:

- `HighestOrderNumber=HighestOrderNumber+1` — Increment the model state HighestOrderNumber to reflect the arrival of the current order

- `ModelEntity.OrderNumber=HighestOrderNumber` — Assigns the current order number to the Customer entity

- `ModelEntity.NumberOrdered=Random.Discrete(1,.5,2,.8,3,.95,4,1)` — Determine the number of pizzas in this order and store the value in the entity state NumberOrdered

---

[5]Note that we do not illustrate the creation of standard states as we have done this in previous chapters.

Figure 9.35: Properties for the NumberOrdered model state.

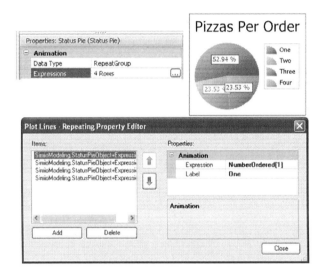

Figure 9.36: Properties for the NumberOrdered model state.

- `NumberOrdered[ModelEntity.NumberOrdered]=`
  `NumberOrdered[ModelEntity.NumberOrdered]+1` — This assignment increments a model state that tracks the number of orders at each size (number of pizzas in the order). Figure 9.35 shows the properties for the NumberOrdered model state. Note that it is a vector of size 4 – one state for each of the four potential order sizes. At any point in simulation time, the values will be the numbers of orders taken at each order size. Figure 9.36 shows the Status Pie chart along with the properties and repeat group used for the chart.

The Create step creates the Pizza entity objects using the *NumberOrdered* state to determine how many entities to create (see Figure 9.37 for the Create step properties). Leaving the Create step, the original entity will move to the Set Node step and the newly created Pizza objects will move to the Assign step. The Set Node step assigns the destination Node Name property to `ParentInput@Register`, so that when the entity object leaves the `TakeOrder` object, it will be transferred to the `Register` object to wait for the `Pizza` entity objects that belong to the order (the Customer object represents the "customer ticket" described above). The Assign step assigns the *ModelEntity.OrderNumber* state the value `HighestOrderNumber`. Recall that

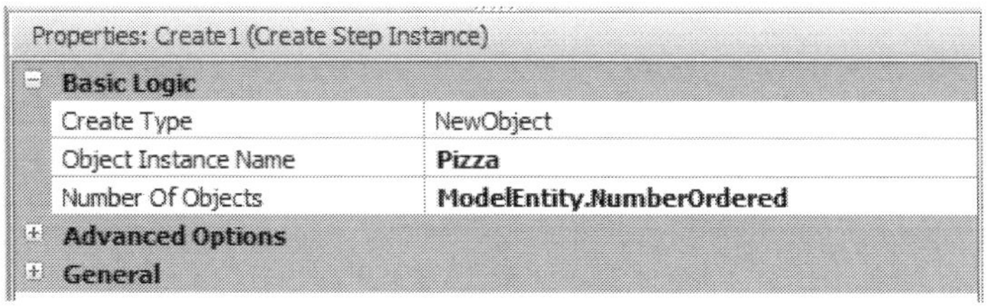

Figure 9.37: Properties for the *Create* step.

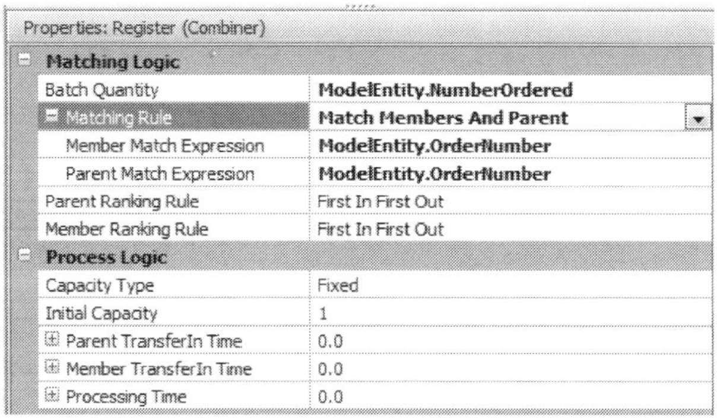

Figure 9.38: Properties for the *Register* Combiner object.

the `Customer` object entity state OrderNumber was also assigned this value
(see above). So the OrderNumber state can be used to match-up the `Customer`
object with the `Pizza` objects representing pizzas in the customer order. The
Set Node step for the newly created Pizza entity objects sets the destination
Node Name property to `Input@Make` so that the objects will be transferred to
the `Make` object. Finally, we have to tell Simio where to put the newly created
objects (the Create step simply creates the objects in free space). The Transfer
step will immediately transfer the object to the node specified in the Node Name
property (`Output@TakeOrder`, in this case). So at this point, the Customer object will be sent to the register to wait for Pizza objects, the Pizza objects will
be sent to the make station and all of the objects associated with the customer
order can be identified and matched using the OrderNumber respective entity
states.

The `Register` Combiner object matches Customer objects representing customer orders with Pizza objects representing the pizzas that comprise the
customer order. Figure 9.38 shows the properties for the Combiner object.
The Batch Quantity property tells the Combiner object how many "members"
should be matched to the "parent." In our case, the parent is the `Customer`
object and the members are the corresponding `Pizza` objects. As such, the
`ModelEntity.NumberOrdered` entity state specifies how many pizzas belong to
the order (recall that we used this same entity state to specify the number of
Pizza objects to create above). The Member Match Expression and Parent
Match Expression properties indicate the expressions associated with the mem-

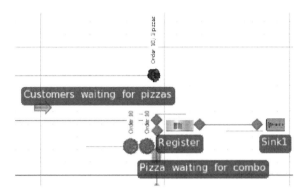

Figure 9.39: Combiner object with one parent and two member objects in queue.

ber and parent objects to match the objects. Figure 9.39 shows the Combiner object with a `Customer` object with label "Order 10; 3 pizzas" waiting in the parent queue and two Pizza objects with label "Order 10" waiting in the member queue. Note how the Floor Labels help us identify the situation – 3 pizzas are in customer order 10, 2 of the pizzas are boxed and waiting. When the final Pizza object arrives to the Combiner object, it will be combined with the other two Pizza objects and the Customer object and the order will be complete. When this happens, the Customer object will leave the Combiner object and be transferred to the Sink object.

The `Make` Server object models the pizza making station. Since the make station has room for three employees to be making pizzas simultaneously, we set the Initial Capacity property to 3. We also set the Processing Time property to `Random.Triangular(2,3,4)` minutes. Since the make process also requires an employee, we will use the Processing add-on process to seize the Worker object and the Processed add-on process to release the worker as we have done previously with the `TakeOrder` object. The difference here is that the make process requires that the employee be physically present at the make station in order to make the pizza. As such, we need to request that the Worker object travel to BasicNode2 (the node in the `WorkerNetwork` designated for the make station). Luckily we can make this request directly when we seize the Worker object. Figure 9.40 shows the Seizes property editor associated with the Seize step in the Make_Processing add-on process. Here, in addition to specifying the Object Name property (`Worker1`), we also set the Request Visit property to `ToNode` and the Node Name property to `BasicNode2` indicating that when we seize the object, the object should be sent to the `BasicNode2` node. When the add-on process executes, the token will remain in the Seize step until the Worker object arrives at `BasicNode2`. Finally, we set the `Box` Server object up similarly (except we set the Processing Time property to `Random.Triangular(0.5,0.8,1)` and specify that the Worker object should visit `BasicNode1`).

So our model is now complete — if a phone line is available when a customer call arrives, the first available employee takes the call and records the order. The corresponding Customer Order entity then creates the Pizza entities asso-

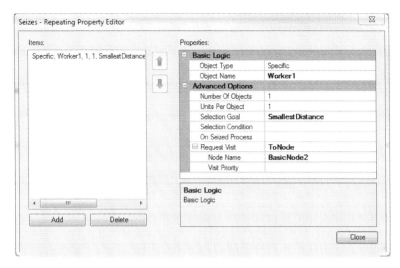

Figure 9.40: Seizes property editor for the *Seize* step in the *Make_Processing* add-on process.

ciated with the orders. The Customer Order entity is then sent to wait for the Pizza orders that comprise the order and the Pizza entities are sent for processing at the Make, Oven, and Box stations. Once all Pizza entities complete processing at the Box station, the Customer Order and Pizza entities for an order are combined and the order is complete. To support our experimentation, we used referenced properties to specify the number of phone lines (`NumLines`) and number of employees (`NumWorkers`).

9.2.1 Experimentation With Model 9-2

Figure 9.41 shows a simple 15-scenario experiment that we set up and ran for Model 9-2 (the run length was 1000 hours and there was no warmup – see Problem 3 at the end of the chapter to explore these choices). The number of phone lines (`NumLines`) and number of workers (`NumWorkers`) are the controls for the experiment. We have also defined three responses:

- `CustTIS` - the time (in minutes) that customer orders spend in the system. Defined as `Customer.Population.TimeInSystem.Average*60`;

- `Oven Util` - the oven's scheduled utilization. Defined as `PizzaOven.Capacity.ScheduledUtilization`; and

- `RejectionsPerHr` - customer rejections (due to phone line capacity) per hour. Defined using the user-defined output statistic `NumRejections`. See Figure 9.42 for the properties of `NumRejections`. Note that since an output statistic is evaluated at the end of a replication, dividing the total number of phone rejections (stored in the model state `PhoneRejections`) by the current time (`TimeNow`) results in the desired metric.

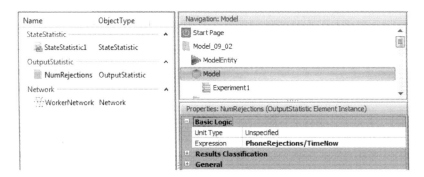

Scenario		Replications		Controls		Responses		
Name	Status	Required	Completed	Num Lines	NumWorkers	Cust TIS	Oven Util	Rejections/Hr
Scenario13	Comple...	10	10 of 10	1	1	13.4015	17.4703	6.0498
Scenario14	Comple...	10	10 of 10	1	2	13.8122	26.833	2.8637
Scenario15	Comple...	10	10 of 10	1	3	13.9567	30.5717	1.4925
Scenario10	Comple...	10	10 of 10	2	1	18.9411	18.3952	5.7273
Scenario11	Comple...	10	10 of 10	2	2	17.7113	29.7542	1.7706
Scenario12	Comple...	10	10 of 10	2	3	16.1282	33.5063	0.5536
Scenario1	Comple...	20	20 of 20	3	1	23.9739	18.602	5.62305
Scenario2	Comple...	20	20 of 20	3	2	20.6896	31.3109	1.28965
Scenario3	Comple...	20	20 of 20	3	3	17.1927	34.523	0.23815
Scenario4	Comple...	10	10 of 10	4	1	28.9758	18.6493	5.5715
Scenario5	Comple...	10	10 of 10	4	2	23.2022	32.1114	1.0026
Scenario6	Comple...	10	10 of 10	4	3	17.6684	34.6456	0.1006
Scenario7	Comple...	10	10 of 10	5	1	34.0238	18.6817	5.5496
Scenario8	Comple...	10	10 of 10	5	2	25.4529	32.6309	0.7896
Scenario9	Comple...	10	10 of 10	5	3	17.9297	34.722	0.0459

Figure 9.41: Simple experiment results for Model 9-2.

Name	ObjectType
StateStatistic	⌄
StateStatistic1	StateStatistic
OutputStatistic	⌄
NumRejections	OutputStatistic
Network	⌄
WorkerNetwork	Network

Navigation: Model

- Start Page
- Model_09_02
  - ModelEntity
  - Model
    - Experiment1

Properties: NumRejections (OutputStatistic Element Instance)

Basic Logic
Unit Type	Unspecified
Expression	**PhoneRejections/TimeNow**

Results Classification
General

Figure 9.42: User-defined output statistic for phone rejections per hour.

As set up, our experiment evaluates scenarios containing between 1 and 5 phone lines and between 1 and 3 workers (for a total of 15 scenarios). The scenarios are sorted in order of increasing value of CustTIS — resulting in the "best" configuration(s) in terms of the time that customers spend in the system being at the top of the table. Rather than manually enumerating all scenarios within a region (as we do in Figure 9.42), we could also have used the OptQuest add-in to assist us with the identification of the "best" configuration(s) (as described in Section 9.1.1). Figure 9.43 shows the results of using OptQuest with Model 9-2. For this optimization, we used two controls and two responses to control the experiment (or, more precisely, to tell OptQuest how to control the experiment). These controls and responses are defined as follows:

- NumLines (control) - the number of phone lines used. We specify a minimum of 1 line and a maximum of 5 lines.

- NumWorkers (control) - the number of workers. We specify a minimum of 1 worker and a maximum of 3 workers.

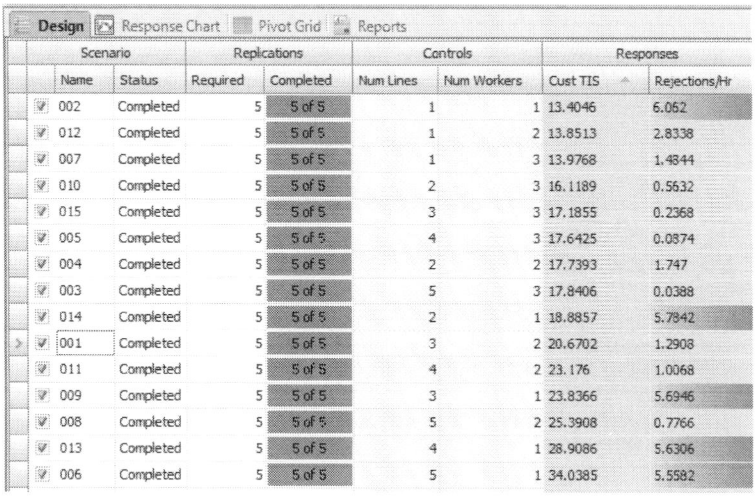

Design	Response Chart	Pivot Grid	Reports					
	Scenario		Replications		Controls		Responses	
	Name	Status	Required	Completed	Num Lines	Num Workers	Cust TIS	Rejections/Hr
☑	002	Completed	5	5 of 5	1	1	13.4046	6.062
☑	012	Completed	5	5 of 5	1	2	13.8513	2.8338
☑	007	Completed	5	5 of 5	1	3	13.9768	1.4344
☑	010	Completed	5	5 of 5	2	3	16.1189	0.5632
☑	015	Completed	5	5 of 5	3	3	17.1855	0.2368
☑	005	Completed	5	5 of 5	4	3	17.6425	0.0874
☑	004	Completed	5	5 of 5	2	2	17.7393	1.747
☑	003	Completed	5	5 of 5	5	3	17.8406	0.0388
☑	014	Completed	5	5 of 5	2	1	18.8857	5.7842
☑	001	Completed	5	5 of 5	3	2	20.6702	1.2908
☑	011	Completed	5	5 of 5	4	2	23.176	1.0068
☑	009	Completed	5	5 of 5	3	1	23.8366	5.6946
☑	008	Completed	5	5 of 5	5	2	25.3908	0.7766
☑	013	Completed	5	5 of 5	4	1	28.9086	5.6306
☑	006	Completed	5	5 of 5	5	1	34.0385	5.5582

Figure 9.43: Result of using OptQuest with Model 9-2.

- CustTIS (response) - the time that customers spend in the system. We would like to minimize this output value (so that we can make our customers happy).

- RejectionsPerHr (response) - the number of customer call rejections per hour. We would like to keep this output value under 4 (we arbitrarily chose this value). Without this service-level constraint, the optimization could minimize the customer time in system by simply reducing the number of phone lines, thereby restricting the number of customers who actually get to place an order (something the pizza shop manager would likely frown on!).

Figure 9.44 shows the control and response properties for the OptQuest-based experiment with Model 9-2. Note that we set the minimum and maximum values for the controls (NumLines and NumWorkers), we set the objective (Minimize) for the Customer TIS response and set the maximum value (4) for the RejectionsPerHr response. The results (Figure 9.43) indicate that the 1-line, 1-worker configuration results in the minimum customer time-in-system (13.4046), but that the average number of rejections per hour violates our constraint (6.062 > 4) — note that the Rejections/Hr cell for this configuration is highlighted in red, indicating the infeasibility. The next best configuration is the 1-line, 2-worker configuration with a slightly higher customer time-in-system (13.8513), but an acceptable average number of rejections per hour (2.8338). The next logical step in our analysis would be to use the Subset Selection Analysis and/or the Select Best Scenario using KN to identify the "best" configuration – we will leave this as an exercise for the reader (see Problem 4).

Properties: NumLines (Control)	
OptQuest for Simio - Parameters	
Include in Optimization	**Yes**
Minimum Value	1
Maximum Value	5
Increment	1

Properties: NumWorkers (Control)	
OptQuest for Simio - Parameters	
Include in Optimization	**Yes**
Minimum Value	1
Maximum Value	3
Increment	1

Properties: Cust TIS (Response)	
General	
Name	**Cust TIS**
Expression	**Customer.TimeInSystem.Average*60**
Objective	**Minimize**
Lower Bound	
Upper Bound	

Properties: Rejections/Hr (Response)	
General	
Name	**Rejections/Hr**
Expression	**NumRejections.Value**
Objective	None
Lower Bound	0
Upper Bound	4

Figure 9.44: Control and response properties for the OptQuest experiment with Model 9-2.

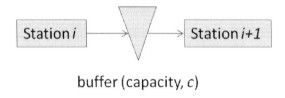

buffer (capacity, c)

Figure 9.45: Two adjacent stations separated by a buffer.

9.3   Model 9-3: Fixed-Capacity Buffers

The final model for this chapter involves modeling an assembly line with fixed-capacity buffers between the stations. In an *assembly line*, items move sequentially from station to station starting at station 1 and ending at station n, where n is the length of the line. Adjacent stations in the line are separated by a *buffer* that can hold a fixed number of items (defined as the buffer *capacity*). If there is no buffer between two stations, we simply assign a zero-capacity buffer (for consistency in notation). Consider two adjacent stations, i and $i+1$ (see Figure 9.45). If station i is finished working and the buffer between the stations is full (or zero-capacity), station i is *blocked* and cannot begin pro-

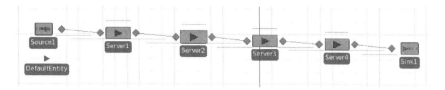

Figure 9.46: Facility view for Model 9-3.

cessing the next item. On the other hand, if station $i + 1$ is finished and the buffer is empty, station $i + 1$ is *starved* and similarly cannot start on processing the next item. Both cases result in lost production in the line. Clearly in cases where the processing times are highly variable and/or where the stations are unreliable, adding buffer capacity can improve the process throughput by reducing blocking and starving in the station. The *buffer allocation problem* involves determining where buffers should be placed and the capacities of the buffers. Model 9-3 will allow us to solve the buffer allocation problem (or, more specifically, will provide throughput analysis that can be used to solve the buffer allocation problem). A more detailed description of serial production lines and their analysis can be found in [2], but our brief description is sufficient for our needs.

Our goal for Model 9-3 is to be able to estimate the line's maximum *throughput* for a given buffer configuration (the set of the sizes of buffers between the stations). *Throughput* in this context is defined as the number of items produced per unit time. For simplicity, our model will consider the case where stations are identical (i.e., the distributions of processing times are the same) and the buffers between adjacent stations are all the same size so that the buffer configuration is defined by a single number — the capacity of all of the buffers. Model 9-3 can be used to replicate the analysis originally presented in [7] and described in [2]. Figure 9.46 shows the Facility view of Model 9-3.

From the Facility view, Model 9-3 looks much like Model 5-1 with three additional stations added between the Source and Sink objects. There are, however some subtle differences required to model our fixed-capacity buffer assembly line. We will discuss these changes in the following paragraphs. In addition, we will also demonstrate the use of referenced properties and the Simio Response Charts as tools for experimentation and scenario comparison.

By default, when we connect two Server objects with a Connector or Path object, entities are automatically transferred from the output node of one station to the input node of the next station once processing is complete at the first station. If the second station is busy, the transferred entity waits in a queue. When the second server is available it removes the first entity waiting in the queue and begins processing. This essentially models an infinite capacity buffer between two stations and does not model the blocking effect associated with finite-capacity buffers. Consider the simple case where we have a zero-capacity buffer (e.g., no buffer) between two stations that we model as resources. What we need instead of the default behavior is to ensure that the subsequent resource has available capacity *before* we release the current resource (i.e., making sure that the entity has somewhere to go before it starts going). This is described as

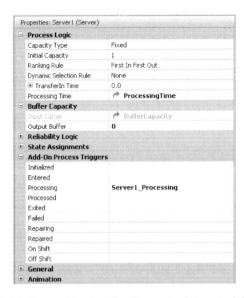

Figure 9.47: Properties for the Server1 object in Model 9-3.

overlapping resources in [25]. Clearly if we have a finite-capacity buffer between the two stations, we have an analogous problem — we just need to make sure that the buffer instead of the subsequent resource has remaining capacity.

Luckily, the Simio Server object makes it fairly easy to model this type of overlapping resources situation. Figure 9.47 shows the properties for the Server1 object in Model 9-3. Our interest here is in the Buffer Capacity section of the properties and the Input Buffer and Output Buffer properties in particular. The Input Buffer for a Server object is used to hold entities that are entering the server when there is not sufficient capacity on the server resource. Similarly, the Output Buffer is used to hold entities that have completed processing on the server, but do not yet have anywhere to go. Whereas the default value for these properties is `Infinite`, we use a referenced property named `BufferCapacity` to specify the Input Buffer size and set the Output Buffer to 0. With this configuration for all of our stations, an entity can only be transferred from one Server to the next if the number of entities in the internal queue representing the input buffer of the subsequent station is less than the value of the referenced property `BufferCapacity`. Note also that we specified the Processing Time property using another referenced property, `ProcessingTime`. Using the referenced properties for the buffer capacity and processing time distributions will simplify experimentation with different line configurations (as we will illustrate below).

Since we are using the model to estimate the maximum throughput for a line, we also need to modify the entity arrival process. As we have seen, the standard Source object will create entities using an interarrival time distribution or an arrival schedule. For our maximum throughput model, we need to make sure that there is always an entity waiting for the first station (`Server1`). This way, the first station will never be starved (this is equivalent to saying that there is

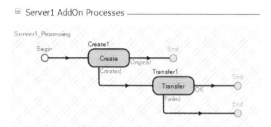

Figure 9.48: Server1_Processing add-on process for Server1.

Scenario		Replications		Controls		Responses
Name	Status	Required	Completed	Processing Time	Buffer Capacity	Throughput
☑ Scenario3	Idle	50	50 of 50	Random.Exponential(10)	0	515.82
☑ Scenario2	Idle	50	50 of 50	Random.Exponential(10)	1	629.56
☑ Scenario4	Idle	50	50 of 50	Random.Exponential(10)	2	700.48
☑ Scenario5	Idle	50	50 of 50	Random.Exponential(10)	3	743.8
☑ Scenario6	Idle	50	50 of 50	Random.Exponential(10)	4	776.5
☑ Scenario7	Idle	50	50 of 50	Random.Exponential(10)	5	801.76
☑ Scenario8	Idle	50	50 of 50	Random.Exponential(10)	6	824.42
☑ Scenario9	Idle	50	50 of 50	Random.Exponential(10)	7	842.28
☑ Scenario10	Idle	50	50 of 50	Random.Exponential(10)	8	856.76
☑ Scenario11	Idle	50	50 of 50	Random.Exponential(10)	9	867.96
☑ Scenario12	Idle	50	50 of 50	Random.Exponential(10)	10	873.9

Figure 9.49: Experiment with exponentially distributed processing times for Model 9-3.

an infinite supply of raw materials for the line). To implement this logic, we simply set the Maximum Arrivals property on Source1 to 1 (so that we will create a single entity at time zero) and we duplicate the entity when it finished processing on Server1 and send the duplicate back to the input of Server1. We accomplish this duplication and routing using the Server1_Processing add-on process for Server1. (see Figure 9.48). The Create step duplicates the entity and the Transfer step connected to the Created output of the Create step transfers the entity to the node Input@server1. Since the Original output goes to the End of the process, the original entity will follow the normal path and either be transferred to the second station if there is available capacity either in Server2 or in the buffer between Server1 and Server2 or will remain on Server1 since the Output Buffer capacity is set to 0. See Problem 8 at the end of the chapter for an alternate method of maintaining the infinite supply of raw materials at the first station.

Now that the initial model is complete, we can use the model to experiment with different configurations. Figure 9.49 shows an experiment for Model 9-3. Note that since we defined referenced properties for the station processing times (ProcessingTime) and the buffer capacities (BufferCapacity), these show up

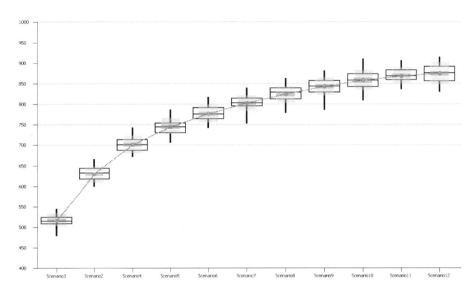

Figure 9.50: Response chart for exponentially distributed processing times for Model 9-3.

as Controls in the experiment. In this experiment we defined eleven scenarios where the processing times are exponentially distributed with mean 10 minutes and the buffer capacity goes from 0 to 10 in steps of 1. For this experiment, we set the run length to 30,000 minutes and the warm-up period to 20,000 minutes. With 10,000 minutes of run time and mean processing times of 10 minutes at each station, we would expect 1,000 items to be produced in the absence of processing time variation. For the scenarios tested the average throughput for 50 replications ranged from 515.82 with no buffers to 873.9 with 10 buffers. Figure 9.50 shows the Response Chart for the experiment in Figure 9.49. The chart clearly shows the expected trend of increasing throughput with increasing buffer capacity, but with diminishing returns.

As we have previously discussed, Simio automatically tracks the busy/idle state for resources and Server objects. In fact, by default, Simio tracks the following states for Server objects:

- Starved - The server is idle and waiting for an entity

- Processing - The server is processing one or more entities

- Blocked - The server has completed processing, but the entity cannot release the server

- Failed - The server has failed

- Offshift - The server uses a schedule and is off-shift

For our serial production line with an infinite supply of items at the first server, no server failures, and no work schedules, the servers can either be processing an item (busy), starved (idle), or blocked. We can use Simio's Status Pie chart

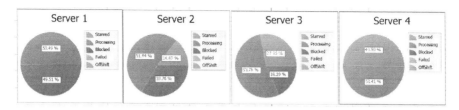

Figure 9.51: Server state status pie charts for Model 9-3.

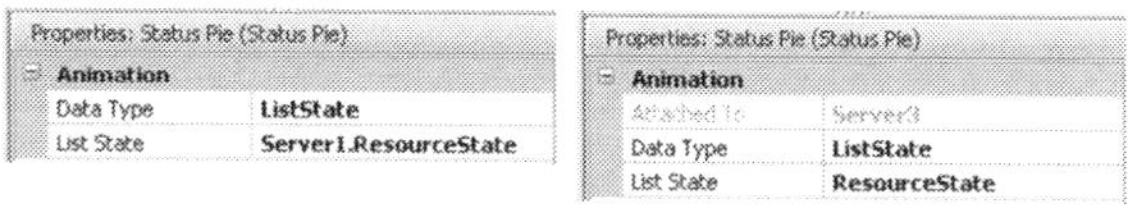

Figure 9.52: Properties for the Status Pie Charts for Server1 (left) and Server3 (right).

feature to track the states of the four servers in our serial line (see Figure 9.51). The Status Pie charts show the percentage of time that the given server was in the respective states. Figure 9.52 shows the properties for the Status Pie charts for `Server1` and `Server3`. In both charts, the Data Type property is set to `ListState` and the List State property points to the server's `ResourceState`. The difference between the two charts is that the `Server3` chart is attached to `Server3` and so the List State property value does not need to reference `Server3` whereas the `Server1` chart is not attached to the server and so the property must explicitly reference the server object (`Server1.ResourceState`). In addition, since the `Server3` chart is attached to the Server object, the chart will "move" with the Server object in the Facility view. Notice that the first server in the line is never starved and the last server is never blocked. This is exactly as we would expect in a model designed for assessing maximum throughput. Comparing the second and third stations, the second station has more blocking and the third station has more starving. If we had a longer line, this pattern would continue — stations nearer the start of the line experience more blocking and stations near the end of the line experience more starving.

Since we used a referenced property to specify the processing time distributions, we can easily change these in the experiment (no need to go back to the model). Figure 9.53 shows the Response Chart for an experiment similar to the one shown in Figure 9.49 except with triangularly distributed processing times with parameters (5, 10, 15) — we simply changed the referenced property values in the Controls section of the experiment. Since the variance is significantly less with the triangular distribution than with the exponential distribution, we expect the blocking and starving effect to be less for the triangular case, and we expect the throughput recovery by adding buffers to be faster (i.e., more throughput recovery with fewer buffers). All of these analysis results match our expectation as well as the results of using the methods described in [2] and [7].

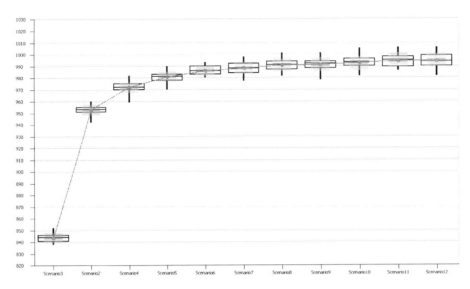

Figure 9.53: Response chart for triangularly distributed processing times for Model 9-3.

9.4   Summary

In this chapter we've looked at Simio models of three systems — each involving one or more advanced simulation concepts. In addition to the modeling aspects, we also focused on the experimentation with each of these models and introduced the OptQuest simulation-based optimization add-in. Once the user defines the appropriate objective, constraints, and requirements, OptQuest "drives" the Simio experiment towards an optimal configuration of the system. We also described the ranking and selection process, by which we try to identify the "best" solution from those scenarios provided by OptQuest[6] and demonstrated the use of Simio's Subset Selection Analysis function and the Select Best Scenario using the KN add-in. These tools significantly simplify the general output-analysis process.

Of course we have merely scratched the surface of the applications of simulation in general and the capabilities of Simio in particular, so you should definitely consider this as the beginning rather than the end of your quest to master the use of simulation and Simio. Further, our coverage of output analysis in general, and simulation-based optimization in particular, has been extremely introductory in nature, and there is much more to be learned from other sources and experimentation.

---

[6]Note that the ranking-and-selection process that we've described will work with any set of scenarios regardless of how they were identified — there is no requirement that the scenarios be generated by OptQuest.

Table 9.3: Zone delivery time distributions and proportions of customers from each zone for Problem 3. The delivery times include round-trip travel and all times are in minutes.

Zone	Proportion	Delivery Time Distribution
1	12%	Triangular(5, 12, 20)
2	30%	Triangular(2, 7, 10)
3	18%	Triangular(8, 15, 22)
4	40%	Triangular(6, 10, 15)

9.5   Problems

1. Modify Model 9-1 so that it includes the patient physical movement and Nurse/Aide escort functionality from Model 8-3. Compare the resource requirements with those of Model 9-1. Does including this level of detail "improve" the validity of the results? If so, why? If not, why not?

2. Modify Model 9-1 to support preemption of the Doctor resource for Urgent patients. In particular, if all doctors are busy when an Urgent patient arrives, the Urgent patient should preempt a doctor from one of the other non-Urgent patients. Compare the results of the models. Would you advise that the emergency department make the corresponding change in the real system?

3. In Model 9-2, we used a 1000 hour run length with no warm-up period (and gave no explanation for these settings). Do they appear reasonable? Why or why not? Could you achieve the same results with a shorter run length? Would a longer run length improve the results?

4. In Section 9.2.1 we experimented with model 9-2 (manually and with OptQuest), but we did not use the Subset Selection Analysis function and/or Select Best Scenario using KN add-in to identify the "best" configuration(s) from the identified set. Do this and make any necessary adjustments to the numbers of replications required to identify the "best" configuration.

5. Many pizza take-out places around these days also deliver. Modify Model 9-2 to support the delivery of customer orders. Assume that 90% of customers would like their pizzas delivered (the remaining 10% will pick them up as we assumed in Model 9-2). The pizza shop has divided the customer delivery area into four *zones* and a delivery person can take multiple orders on the same delivery trip if the orders come from the same zone. Table 9.3 gives the delivery time distributions for each zone and the proportion of customers from each zone. For simplicity, if a delivery includes more than one order, simply add 4 minutes to the sampled delivery time for each extra order. For example, if a delivery includes 3 orders and the originally sampled delivery time is 8 minutes, you would use $8+4+4 = 12$ minutes for the total delivery time instead. For customer arrivals, processing times, and worker characteristics, use the same data as was given

Table 9.4: Processing time and failure/repair distributions for Problem 7. Note that all times are in minutes and the failures are calendar-based.

Num	Processing Time	Uptime	Time to Repair
1	Triangular(3,6,9)	Exponential(360)	Exponential(20)
2	Exponential(3)	Exponential(120)	Triangular(10,15,20)
3	Erlang(6,3)	Exponential(300)	Exponential(5)
4	Triangular(2,7,12)	N/A	N/A
5	Lognormal(4,3)	Exponential(170)	Triangular(8,10,12)
6	Exponential(6)	N/A	N/A

for Model 9-2. Develop an experiment to determine how many delivery people should be hired. Be sure to justify your solution (i.e., how you decided the appropriate number of delivery people).

6. Modify the model from Problem 3 so that delivery people wait until there are *at least* 2 orders waiting to be delivered to a given zone before they start a delivery. Compare the results to the previous model. Is this a good strategy? What about requiring at least 3 orders?

7. Consider a six-station serial manufacturing line similar to the assembly line in Model 9-3. As with Model 9-3, there are finite-capacity buffers between each pair of stations (assume that the first station is never starved and the last station is never blocked). The difference in the current system is that the stations are not identical in terms of the processing time distributions and some are subject to random failures. The processing time distributions and failure data are given in Table 9.3. Develop a Simio model of the serial line and create an experiment that will find a "good" allocation of 30 buffer slots. For example, one allocation would be $[6, 6, 6, 6, 6]$ (6 slots between each pair of stations). The objective is to maximize the line throughput. Does your experiment find the guaranteed optimal buffer allocation? If so, why? If not, why not?

8. Modify Model 9-3 to use the `On Event` Arrival Mode property for the Source object to create new entities for the first station rather than using the add-on process associated with the Server1 object as described in Section 9.3.

9. In the OptQuest-based experiment described in Section 9.2.1, we minimized the customer time in system subject to a service level constraint. However, we didn't consider staffing costs. As an alternative, develop a single "total cost" metric that incorporates customer waiting time, worker costs, and customer service. Add the total cost metric as an output statistic and use OptQuest to find the optimal configuration(s).

10. Model 9-2 uses a *pooled* employee strategy for taking orders, making pizzas and boxing pizzas. Develop a model that uses a dedicated employee strategy (i.e., each employee is dedicated to a task) and compare the results with those of Model 9-2. Under what conditions is one strategy preferred over the other? Justify your answer with experimentation.

Chapter 10

Customizing and Extending Simio

You may be able to solve many of your simulation problems using the Simio Express Edition which is limited to using only the Simio Standard Library. Or you can solve many problems by supplementing with the add-on processes that are available in the Simio Design Edition and above. But in some cases you may need or want more capability than is offered by that approach. For example:

- You may find that you need more object customization than is easily available using add-on processes.

- You might make repeated use of the same object enhancements and prefer to create your own reusable objects.

- You might have a team of more advanced modelers who support the use of simulation by occasional, less skilled users who merit their own customized tool.

- You may desire to build commercial libraries that you market and sell to others.

All of these may be accomplished by creating custom Simio objects and libraries. If you have read the initial 9 chapters and worked along with the examples, you actually know most of what you need to know in order to define custom objects – the first part of this chapter will bring together concepts you already know and fill in a few missing concepts.

Even though custom objects are quite flexible and powerful, a few motivated modelers may desire to go beyond what can be done with objects. Simio also provides additional capabilities for more advanced users. If you have some programming background (like Visual Basic, C++, C#, or any .NET language), you can:

- Write your own design-time add-ons for example to automatically build a model from external data.

- Add new dynamic selection rules to those already available within Simio, for example a comprehensive work selection rule for a Vehicle, Worker, or Server.

- Add custom routines for importing and exporting tables.

- Add custom algorithms for designing (e.g., setting up or analyzing scenarios) or running (e.g., optimizing) experiments.

- Add custom Steps and Elements to extend the capabilities of the Simio engine.

An overview of these capabilities is provided near the end of this chapter.

Some of the material in this chapter has been adapted from Introduction to Simio [45] (the e-book included with the Simio software) and used with permission.

10.1   Basic Concepts of Defining Objects

The Standard Library that we have been using up to this point is just one of many possible libraries that you can use to build your models. You can build libraries of objects for your own purposes or build libraries to share across your enterprise. One of the basic principles of Simio is the notion that any model can be an object definition. Models are used to define the basic behavior of an object. It is fairly easy to take a Simio model that you have developed and then "package" it up for someone else to use as a building block for their models. It is also fairly easy to build sub-models within one project and then build your model using these sub-models as building blocks.

The material we will be discussing here is an extension of the Simio process-related ideas we presented earlier in Chapters 4, 5, and 9. In those chapters, we have already learned how to use processes to extend the basic functionality of a specific object instance without altering the core behavior of the object definition. In the next sections our focus will be either on modifying or extending the core behavior of an object definition — thereby changing the behavior of all instances of that object — or on building new objects from scratch.

An object definition has five primary components: properties, states, events, external view, and logic (see Figure 10.1) The properties, states, events, and external view are all defined in the Definitions window for a model. The logic is defined in the combination of the Facility and Processes windows for the model.

We initially used model properties to create properties that were referenced in our model and could be used for experimentation. These same properties are used to define inputs for a model when used as an object definition. *Model properties* provide inputs to model logic and remain static during the run. *States* are used to define the current state of the object. Anything that can change during the execution of the simulation is represented in the object as a state. *Events* are logical occurrences at an instant in time such as an entity entering a station or departing a node. Objects may define and fire their own events to notify other objects that something of interest has happened. The *External View* is a graphical representation for instances of the object. The external view

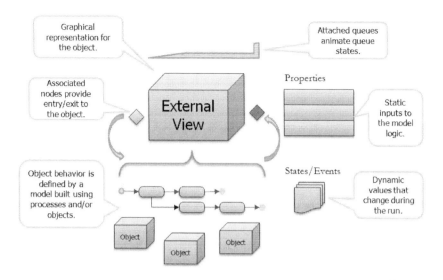

Figure 10.1: Anatomy of an object.

is what someone sees when they place the object in their model. The *Logic* for the object is simply a model that defines how the object reacts to events. The logic gives the object its behavior. This logic can be built hierarchically using existing objects, or can be defined with graphical process flows.

10.1.1 Model Logic

The logic for a fixed object definition is defined using a facility and/or process model, as determined by the Input Logic Type on the node symbols. In the case of entities, transporters, nodes and links there are no node symbols and hence these models are typically built using process logic.

Whenever you have built a model of a system it represents the logic component of an object definition. You can typically turn any model into a useable object definition by simply adding some properties to supply inputs to the model, along with an external view to provide a graphical representation for the object along with nodes that allow entities to enter and exit the model.

There are three approaches to defining the model logic for an object:

- The first approach is to create your model hierarchically using a facility model. This approach can also be combined with the use of add-on processes to define custom behavior within the facility objects. This approach is typically used for building up higher level facility components such a workcenter comprised of two machines, a worker, and tooling.

- The second and most flexible approach is to create the object definition behavior from scratch using a process model. This is the approach that was used to create the objects in the Standard Library.

- The third approach is to *sub-class* an existing object definition and then change/extend the behavior of the new object using processes. This ap-

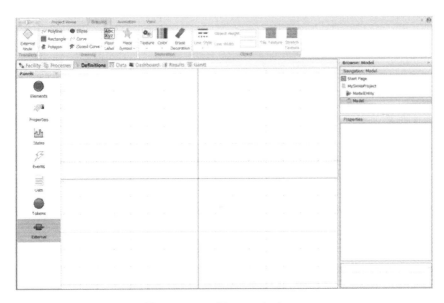

Figure 10.2: External view.

proach is typically used when there is an existing object that has behavior similar to the desired object, and can be "tweaked" in terms of property names, descriptions, and behavior to meet the needs of the new object.

10.1.2   External View

The external view is a graphical representation for instances of a fixed or dynamic object. The external view is not used for nodes or links. The external view is defined by clicking on the External panel in the Definitions window for a model as shown in Figure 10.2.

The external view is what a user sees when they place an instance of the object in their model. The graphics for the external view can be composed of symbols from the symbol library or downloaded from Trimble 3D Warehouse[1], as well as static graphics drawn using the drawing tools on the Drawing ribbon. In addition the view may contain attached animation components that are placed into the view from the Animation window. These animation components include animated queues, status labels, plots, pie charts, linear/circular gauges, and buttons. Note that in the case of dynamic objects (i.e. entities) these animated objects are carried by each dynamic object in the system. For example an entity could carry a status label showing the value of one of its states, or a button that when clicked by the user causes some action to take place (as illustrated in Model 9-2).

When you build a new fixed object you typically provide associated input/output nodes for the object so that dynamic entities may enter and/or leave your fixed object. For example a Server object has associated node objects for input and output. The characteristics of these associated objects are

---

[1]Formerly Google 3D Warehouse.

also defined in the external view of the object by placing external node symbols in the external view using the Drawing ribbon. Note that these are not node objects but symbols identifying the location where associated node objects will be created when this object is placed.

When you place a node symbol in the external view you must also define its properties. The *Node Class* specifies the type of node that should be created. The drop list will give you a choice from all available node definitions which will include `Node` (empty model), `BasicNode` (simple intersection), and `TransferNode` (intersection with support for setting destination and selecting transporters on which entities will ride), as well as any other node definitions that you have loaded as libraries or are defined in your project. In the case of the Standard Library the BasicNode class is always used for input nodes and the TransferNode is used for output nodes. The *Input Logic Type* specifies how arriving entities attempting to enter this object are to be handled by this object. The `None` option specifies that entities are not permitted to enter the object. The `ProcessStation` option specifies that the arriving entity may enter at a specified station, which fires a station-entered event that can be used to trigger process logic. This is used when defining the object logic using a process model. The `FacilityNode` option specifies that the arriving entity is to be sent to a node inside the facility model for the object. This option is used for defining object logic using a facility model. In the General section the *Name* for the node symbol is specified. This node symbol name is used to create the name of the associated node object that is created at this symbol location when this object is instantiated using the format *NodeSymbolName@ObjectName*. For example if the symbol name is `Input` and the object name is `Lathe`, the associated object that is automatically created is named `Input@Lathe`. In the Standard Library the input node symbols are named Input and output node symbols are named Output.

10.1.3   Sub-classing an Object Definition

The basic idea of sub-classing an object definition is that the new object definition that you are building inherits its properties, states, events, and logic from an existing object definition. The sub-classed object will initially have the same properties and behavior of the original object, and if the original object definition is updated, then the sub-classed object will inherit the new behavior. However after sub-classing you can then hide or rename the inherited properties, selectively override portions of the process logic that is inherited, or add additional processes to extend the behavior. For example you might create a new object definition named MRI that is sub-classed from Server and incorporated into a library for modeling health care systems. You might hide some of the normal Server properties (e.g., Capacity Type), and rename others (e.g., rename Process Time to Treatment Time). You might also replace the normal internal process that models a failure by a new process that models the failure pattern for an MRI, while continuing to inherit the other processes from the Server.

To sub-class an object definition, select the Project in the navigation window, and then select the Models panel. The Edit ribbon (Figure 10.3) lets you

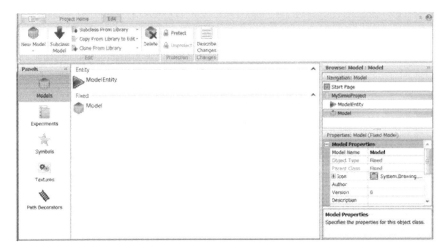

Figure 10.3: Edit ribbon.

add a new blank model, create a sub-class of the selected model, or create a sub-class from an object definition from a library. You can also sub-class a library object using the right-click menu on the library object definition from within the Facility window. There are actually three ways to create new objects based on Library Objects:

- `SubClass From Library` subclasses (derives) an object from the library object as described above. Changing the original will change the behavior of the sub-classed object since it inherits process logic from its base object definition

- `Copy From Library for Edit` creates a copy of the object without sub-classing. The newly created object definition is a copy of the library object, but does not maintain an inherited relationship; hence if the original object definition is changed the copy is not affected.

- `Clone From Library` creates an object that is identical to the original (from Simio's standpoint actually indistinguishable from the original). This should be used when you want to rearrange how models are organized in different libraries.

You can also protect an object definition with a password. In this case the password is required to view or change the internal model for the object.

The model properties include the *Model Name*, *Object Type* (fixed, link, node, entity, or transporter), *Parent Class* (from which this object was sub-classed), *Icon* for displaying the model, along with the *Author*, *Version number*, and *Description*. The properties also include *Resource Object* for specifying if the object can be seized and released as a resource, and *Runnable* to specify if the object can be run as a model or only used as a sub-model within other models.

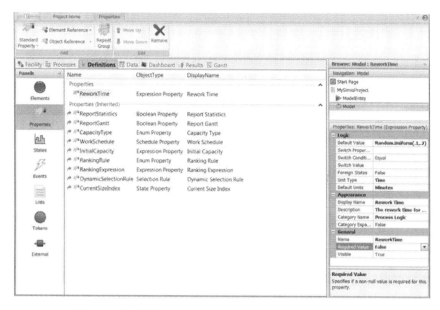

Figure 10.4: Properties panel in Definitions window.

10.1.4   Properties, States, and Events

Whenever you create an object definition you inherit properties, states, and events from the base object class and you can add new members as well. You can view the inherited properties, states, and events in the Definitions window for the model. The inherited and new members are placed in separate categories that can be independently expanded and collapsed. Figure 10.4 shows the Properties panel for a model type Fixed with a single new property named ReworkTime. Note that this property can be referenced by both the objects and process flows within the model.

The characteristics of ReworkTime are shown in the Property window, and include the Default Value, Unit Type/Default Units, Display Name, Description, Category Name, and flags specifying if this is a Required Value and if this property is Visible to the user. The characteristics also include a Switch Property Name, Switch Condition, and Switch Value. These can be used to hide or show a property based on the value the user specifies for another property. For example you might have property Y show only if the user sets property X to value greater than 0. This lets you dynamically configure the Property window based on user inputs.

In the case of an inherited property you can change the *Visible* flag along with the *Default Value, Display Name, Description,* and *Category.* Hence you can hide properties or change their general appearance. Although you can add additional states and events to a model you cannot rename or hide inherited states and events.

10.2   Model 10-1: Building a Hierarchical Object

In this example we are going to build a new object definition for a tandem server comprised of two Standard Library Servers in series, with a Connector in between. The first Server has a capacity of one and has no output buffer. The second Server also has a capacity of one and has no input buffer; hence the second Server will block the first Server whenever it is busy. The object has two properties that specify the processing time on each of the two servers. The object also animates the entity in process at each Server, as well as animating pie charts for the resource state of each Server.

10.2.1   Model Logic

Since we are building this object hierarchically (i.e., from other objects in the Facility Window), we will begin by creating an object and defining its logic in the Facility Window.

- Start with a new project. Add a new fixed model from the Project Home ribbon and rename this new model to TandemServer.

- Select the TandemServer as the active model and place two Servers connected by a Connector its Facility window. Set the output buffer for Server1 to 0, and the input buffer for Server2 to 0.

Next we will define the properties of our Tandem Server and customize the display of its Properties Window:

- In the Properties panel of the Definitions window add two new Standard Properties of type Expression. Name them `ProcessTimeOne` and `ProcessTimeTwo`, and set their Default Value, UnitType/Default Units, Display Name, and Category Name as shown in Figure 10.5.

- Expand the inherited properties. Since the inherited properties named CapacityType, WorkSchedule, and InitialCapacity are not relevant for our TandemServer object, set the Visible characteristic in the General category to `False` to hide each of these as shown in Figure 10.6.

10.2.2   External View

Next we will use the External view to define what we want users to see when they place the object. Starting with Simio version 4, objects, nodes, links, and status animation that are defined in the Facility Window are automatically added to the external view. This is particularly handy if you want to show entity movement on links that are internal to your object. In our case we want the finer control (and learning experience) of creating our user view manually, so we will start by turning off the default user view.

- Select the Facility window. To suppress the automatic inclusion of these items in the external view right-click on each item in the Facility View and deselect the `Externally Visible` option. Rather than change each item individually, you can also use the Control-Drag to highlight all of

Properties: ProcessTimeOne (Expression Property)	
Logic	
Default Value	**Random.Triangular(.1,.2,.3)**
Switch Property Name	
Switch Condition	Equal
Switch Value	
Foreign States	False
Unit Type	**Time**
Default Units	**Minutes**
Appearance	
Display Name	**Process Time One**
Description	**Processing time on first server**
Category Name	**Process Logic**
Category Expanded	False
General	
Name	**ProcessTimeOne**
Required Value	True
Visible	True

Figure 10.5: Characteristics (properties) of Tandem Server process time one.

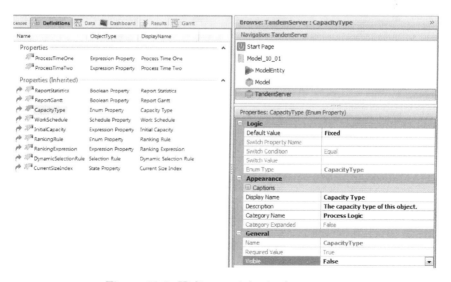

Figure 10.6: Hiding an inherited property.

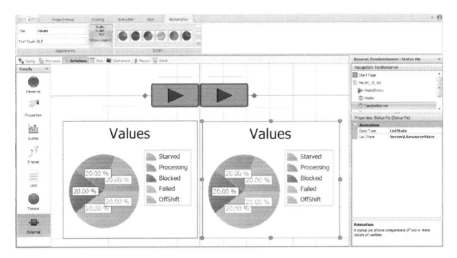

Figure 10.7: Defining the external view for Tandem Server.

the items and deselect the `Externally Visible` option for all of them at once.

- Select the External panel and place two Server symbols back to back using the Place Symbol Ribbon button.

- Place an External Node on the input side, with Node Class specified as `BasicNode`, Input Logic Type set to `FacilityNode`, Node specified as `Input@Server1` and give it a base Name of `Input`.

- Place a second External Node on the output side with Node Class specified as `TransferNode`, Input Logic Type set to `None` (no entry allowed on this side of the tandem server), and give it a base Name of `Output`.

- Using `Queue` button on the Animation ribbon, draw animated queues for the input buffer, the two processing stations, and the output buffer. As you place each queue select the Alignment of `None` on the Appearance ribbon. Specify the Queue States as `Server1.InputBuffer.Contents`, `Server1.Processing.Contents`, `Server2.Processing.Contents`, and `Server2.OutputBuffer.Contents` respectively.

- Using `Status Pie` button on the Animation ribbon, add two animated pie charts sized approximately 5 by 6 meters each. Specify the Data Type as ListState and List State as `Server1.ResourceState` and `Server2.ResourceState`, respectively. It should look something like shown in Figure 10.7.

Note that in our external view we are sending entities arriving to the associated node that is created by the Input node symbol to the node named Input@Server1 in our facility model of the TandemServer. Here they will proceed through the two Servers until they reach the output node of Server2, where we want to send them back up to the associated node defined by the external

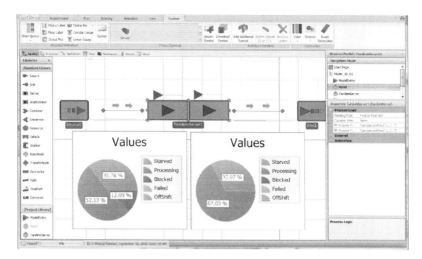

Figure 10.8: Completed Tandem Server in operation.

node symbol named Output. To inform Simio about the external output node, return to the Facility window for TandemServer and right-click on the output node for Server2. From the drop-down menu choose Bind to New External Output Node and specify the External Node Name as Output. This will cause the departing entities to be sent out from the associated output node that is created for the Output.

We will also edit the processing time for Server1 to be ProcessTimeOne and the processing time for Server2 to be ProcessTimeTwo. Do this by right clicking on each Processing Time property, select Set Referenced Property, then select the appropriate property that you defined earlier. Now these processing times will be specified by our newly created properties for TandemServer.

Summary

In this section we built an object using the hierarchical approach. We specified the object's logic in the Facility window. We defined the properties that we wanted users to see as well as hid some properties that we did not want users to see. Then we used the External view to define what we want users to see when they place our new object in their Facility window.

We are now ready to use our new object definition. Click on our main model (Model) to make it the active model, and then place a Source, TandemServer (selected from the Project Library on the lower left), and Sink, and connect them with Paths. If you click on the TandemServer you will now see our custom properties displayed in the Property window. Figure 10.8 shows our simple model employing our new TandemServer object in operation.

10.3   Model 10-2: Building a Base Object

In our previous example we built a new object definition using a facility model. We will now build a new object definition using a process model. Our new object will be a lathe (a type of machine tool) that can process one part at a time. Our lathe has an input buffer for holding waiting parts and an output buffer for parts waiting to exit to their next location.

10.3.1   Model Logic

Start with a new project (ModelEntity and Model). Add a new Fixed Class[2] model from the Project Home ribbon and rename it `Lathe`. Make Lathe the active model and click on the Properties panel in the Definitions window. We will define four new properties for our new object and hide inherited properties that we do not want the user to see:

- Add a new Standard Property of type Expression. Name it `TransferInTime`, and set its Default Value to `0.0`, UnitType to `Time`, Default Units to `Minutes`, Display Name to `Transfer In Time`, and Category Name to `Process Logic`.

- Add another new Standard Property of type Expression. Name it `ProcessingTime`, and set its Default Value to `Random.Triangular( 0.1, 0.2, 0.3)`, UnitType to `Time`, Default Units to `Minutes`, Display Name to `Processing Time`, and Category Name to `Process Logic`.

- Add two new Standard Properties of type Integer for specifying the buffer sizes. Name them `InputBufferCapacity` (with display name `Input Buffer`) and `OutputBufferCapacity` (with display name `Output Buffer`). Set their Default Values to `Infinity` and their Category Names to `Buffer Capacity`.

- Expand the inherited properties. Since the inherited properties named CapacityType, WorkSchedule, and InitialCapacity are not relevant for our Lathe object, set the Visible characteristic in the General category to `False` to hide each of these just as we did in the previous example (Figure 10.6).

To support building the logic in the Process window, we will first add the stations that we will be using. While still in the Definitions window, click on the Elements panel and add three station elements. Name them `InputBuffer`, `Processing`, and `OutputBuffer`. Use Specify the Initial Capacity for the `InputBuffer` station as the property named `InputBufferCapacity` (using the right click, Set Referenced Property approach as we did earlier) and the Initial Capacity for the `OutputBuffer` station as the property named

---

[2]You may note that there is a Processor option under Fixed Class. When building base objects, using this feature is often much simpler because it creates some framework for you automatically. So you can better understand how that framework is created, we are not using Processor in this example.

OutputBufferCapacity. Specify the Initial Capacity for the Processing station as 1.

Navigate to the Process window (still in the Lathe object). Now we will build process flows for moving through each of the three stations we have defined. Entities will enter our Lathe object and the InputBuffer station by executing the process we will name InputBufferTransferIn.

- Click on the Create Process button to create our first process. Name it InputBufferTransferIn. Set the Description to something like "Transfer in from outside the object. Hold the entity until the processing capacity is available."

- The token that executes this process is triggered by the InputBuffer.Entered event that is automatically generated each time an entity initiates entry into the station. Set the Triggering Event for the process to InputBuffer.Entered.

- Click and drag to add a Delay step, an EndTransfer step and a Transfer step to the process.

- Set the DelayTime to TransferInTime, the time each entity requires before its transfer into the station is considered complete. Using *F2*, rename[3] the Delay step to TransferringIn. The EndTransfer step signals to the outside world (perhaps a conveyor, robot, person, etc.) that the transfer is now complete. Once the transfer is complete a new transfer can begin as long as there is remaining space in the InputBuffer station.

- After the transfer in is complete, the token then initiates a transfer of the entity from the InputBuffer station to the Processing station. Set the Transfer step From property to CurrentStation, the To property to Station, and the Station Name property to Processing.

The completed InputBufferTransferIn process is shown at the top of Figure 10.9. The Processing station executes slightly different logic because it has no transfer in time, but does have a processing time:

- Click on the Create Process button to create our next process. Name it ProcessingTransferIn. Set the Description to something like Delay for processing.

- The token that executes this process is triggered by the Processing.Entered event that is automatically generated each time an entity initiates entry into the station. Set the Triggering Event for the process to Processing.Entered.

- Click and drag to add an EndTransfer step, a Delay step, and a Transfer step to the process.

---

[3]The step "name" you are changing is the unique label given to that step in that process. Using meaningful names not only makes the process easier to understand, but also makes the model trace easier to understand. You should use this feature routinely.

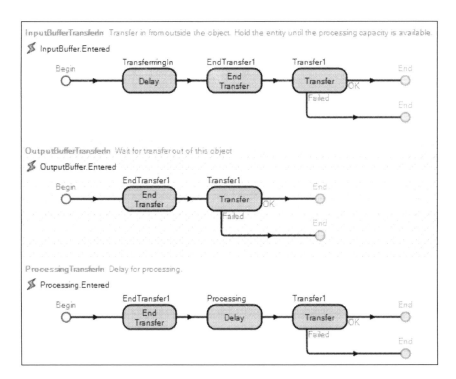

Figure 10.9: Three processes that define the behavior of the Lathe object.

- The EndTransfer step immediately ends the incoming transfer (there is no transfer-in delay). Once the transfer is complete a new transfer could potentially begin if there were any remaining space in the Processing station. Since the capacity is fixed at one in our object, the next transfer in cannot occur until the current entity has transferred out.

- Set the DelayTime to ProcessingTime, the time the user has specified for processing. Using *F2*, rename the Delay step to Processing.

- After the processing delay is complete, the token then initiates a transfer of the entity from the Processing station to the OutputBuffer station. Set the Transfer step From property to CurrentStation, the To property to Station, and the Station Name property to OutputBuffer.

The completed ProcessingTransferIn process is shown in bottom of Figure 10.9.

Our shortest and simplest process is for the OutputBuffer station which simply passes the outgoing entity out of the object:

- Click on the Create Process button to create our last process. Name it OutputBufferTransferIn. Set the Description to something like Wait for transferring out.

- The token that executes this process is triggered by the OutputBuffer.Entered event that is automatically generated each time an entity initiates entry into the station. Set the Triggering Event for the process to `OutputBuffer.Entered`.

- Click and drag to add an EndTransfer step and a Transfer step to the process.

- The EndTransfer step immediately ends the incoming transfer (there is no transfer-in delay).

- The token initiates a transfer of the entity from the OutputBuffer out through the associated node object that is defined by the ParentExternalNode named Output. Set the Transfer step From property to `CurrentStation` and the To property to `ParentExternalNode`. (Note: We will not be able to specify the External Node Name property until we have first defined it in the external view.)

The completed `OutputBufferTransferIn` process is shown in the middle (shaded) section of Figure 10.9.

10.3.2   External View

Next we will define the external view. Click on the External panel in the Definitions window.

- Place a lathe symbol from the symbol library in the center of the external view.

- Add an external node from the Draw ribbon to the left side of the lathe. Select Node Class of `BasicNode` because this will be for input only. Name this node `Input`.

- Add an external node to the right side of the lathe. Select Node Class of `TransferNode` because we want the extended capabilities that offers on the output side. Name it `Output`.

- Specify the Input Logic Type on the Input external node as `ProcessStation`, and then select `InputBuffer` from the drop list. Note that arriving entities will transfer to the InputBuffer station upon arrival to the associated node object corresponding to this external node symbol.

- Add an animated queue to the left of the lathe symbol for animating the InputBuffer station. Specify the Queue State of `InputBuffer.Contents`.

- Add an animated queue to the right of the lathe for animating the OutputBuffer station. Specify the Queue State of `OutputBuffer.Contents`.

- Add an animated queue adjacent to the lathe for animating the Processing station. Specify the Queue State of `Processing.Contents`. Use the Shift-Drag technique we learned in Chapter 8 to raise this queue and place it in the throat of the lathe.

Our external view for the Lathe is shown in Figure 10.10. Since we have defined our external nodes, we can now go back to the process window and specify our ParentExternalNode named `Output` on our OutputBuffer Transfer step).

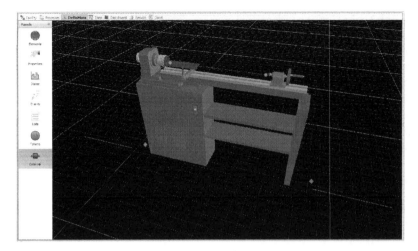

Figure 10.10: External view of lathe object.

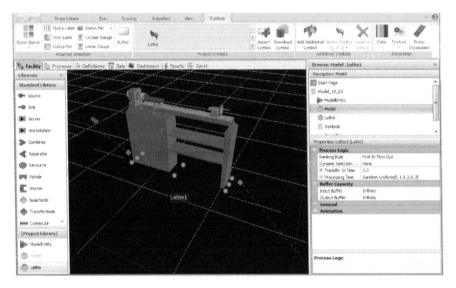

Figure 10.11: Completed lathe object used in a model.

Summary

We are now ready to use our new object definition in a model. Click on Model in the navigation window to make it our active model and drag out a Source, Lathe (from the Project Library), and Sink, and connect them with Paths. Click on Lathe to edit its properties. You can now run the model to see something like illustrated in Figure 10.11.

10.4   Model 10-3: Sub-Classing an Object

In this example we are going to build a new object definition named MRI that represents a medical device (a magnetic resonance imaging device). Since this device has many behaviors in common with the Standard Library Server, we will subclass it from Server. We will hide and rename some properties, and add a new property to specify an optional repair person that is required to perform a repair on the MRI. We will then modify the repair logic to seize and release this repair person.

10.4.1   Model Logic

Beginning with a new project we right-click on Server in the Facility window and select `Subclass` (we could do this same operation in the Project window). This adds a new model to our project named MyServer, which we will rename MRI.

We need to add a new property so that the user can specify which repair person to use:

- Click on MRI to make it the active model, and then click on the Properties panel in the Definitions window.

- Add a new object reference property, name it `RepairPerson`, and specify its category as `Reliability Logic`.

- Specify the Switch Property Name as `Failure Type`, the Switch Condition as `NotEqual`, and the Switch Value as `NoFailures`. Hence this property will not be visible if the `NoFailures` option is selected by the user.

- We will also set the Required Value flag to `False` for this property.

This property definition is shown in Figure 10.12.

We want to change the terminology used in this object from generic to medical. Change the Display Name of the ProcessingTime property to `Treatment Time`, and set the description for this property to `The time required to process each patient`. The ProcessingTime property should now look like Figure 10.13. We will also hide the inherited properties named Capacity Type, WorkSchedule, and InitialCapacity so they do not clutter the object.

Next we will modify the process logic that we have inherited to make use of the repair person during maintenance. Click on the Processes window to display the inherited processes from the base Server definition. Note that you cannot edit any of the processes at this point because they are owned by the Server and are inherited for use by the MRI. This inheritance is indicated by the green arrow icon on the left that points up. We will select the first inherited process in MRI named FailureOccurrenceLogic and then click on Override in the Process ribbon. This will make a copy of the inherited process from the Server for use by the MRI in place of the same process that is owned by the Server. We can then edit this overridden process in any way that we wish. This override is indicated by the arrow icon on the left that is now pointed down.

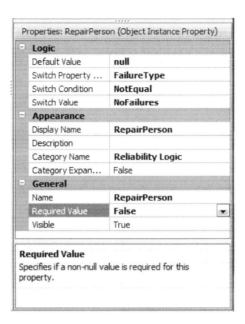

Figure 10.12: Property definition of RepairPerson.

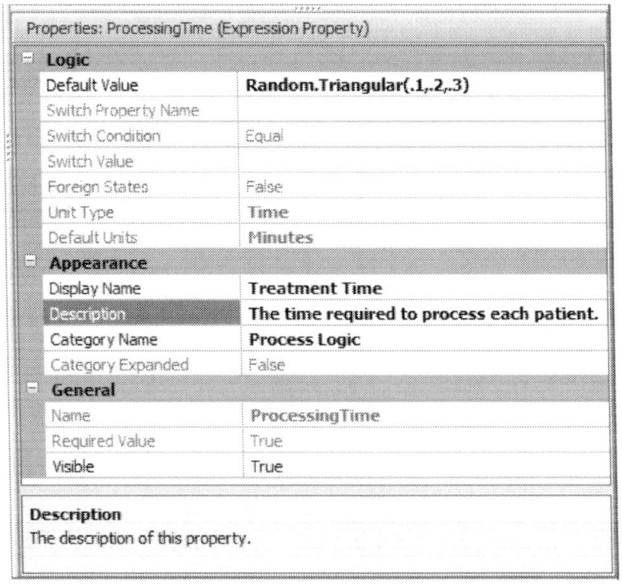

Figure 10.13: Property definition of ProcessingTime to appear as Treatment Time.

Figure 10.14: FailureOccurrenceLogic overridden to add repair person.

We could also restore back to the original process by clicking on Restore in the Process ribbon.

Now that we can change the logic in this process, we will add a Seize step immediately before the TimeToRepair delay, and then a Release step immediately after this delay. In both steps we will specify the RepairPerson property as the object to seize/release (use the right-click and Set Reference Property). The beginning of our revised process is shown in Figure 10.14.

10.4.2 External View

Next we will customize the external view for the MRI. We need to customize the animation in our user view similarly to what we did in previous objects. But we do not need to worry about the nodes and framework because it was inherited from the server object.

Click on the External panel in the Definitions window. In the Place Symbol drop list select Download Symbol to search for and download an MRI graphic from Google Warehouse. Add animated queues for the queue states Input-Buffer.Contents, Processing.Contents, and OutputBuffer.Contents. Note that the two external node symbols are inherited from the Server and cannot be modified beyond adjusting their positions.

Summary

We are now ready to use our new MRI object definition in a model. Click on Model to return to our main model, and drag out a Source, MRI (from the Project Library), and a Sink, and connect them with Paths. Select MRI to edit its properties and then run. Figure 10.15 shows this simple model in operation.

10.5 Working With User Extensions

You have already learned to build models using the Standard Library, extend the Standard Library using add-on processes, and (earlier in this chapter) to build your own custom objects. If you have a moderate level of programming skills, Simio goes one step beyond that to allow you to actually extend the underlying simulation engine itself. Simio's architecture provides many points where users can integrate their own custom functionality written in a .NET language such as Visual C# or Visual Basic .NET.

The types of user extensions supported include:

- User Defined Steps
- User Defined Elements

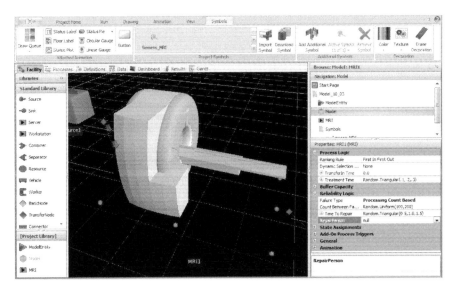

Figure 10.15: New MRI object in use and illustrating custom properties.

- User Defined Selection Rules

- Design Time Add-Ins

- Table Imports and Binding

- Design or Run Experiments

Simio's extension points have been exposed as a set of interfaces that describe the methods and calling conventions to be implemented by any user-extended components. This set of interfaces is referred to as the Simio application programming interface (API). For detailed information on the Simio API, see the API Reference.

While you can create extensions using any .NET programming language, Simio provides extra support for C# users. To help you get started creating Simio user extensions, a number of predefined Simio project and project item templates for Microsoft Visual Studio® 2008 are available. These templates provide reusable and customizable project and item stubs that may be used to accelerate the development process, removing the need to create new projects and items from scratch.

In addition, several user extension examples are included with the Simio installation. You can explore these examples and possibly even customize them to solve your own problems. The examples include:

- Binary Gate - An element and three steps for controlling flow through a gate.

- TextFileReadWrite - An element and two steps for reading and writing text files.

- CSVGridDataProvider - Supports import and export between tables and text files.

- ExcelGridDataProvider - Supports import and export between tables and Excel files.

- SelectBestScenario - Illustrates an experiment data analysis add-in.

- SimioSelectionRules - Contains the implementation of all of Simio's dynamic selection rules.

- SourceServerSink - Illustrates a design time add-in that builds a simple facility model from code.

These examples can be found in a UserExtensions subfolder of the Simio example models - typically under a Public or All Users folder.

10.5.1 How to Create and Deploy a User Extension

The general recommended steps to create and deploy a user extension are as follows:

- Create a new .NET project in Visual Studio 2008, or add an item to an existing project, using one of the Simio Visual Studio templates. Note that, in addition to the commercial versions of Visual Studio, Microsoft also offers Express editions which are available as free downloads from *www.microsoft.com/express/Windows*.

- Complete the implementation of the user extension and then build the .NET assembly (.dll) file.

- To deploy the extension, copy the .dll file into the UserExtensions folder located in the Simio install directory (typically under Program Files).

A correctly deployed extension will automatically appear at the appropriate location in Simio. In some cases these are clearly identified as user add-ins, for example:

- In a model's Processes Window, all user defined steps will be available from the left hand steps panel under User Defined.

- In a model's Definitions Window, when defining elements, all user defined elements will be available via the User Defined button in the Elements tab of the ribbon interface.

In other cases it appears as though Simio has new features, for example:

- User defined selection rules are available for use in a model as dynamic selection rules.

- Application add-ins are available for use in a project via the Select Add-In button in the Project Home tab of the ribbon.

You can find additional information on this topic by searching the main Simio help for "extensions" or "API". These topics provide a general overview and introduction to the features. More detailed information is available in the help file `Simio API Reference Guide.chm` that can be found in the Simio folder under Program Files. This provides over 500 pages of very detailed technical information. Although as of this writing there is no training course provided for creating Simio user extensions, an appendix in the `Learning Simio` slide set does provide additional step by step instructions. If you are a *Simio Insider*[4] (which we strongly encourage) you can find additional examples and discussions in the topics *Shared Items* and *API*.

10.6 Summary

In this chapter we have reviewed the basic concepts for building object definitions in Simio. This is a key feature of Simio because it allows non-programmers to build custom objects that are focused on a specific model or application area.

In our examples here we placed our objects from the Project Library. The other alternative is to build the object definitions in their own project and then load that project as a library from the Project Home ribbon.

With Simio you often have the choice to make between working with the Standard Library embellished with add-on processes, and creating a new library with custom logic. If you encounter the same applications over and over again it is more convenient to build custom objects and thereby avoid the repeated need for add-on processes. Of course you can also provide support for add-on processes in your own objects by simply defining a property for the process name and passing that property to an Execute step within your process.

With a bit of experience you will find object building to be a simple and powerful capability for your modeling activities.

10.7 Problems

1. Compare and contrast the three ways of creating model logic. What determines your choice when creating a specific object?

2. What is the difference between the *External View* and the *Dashboard*[5] and what are the limitations of each?

3. When you have built a model, say a paint line, what are the typical changes required to make it suitable for reuse as an object in a larger model (e.g., a car plant containing many similar but not identical paint lines)?

4. Reproduce Model 10-2, but this time start with a *Fixed Class - Processor*. What advantages and disadvantages does that approach provide?

---

[4]You can become a Simio Insider at `www.simio.com/forums/`

[5]Renamed Console in Simio version 6.

5. Add a time-based failure to Model 10-2. Provide options for the Time Between Failures, Time To Repair, and optional Repair Resource. Include an external node where the repair operator will come to do the repair.

   Use this object in a model demonstrating its full capability.

6. Create a Palletizer object with one input node for incoming boxes and one output node for the outgoing palletized boxes. Pallets are created on demand within the Palletizer. Boxes can be added to the pallet in a user-specified quantity.

   Demonstrate your Palletizer object in a model with two types of boxes. BoxA can be stacked 5 to a pallet, the smaller BoxB can be stacked 10 to a pallet. The full pallets are transported to shipping where the boxes are removed from the pallets and all boxes and pallets are counted.

7. Create your own add-in for use in the experiment window. Your add-in should read scenario data from an external data file and then create appropriate scenarios in the Experiment Design view.

Chapter 11

Case Studies Using Simio

This chapter includes four "introductory" and two "advanced" case studies involving the development and use of Simio models to analyze systems. These problems are larger in scope and are not as well-defined as the Problems in previous chapters. For the first two cases, we've also provided some analysis results based on our sample models. However, unlike the models in previous chapters, we do *not* provide detailed descriptions of the models themselves.

The first case (Section 11.1) describes a manufacturing system with machining, inspection, and cleaning operations, and involves analyzing the current system configuration along with two proposed improvements. The second case (Section 11.2) analyzes an amusement park and considers a "Fast-Pass" ticket option and the impact it might have on waiting times. The third case (Section 11.3) models a restaurant and asks about the important issues of staffing and capacity. The fourth case (Section 11.4) considers a branch office of a bank, with questions about how scheduling and equipment choice affect costs.

The two advanced case studies are intended to represent larger realistic problems that might be encountered in a non-academic setting. In the "real world," problems are typically not as well-formed, complete, or tidy as they are in typical homework Problems. You often face missing or ambiguous information, tricky situations, and multiple solutions based on problem interpretation. You may expect the same from these cases. As these situations are encountered, you should make and document reasonable assumptions.

11.1 Machining-Inspection Operation

11.1.1 Problem Description

You've been asked to evaluate your company's current machining and inspection operation (see Figure 11.1) and investigate two potential modifications, both of which involve replacing the inspection station and its single worker with an automated inspection process. The goal of the proposed modification is to improve system performance.

In the current system, four different part types arrive to an automated machining center, where they're processed one part at a time. After completing machining, parts are individually inspected by a human inspector. The inspec-

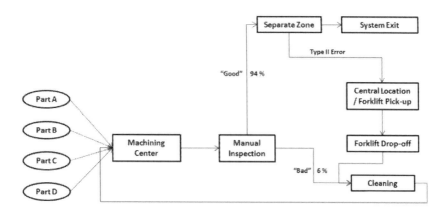

Figure 11.1: Machining-inspection operation layout.

Table 11.1: Arrival and machining data.

Type	Arrival Rate (parts/hour)	Machining-Time Distribution
A	5.0	tria(2.0, 3.0, 4.0) minutes
B	6.0	tria(1.5, 2.0, 2.5) minutes
C	7.5	expo(1.5) minutes
D	5.0	tria(1.0, 2.0, 3.0) minutes

tion process is subject to error — classifying "good" parts as "bad" (type I error) or classifying "bad" parts as "good" (type II error). After inspection, parts classified as "good" are sent to another zone in the plant for further processing, while parts classified as "bad" must be cleaned in an automatic cleaner (with a capacity of one part) and then re-processed by the machining center (using the same processing parameters as "new" parts).

Additional details of the system include:

- Part arrival, machining time, and cleaning time process details are given in Table 11.1. All four arrival processes are stationary Poisson processes, and independent of each other.

- All parts go to inspection immediately following machining. The time to inspect a part is independent of the part type and follows an exponential distribution with mean 2 minutes. 6% of inspected parts are found to be "bad" and must be cleaned and then re-processed (this is independent of the number of times a part has been cleaned and re-processed). Parts that fail inspection have priority over the "new" parts in the machining queue after they're cleaned. It takes 12 minutes to clean each part (the cleaning time is also independent of part type).

- Out of all the parts that are deemed "good" and sent to another zone, 3% are later found to be "bad" (type II error). These parts are put on a pallet in a central location where a forklift will periodically pick up and deliver the pallet to the cleaning machine. These parts are given priority

Table 11.2: System and operation costs.

Item	Cost
Unnecessary cleaning due to type I error	$5/part
Simple robot	$1,200/week each
Complex robot	$5,000/week
Holding cost per part	$0.75/hour spent in system
Inspection worker	$50/hour

in the automatic cleaner queue. The time between forklift pickups is exponentially distributed with mean 3 hours.

- Out of all the parts that are deemed "bad" and sent to the automatic cleaner, 7% were actually good and did not require cleaning and re-machining (type I error). The per-part cost associated with the unnecessary cleaning and processing is given in Table 11.2.

- The machining center and cleaning machine are both subject to random, time-dependent failures. For both machines, the up times follow an exponential distribution with mean 180 minutes, and the repair times follow an exponential distribution with mean 15 minutes. If a part is being processed when either of the machines fails, processing is interrupted but the part is not destroyed.

- The system operates for two shifts per day, six days per week. Each shift is eight hours long and work picks up where it left off at the end of the previous shift (i.e., there's no startup effect at the beginning of a shift).

The two scenarios you've been asked to evaluate both concern possible automation of the inspection operation:

1. Installation and use of four "simple" robots, one for each part type. These robots inspect one part at a time. The probabilities of type I and type II errors are 0.5% and 0.1%, respectively. The time for each "simple" robot to inspect a part is independent of the part type and follows a triangular distribution with parameters (5.5, 6.0, 6.5) minutes.

2. Installation and use of one "complex" robot that can inspect all four different part types, but can inspect only one part at a time. The probabilities of type I and type II errors are 0.5% and 0.1% respectively. The time for the "complex" robot to inspect a part is independent of the part type and follows the a triangular distribution with parameters (1.0, 2.0, 2.5) minutes.

Since these robots are manufactured by the same company as the machining center and automatic cleaner, the robots are assumed also to be subject to random failures with the same probability distributions as given above for the machining and cleaning machines. The cost data to be used in your analysis are in Table 11.2. Economic analysis for purchasing the robots has already been developed, and the associated costs (purchase price, installation, operation and maintenance, etc.) are given as a single weekly cost in Table 11.2.

Develop a Simio simulation to analyze and determine which inspection process your company should use. Your results should include estimates of the following for each scenario:

- Average total cost incurred each week.

- Average numbers of type I and type II errors each week.

- Average total number of parts that finish processing (exit the system) each week.

Using your Simio model, answer the following questions. Your answers should be supported by results and analysis from your simulation experiment(s).

1. After looking at the specifications for the robots, the manager has noticed that the average time of inspection for a "simple" robot is much higher than that of the current process. He is nervous that changing the current inspection process might actually *decrease* overall production (i.e., number of completed parts per week). Will changing the method of inspection change the number of parts that are completed each week? If so, describe how the number of completed parts changes with respect to the current system (i.e., amount of increase/decrease).

2. Given the three methods of inspection, which operation do you recommend to minimize costs?

3. If you find that one or both of the automatic inspection processes costs the same as, or more than, manual inspection, determine what would enable that inspection process to operate at a weekly cost that's lower than that with manual inspection. Assume that all costs are fixed and can't be adjusted.

11.1.2 Sample Model and Results

This section gives the results for a sample model that the authors built. Note that the results depend on a number of assumptions, so your results might not match exactly. The Facility View of the sample model is shown in Figure 11.2. For our experimentation, we used the following controls and parameters:

- Run length: 192 hours

- Warm-up period: 96 hours

- Number of replications: 2,000

- Experiment Controls:

    - Automatic Inspection (boolean operator)
    - One Inspection Machine (boolean operator)

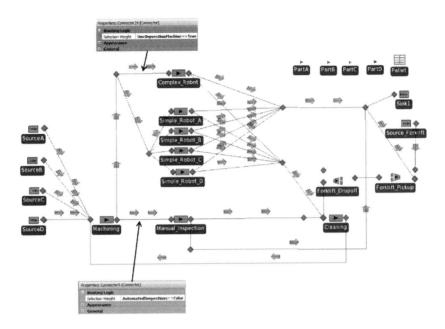

Figure 11.2: Facility View of the sample machining-inspection model.

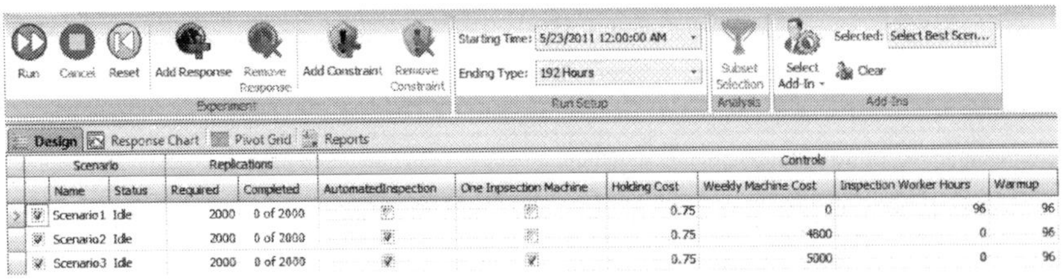

Figure 11.3: Snapshot of the machining-inspection Experiment Design.

  – Weekly Machine Cost

  – Inspection Worker Hours

  – Warm-up

Figure 11.3 shows how the different controls were used in the Simio Experiment to govern a given scenario's activity and total cost. We essentially coded all three options into a single model so that we could use Simio's scenario-comparison capabilities in our analysis. In order to compare the different scenarios in the same experiment, the boolean operators `Automatic Inspection` and `One Inspection Machine` were used. When the box associated with a control is checked, the property is given the value of *True*. The model references these values in the selection-weight property of the connectors before the different inspection processes. Two examples of this are called out in Figure 11.2.

Table 11.3: Average total costs for each scenario.

Inspection scenario	Cost	Half width
Current manual	$7,303.31	$63.20
Automated: four "simple" robots	$7,268.96	$58.13
Automated: one "complex" robot	$7,119.54	$56.37

The expression that we use for the total weekly cost is:

$$
\begin{aligned}
TotalCost = \ & WeeklyMachineCosts \\
& + \ WeeklyHoldingCost.Value \\
& + \ TotalTypeIErrors.Value * 5 \\
& + \ InspectionWorkerHours * 50
\end{aligned}
\tag{11.1}
$$

The methods for calculating each of the four sections of equation (11.1) are:

1. **WeeklyMachineCosts:** These costs were given in the problem description. The property `WeeklyMachineCosts` is a standard numeric property that's controlled in the experiment.

2. **WeeklyHoldingCost.Value:** This is the value of the output statistic that references the model state `HoldCost`. Equation (11.2) is used to increment the state `HoldCost` each time an entity enters the sink (exits the system):

$$
HoldCost = (TimeNow - ModelEntity.ArrToSys) * 0.75 + HoldCost.
\tag{11.2}
$$

   Equation (11.2), in essence, calculates the total time the entity spent in the system and multiplies that value by the holding cost given in the problem description. The state is then incremented using this value. Using this method, the output statistic `WeeklyHoldingCost` will be the sum of the holding costs for all entities.

3. **TotalTypeIErrors.Value*5:** The cost of $5 was given in the problem description as the cost associated with a type I error. `TotalTypeIErrors` is an output statistic that references the model state, `TypeIError`. This state is incremented when a type I error is committed in the inspection process.

4. **InspectionWorkerHours*50:** The cost of $50 per hour was given in the problem description. The property `InspectionWorkerHours` is a standard numeric property that's controlled in the experiment.

Using the expression in equation (11.1) and the experiment parameters listed above, the average weekly cost for each scenario was calculated. These results are shown in Table 11.3 (all half widths are of 95% confidence intervals on the expected value). The average weekly totals of type I and type II errors for each inspection process are given in Table 11.4. The average total number of parts that finish processing (exit the system) each week can be seen in Table 11.5.

Table 11.4: Average type I and type II errors for each scenario.

Inspection scenario	Type I errors	Type II errors
Current manual	10.23	69.93
95% CI	[10.09, 10.37]	[69.55, 70.31]
Automated: four "simple" robots	1.02	2.26
95% CI	[0.97, 1.06]	[2.19, 2.32]
Automated: one "complex" robot	1.04	2.24
95% CI	[1.00, 1.09]	[2.17, 2.30]

Table 11.5: Average number of parts processed weekly for each scenario.

Inspection scenario	Finished parts	Half width
Current manual	2,255.29	2.21
Automated: four "simple" robots	2,253.88	2.13
Automated: one "complex" robot	2,255.95	2.18

Answers to Questions

1. Will changing the method of inspection change the number of parts that are completed each week? If so, describe how the number of completed parts changes with respect to the current system (i.e., amount of increase/decrease).

Although the answer to this question could be obtained through simple queueing analysis[1], we can use the results from our simulation experiment to answer this question as well. As seen in the SMORE plot in Figure 11.4, it appears that the average total number of parts processed per week does not vary much across the scenarios. To verify that there is no *statistically significant* difference between the mean number of completed parts per week in scenario 1 (current manual inspection) on the one hand, vs. each of scenario 2 (automated inspection using four "simple" robots) and scenario 3 (automated inspection using one "complex" robot), two different paired t tests were conducted. The results from these tests are in Figures 11.5 and 11.6.

Since the 95% confidence interval of the mean difference, $[-1.54, 4.38]$, contains zero, we conclude that there is no statistically significant difference between scenarios 1 and 2 with respect to the average numbers of parts processed per week. The same determination can be made from Figure 11.6, where scenarios 1 and 3 are compared. Again, there is no statistically significant difference between scenarios 1 and 3 with respect to the average numbers of parts processed per week since the 95% confidence interval of the mean difference, $[-3.70, 2.38]$, contains zero. With this statistical evidence, we can conclude that changing the method of inspection will *not* significantly change the number of parts completed each week.

Another approach to compare these scenarios involves Simio's *Scenario Subset Selection* capability. When the objective functions of the responses are set

---

[1]For any stable system, flow in equals flow out, so the expected number of parts will simply be the arrival rate times the duration.

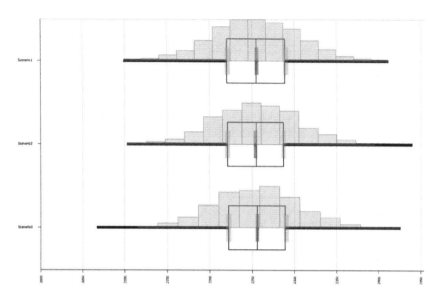

Figure 11.4: SMORE plot of the average number of completed parts per week for each scenario.

Paired T-Test and CI: scenario1, scenario2

```
Paired T for scenario1 - scenario2

                 N      Mean   StDev  SE Mean
scenario1     2000   2255.29   50.35     1.13
scenario2     2000   2253.88   48.54     1.09
Difference    2000      1.42   67.48     1.51


95% CI for mean difference: (-1.54, 4.38)
T-Test of mean difference = 0 (vs not = 0): T-Value = 0.94  P-Value = 0.348
```

Figure 11.5: Minitab results for a paired t test comparing scenarios 1 and 2.

Paired T-Test and CI: scenario1, scenario3

```
Paired T for scenario1 - scenario3

                 N      Mean   StDev  SE Mean
scenario1     2000   2255.29   50.35     1.13
scenario3     2000   2255.95   49.75     1.11
Difference    2000     -0.66   69.33     1.55


95% CI for mean difference: (-3.70, 2.38)
T-Test of mean difference = 0 (vs not = 0): T-Value = -0.43  P-Value = 0.671
```

Figure 11.6: Minitab results for a paired t test comparing scenarios 1 and 3.

Table 11.6: Responses and associated objectives.

Response	Objective
Total processed	Maximize
Type I errors	Minimize
Type II errors	Minimize
Time in system (TIS)	Minimize
Total cost	Minimize
Holding cost per week	Minimize

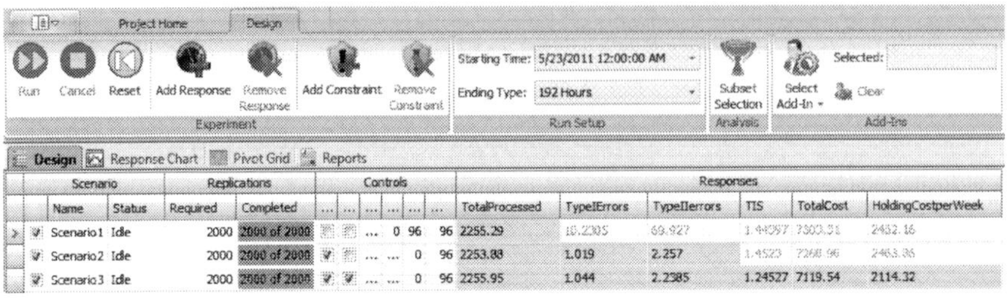

Figure 11.7: Results for analysis of several responses using Simio's Subset Selection.

according to Table 11.6, and the "Subset Analysis" is activated, Simio uses a ranking-and-selection procedure to group the response values into subgroups, "Possible Best" and "Reject," within each response. In the Experiment Window's Design window, the cells in a Response column will be highlighted brown if they're considered to be "Rejects," leaving the "Possible Best" scenarios with their intended coloring. The results obtained through this analysis method are in Figure 11.7. Simio's analysis using Scenario Subset Selection gives us the same results as did our paired t tests. The response, `TotalProcessed`, for each scenario has not been highlighted brown; therefore, Simio's Subset Selection has identified each of the three scenarios as a "Possible Best" for maximizing the average weekly number of parts processed. This analysis verifies our expectation that there is no significant difference in the numbers of parts that are processed each week.

2. *Given the three methods of inspection, which operation do you recommend to minimize costs?*

By examining the SMORE plot shown in Figure 11.8, it's not clear which inspection method would be the least expensive. To determine if there's any statistically significant difference between the weekly costs for the different scenarios, we again use Simio's Scenario Subset Selection. Figure 11.7 shows that the response values associated with the total cost of scenarios 1 and 2 are both highlighted in brown, indicating a "Reject," and the response value associated with the total cost of scenario 3 has been identified as a "Possible Best." Based on this information, and the fact that the overall production does not decrease, we can recommend that the company should change their inspection process to use the one "complex" robot.

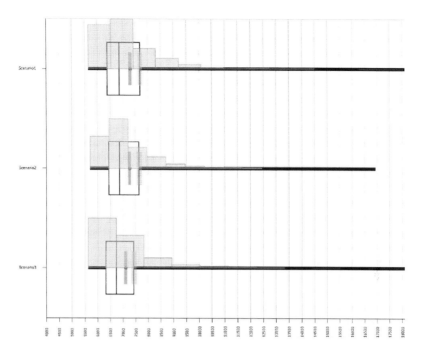

Figure 11.8: SMORE plot of the total cost for each scenario.

3. If you find that one or both of the automatic inspection processes costs no less than manual inspection, determine what would enable that inspection process to operate at a weekly cost that's lower than that with manual inspection. Assume that all costs are fixed and can't be adjusted.

As discussed above, there is no statistically significant difference between the weekly costs of scenarios 1 and 2. To determine what would lower the costs of this inspection process, we consider the elements that comprise the total cost in equation (11.1). The three costs that are driving scenario 2's total cost are the machine cost, the cost associated with type I errors, and the holding cost. Since lowering the fixed machine cost is not an option and the type I errors are only costing about $5 per week, it seems that we must lower the holding cost in order to lower the total cost of scenario 2. Since the holding cost is directly related to the amount of time a part spends in the system, reducing the processing time for the four simple machines may seem like a reasonable solution. This hypothesis is also supported by the results in Figure 11.7, since the scenario with the lowest total cost also possessed the lowest weekly holding cost and lowest average time that an entity spends in the system. To test this hypothesis, we set up an experiment where we systematically reduced the processing time for the "simple" robots (see Table 11.7).

As seen in Figure 11.9, as the average processing time for the "simple" robots decreases, the total cost decreases as well.

Using the Simio Scenario Subset Selection, we can further analyze the effects

Table 11.7: Scenarios for the new experiment to determine the effects of changing the process time of the four simple machines.

Scenario	"Simple" robot inspection time
Original scenario 1 (manual inspection)	N/A
Original scenario 2	tria(5.5, 6.0, 6.5)
Original scenario 3 (automated "complex" inspection)	N/A
Scenario 4	tria(5.4, 5.9, 6.4)
Scenario 5	tria(5.3, 5.8, 6.3)
Scenario 6	tria(5.2, 5.7, 6.2)
Scenario 7	tria(5.1, 5.6, 6.1)
Scenario 8	tria(5.0, 5.5, 6.0)

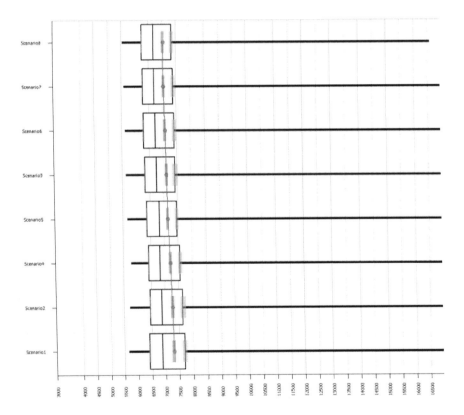

Figure 11.9: SMORE plot for the total cost for each of the eight scenarios. As seen in Table 11.7, scenarios 4–8 use "simple" robot inspection times with decrements of 0.1 minute.

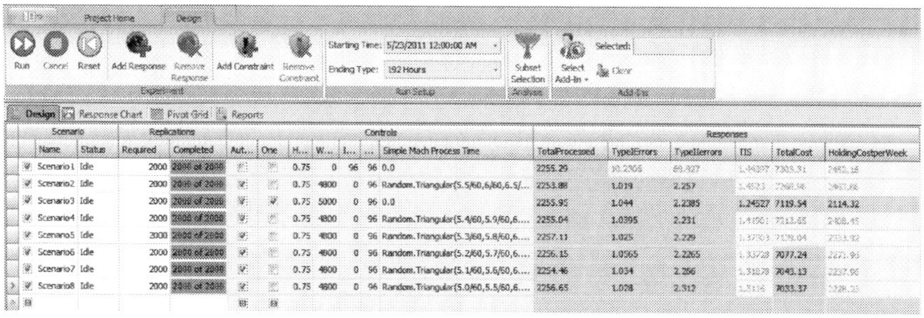

Figure 11.10: Results for analysis of several responses for eight scenarios using Simio's Scenario Subset Selection.

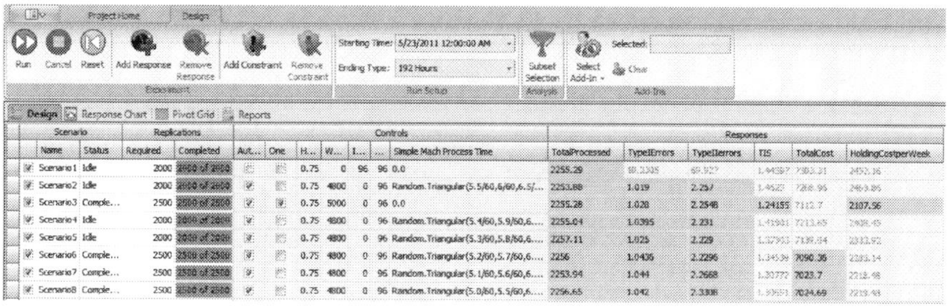

Figure 11.11: Results for analysis of several responses for eight scenarios using Simio's Scenario Subset Selection. The number of completed replications for scenarios 3, 6, 7, and 8 were increased to 2500.

of altering the processing times for the "simple" robot. Figure 11.10 provides the results of the Subset Selection when the objectives of the Responses are set according to Table 11.6.

As shown in Figure 11.10, the average total cost of scenario 1 is statistically significantly different from that of scenarios 6–8. Therefore, if the manufacturer of the robots can reduce the average processing time by 0.3 minute (through the distribution tria(5.2, 5.7, 6.2)), the inspection operation using the four "simple" robots will yield a lower weekly cost than that of manual inspection. Figure 11.10 also shows that there are now four "Possible Best" scenarios: 3, 6, 7, and 8. If we increase the number of replications from 2000 to 2500 for each of these scenarios, and use Scenario Subset Selection, scenario 3 is removed from the subgroup of "Possible Best" (see Figure 11.11). With this information, it's clear that if the manufacturer of the robots can reduce the "simple" robot's average processing time by 0.3 minute (through the distribution tria(5.2, 5.7, 6.2)), we could recommend the use of the four "simple" robots for part inspections.

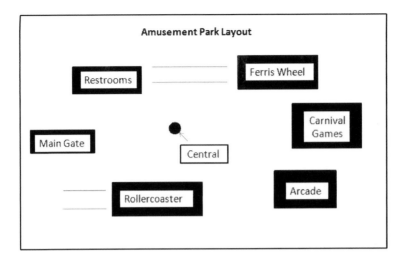

Figure 11.12: Amusement-park layout.

Table 11.8: Amusement park arrival-rate schedule.

Time interval	Customers per hour
10:00 am - 12:00 pm	75
12:00 pm - 2:00 pm	150
2:00 pm - 4:00 pm	100
4:00 pm - 6:00 pm	75

11.2   Amusement Park

11.2.1   Problem Description

An amusement park consists of a roller coaster, a ferris wheel, an arcade, carnival games (one area), concessions, and a bathroom area (see Figure 11.12 for the basic layout of the park). The park's hours are from 10:00 a.m. to 10 p.m. Customers start arriving when the park opens at 10:00 a.m. and stop arriving at 6:00 p.m. Customers arrive according to a non-stationary Poisson process with rate function in Table 11.8.

Once a customer is admitted to the park, he/she has the opportunity to purchase a "Fast Pass," which allows the customer to have priority at the roller coaster and ferris wheel. The manager of the park has had complaints about wait times from customers since the addition of the Fast Pass and has requested assistance with managing the Fast-Pass process. The manager's objective is to maximize overall profits of the Fast Pass, subject to the constraints of wait times for Fast-Pass and regular customers.

Additional Information

- If there are no Fast Passes handed out for the day, customers typically wait about 20 minutes for rides (this is historical performance information

Table 11.9: PMF for customer times in the park.

Hours in park	Probability
0.5	0.01
1.0	0.14
2.0	0.25
4.0	0.40
5.0	0.20

from the current system). While customers are accustomed to waiting at amusement parks, they become extremely dissatisfied after waiting longer than 60 minutes.

- Lines that are "too long" discourage regular customers from entering the respective queue, but Fast-Pass customers have a separate queue and are never discouraged from entering their line. If there are more than 100 people at the roller-coaster line or more than 80 people in the ferris-wheel line, regular customers become discouraged and find something else to do.

- The Fast-Pass customers are willing to wait between 5 and 10 minutes at rides, but become dissatisfied after waiting more than 10 minutes.

- 50% of customers have no interest in going to the arcades.

- 50% of customers will go to the bathroom at most once, and 50% will go at most twice. There are only two bathrooms (capacity of 5 each), currently with an average wait of around 9 minutes over the day.

- Customers do not go to the food area more than once.

- The expected time customers spend in the park varies by customer. Table 11.9 gives the probability mass function (PMF) for times spent in the park, which we treat as a discrete random variable with possible values 0.5, 1, 2, 4, or 5 (in hours).

Customer Travel Times and Area Preferences

Customers choose their next destination based on availability and expected wait times. For the bathroom and concession areas, customers are limited to once or twice per visit (as described above). The park has a centrally located "big board" that shows the current expected waiting time for all rides. All customers visit the central area of the amusement park between each area visit to view the big board to help decide where to go next. All areas that are available for the particular customer are equally probable. For example, if the roller coaster has a very long line and the customer has already eaten, and has no interest in the arcade, then the customer has the same probability of going next to the carnival area, bathroom, or ferris wheel. Additionally, when a customer's time to stay in the park has been surpassed, the path to the exit will become an available option. Customer walk times to and from the central area and all other areas are in Table 11.10. The *offset* mentioned in the table represents the minimum time

Table 11.10: Customer walk times (in minutes).

Area	Travel time (offset, additional exponential mean)
Roller coaster	[1, 1]
Ferris wheel	[2, 1]
Carnival games	[2, 2]
Arcade	[3, 2]
Bathroom	[1, 1]
Entrance/exit	[3, 1]

Table 11.11: Ride-area information (times in minutes).

Ride	Available Cars/ Buckets	Ride Cap.	Queue Cap.	Ride Time	Time to Seat	Time to Unseat
Roller coaster	2	20	100	4	tria(5, 10, 13)	0.5
Ferris wheel	30	60	80	10	1.2	0.5

Table 11.12: Other area information (times in minutes).

Area	Distribution of time spent in area	Capacity	Buffer capacity
Carnival games	tria(5, 10, 30)	40	40
Arcade	tria(1, 5, 20)	30	5
Bathroom	tria(3, 5, 8)	10	30
Concessions	tria(1, 4, 8)	4	25

it takes for customers to reach the destination; the second number represents the mean of the exponentially distributed additional component of the walk time beyond the offset.

Ride Areas

The ride-area information is in Table 11.11. A car denotes the extra roller coaster car available for loading while the first car is in use. A ferris-wheel bucket can seat two people at a time.

Other Areas

The other four areas have unique service times and capacities as shown in Table 11.12. The buffer capacity is the maximum number of people who will wait in that area while not being served (e.g., at most five people will wait for an arcade game when all games are being used).

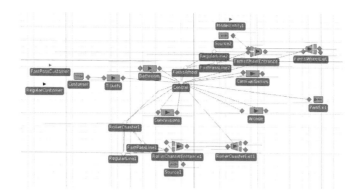

Figure 11.13: Sample Simio model for the amusement park case.

Statistics to Track

- The average and maximum time in queue for both Fast Pass and regular customers for the roller coaster and ferris wheel.

- The average number of Fast Pass and regular customers in the amusement park.

- The average time a Fast Pass and regular customer waits in the queue of a ride.

- The average number of customers waiting and average time waiting at the bathroom.

The manager is interested in the Fast Pass process. Suppose that the park can limit the number of Fast Passes sold (i.e., $x\%$ of customers in the park will have Fast Passes — the manager is confident that s/he can do this through the pricing of the Fast Pass). What proportion of Fast Passes would you recommend to the manager as a limit before satisfaction is lost and customers become unhappy (i.e., what should x be)?

11.2.2 Sample Model and Results

This section gives the results for a sample model that the authors built. As with the previous case study, the specific results depend on a number of assumptions, so your results might not match ours exactly. Here are our model specifications:

- Replication run length: 12 hours

- Base time units: minutes

- 100 Replications

Figure 11.13 shows the Facility View of our sample model. We expect the waiting times for regular and Fast Pass customers to increase as we increase the number of Fast Pass customers in the park. To test this, we added a property that controls the proportion of Fast Pass customers and created an experiment

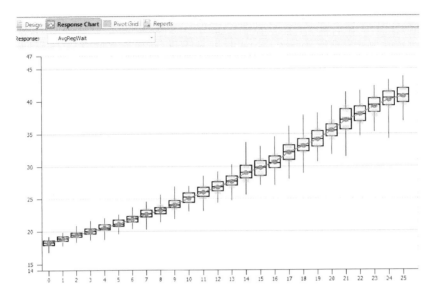

Figure 11.14: SMORE plots of average regular customer wait times by percentage of Fast Pass customers.

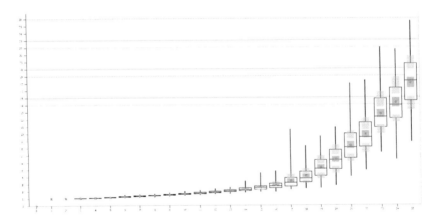

Figure 11.15: SMORE plots of average Fast Pass customer wait times by percentage of Fast Pass customers.

where we varied this property from 0% to 25% in increments of 1%. Figure 11.14 shows the SMORE plot of the response that shows the wait time for regular customers plotted over the percentages of Fast Pass customers. As expected, as the percentage of those who purchase a Fast Pass increases, the wait times increase.

Similarly, Figure 11.15 shows a SMORE plot of the wait time for Fast Pass customers plotted over the percentage of Fast Pass customers. As the percentage of those who purchase a fast pass increases, the wait times also increase.

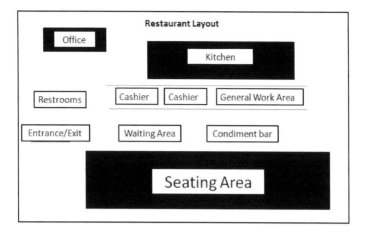

Figure 11.16: Restaurant layout.

In order to determine the "best" proportion of Fast Pass customers, additional cost information for customer waiting would be required. However, the SMORE plots provide relative performance information that would help the manager in making the decision. For example, if 15% of customers have the Fast Pass option, it appears that regular customers would wait between 30 and 35 minutes, and Fast Pass customers would wait approximately 5 minutes. Of course, these estimates are based on a visual inspection of the SMORE plots in Figures 11.14 and 11.15, and additional, more detailed experimentation would be required to get more precise results.

11.3 Simple Restaurant

11.3.1 Problem Description

The layout of the restaurant is shown in Figure 11.16. The restaurant opens at 8:00 a.m., serves breakfast until 11:00 a.m., then serves lunch/dinner until 8:00 p.m. At 8:00 p.m. the doors close, but guests are allowed to stay until they finish their meals. Customers have the option for dine-in or carry-out (To Go). Historically, 10% of customers order To Go and priority is given to their orders while they're being made in the kitchen.

A group enters, is greeted, and is given the estimated wait time (if any). If the group is dining in, they decide whether to stay or to leave (balk). If the group elects to stay, the order is placed and food is prepared in the kitchen. A restaurant staff member delivers the order to the waiting group, then the group either leaves (if the group ordered To Go), or sits down at one of the tables. After a group finishes eating, a restaurant staff member clears and cleans the table so others can sit down.

Customers arrive in groups of size 1, 2, 3, or 4, with probabilities 0.10, 0.30, 0.40, and 0.20 respectively, according to a non-stationary Poisson process with the rate function shown in Table 11.13. The make time for an order depends on

Table 11.13: Restaurant group arrival-rate schedule for an average busy day.

Time interval	Groups per hour
8:00 am - 9:00 am	39
9:00 am - 10:00 am	35
10:00 am - 11:00 am	48
11:00 am - 12:00 pm	44
12:00 pm - 1:00 pm	49
1:00 pm - 2:00 pm	39
2:00 pm - 3:00 pm	26
3:00 pm - 4:00 pm	20
4:00 pm - 5:00 pm	39
5:00 pm - 6:00 pm	48
6:00 pm - 7:00 pm	37
7:00 pm - 8:00 pm	31

Table 11.14: Service-time and worker information; times in minutes.

Service	Service-time distribution	Number of workers
Taking orders	exponential(mean 0.5)	1-2
Making orders	tria(2, 5, 9)	5-10
Cleaning tables	exponential(mean 0.5)	1-2
Distributing	—	1-3

the size of the group (e.g. the make time for a group of three will be triple the make time for one order). See Table 11.14 for the distribution of service times and the possible number of workers performing the respective tasks. Orders sent to the kitchen are grouped accordingly and delivered together.

If the expected wait for a table is longer than 60 minutes or there are more than 5 groups waiting for a table, the group will balk. Currently there are 25 tables in the restaurant. The time it takes for groups to eat has a triangular(25, 45, 60) distribution, in minutes.

Currently there are three types of workers: Kitchen workers, general workers, and managers. The kitchen workers prepare and make the orders sent to the kitchen. General workers can work as a cashier (maximum of 2 registers), distribute completed orders, and clean tables. Managers help out with making and distributing orders, but they do not clean tables or work as cashiers. The manager's priority is making orders in the kitchen. There is at least one manager present at all times. The kitchen gets overcrowded if there are more than ten workers in the kitchen at one time.

Customers waiting longer than 30 minutes (immediately after ordering) become dissatisfied, and food becomes cold eight minutes after being made. Managers are paid $10.00/hour, kitchen workers are paid $7.50/hour, and general workers are paid $6.00/hour. Individual customers spend $9.00 on average.

Current staffing is two managers, eight kitchen workers, and five general workers for a busy day. The owner plans to open a new restaurant with similar expectations and wants to know the following about the current system:

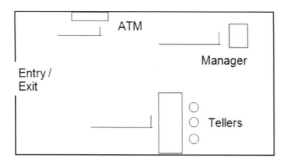

Figure 11.17: Bank layout.

1. How many groups balk due to long waits? What is the average wait for a table?

2. How many managers, kitchen workers, and general workers should be present each day of the week to minimize labor costs and maintain customer satisfaction (regarding wait times and hot food)? What is the total profit (sales minus labor cost) for the scenario that you chose?

3. How many tables are needed to reduce the number of balks due to long waits for a table?

4. Since balks occur frequently due to long waits for tables, the restaurant hasn't experienced its true demand. Create a scenario with the number of tables that you suggested in the previous question. Determine how many workers of each type are needed and the average wait for a table in this system. What is the total profit?

5. If the current rate of arrivals increased by 20%, how many tables would you suggest to reduce the number of balks due to long waits for a table?

6. If the rate of arrivals decreased by 30%, how many workers of each type would you suggest?

11.4 Small Branch Bank

Consider the following setup and operation of a simple bank branch. The bank branch includes an ATM, a bank of tellers, and a manager. The bank branch is open from 9:00 a.m. to 5:00 p.m., five days a week. Customers arriving after 5:00 p.m. will not be allowed to enter, but the bank will stay open long enough to help all customers in the bank when it closes at 5:00 p.m. A block diagram of the bank is given Figure 11.17.

Customers arrive in four independent arrival streams, each of which is a nonstationary Poisson process with rate functions being piecewise-constant as given in Table 11.15. Arriving customers will either go to the bank of tellers (*teller customers*), the ATM (*ATM customers*), or the manager (*manager customers*) for service. Customers who don't care whether they get service from a teller or the ATM (*teller/ATM customers*) will always make their decision

Table 11.15: Arrival rates.

Time period	Arrival rates (customers per hour)			
	Teller	ATM	Teller/ATM	Manager
9:00 – 10:00	20	20	30	4
10:00 – 11:30	40	60	50	6
11:30 – 1:30	25	30	40	3
1:30 – 5:00	35	50	40	7

Table 11.16: Travel times, in seconds.

	Entry/Exit	ATM	Teller	Manager
Entry/Exit	—	10	15	20
ATM	10	—	8	5
Teller	15	8	—	9
Manager	20	5	9	—

based on which line is shortest. If both lines are of equal length, the Teller/ATM customers will always go to the ATM. If the shortest line has more than 4 customers, there is a 70% that customer will leave (balk). *All customers who balk are considered to be "unhappy customers."*

If there are more than 3 people in the ATM line and fewer than 3 people in the teller line when an ATM customer arrives, there is an 85% chance he/she will switch from being an ATM customer to being a teller customer, and will enter the teller queue, but if the teller queue is occupied by 4 or more people there is an 80% chance that this ATM customer will balk. If there are more than 4 people in the teller line when a teller customer arrives, he/she will balk 90% of the time. There has recently been some economic analysis developed that shows that each "unhappy customer" costs the bank approximately $100 for each occurrence.

While most customers are finished and leave the bank after their respective transaction, 2% of the ATM customers and 10% of the teller customers will need to see the manager after completing service at the ATM or teller. In addition, if a customer spends more than five minutes in the teller queue, there is a 50% chance that he or she will get upset and want to see the manager after completing service at the bank of tellers. If a customer spends more than 15 minutes in the manager queue there is a 70% chance that he or she will get upset. All customers who get upset as a result of waiting time should also considered as "unhappy customers." Customer travel times between areas of the bank are given in Table 11.16 and should be included in your model; assume they're all deterministic.

The ATM can serve one customer at a time and has been known to jam at random times, causing delays for customer transactions. The jam has to be fixed by a "hard reset" of the machine. Recently one of the managers had an intern collect data to determine how often the ATM jams and how long it takes to complete the hard reset. These data can be found in the file `ATMdata_bank.xls` in the "students" section of this book's website (see the Preface for instructions on how to access this website). Two histograms in reference to these data for

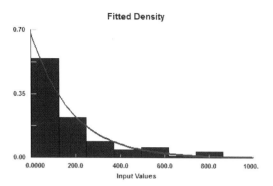

Figure 11.18: A histogram of the data gathered for the ATM up times.

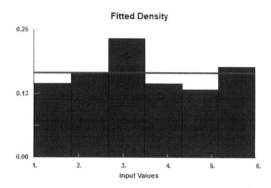

Figure 11.19: A histogram of the data gathered for the time to hard-reset a jammed ATM.

the up times (times from the end of a hard reset to the next time the ATM jams) and time required to reset are shown in Figures 11.18 and 11.19, respectively.

The manager can serve one customer at a time and there is one manager on duty at all times. There are two types of tellers, classified as either *"experienced"* or *"inexperienced."* The bank of tellers has a single queue and each teller can serve only one customer at a time. The hourly costs associated with experienced and inexperienced tellers are $65 and $45, respectively. This cost includes the teller's hourly pay as well as overhead costs (workers compensation, benefits, 401K plans, vacation pay, etc.). Customer service times follow the distributions given in Table 11.17. For the manager, the service times are independent of whether the customer has been to the ATM or the teller prior to visiting the manager.

By convention, the bank divides its days into four different time blocks. The mangers of the bank need your help in determining the number of each type of teller that should be scheduled to work during each time block in order to minimize the costs of operating the bank (for analysis purposes, you should be concerned only with the costs that are provided). Because there are only three

Table 11.17: Service times for the branch bank, in minutes.

Server	Distribution of service time
ATM	tria(0.50, 0.83, 1.33)
Experienced teller	tria(2, 3, 4)
Inexperienced teller	tria(4, 5, 6)
Manager	Exponential (mean 5)

stations in the teller area, there can be no more than three tellers working during any one time block. Each teller must work the entire time block for which s/he is scheduled:

- Time block 1: 9:00 a.m. – 11:00 a.m.

- Time block 2: 11:00 a.m. – 1:00 p.m.

- Time block 3: 1:00 p.m. – 3:00 p.m.

- Time block 4: 3:00 p.m. – 5:00 p.m.

The managers at the bank have also been considering installation of an additional ATM. The ATM supplier has given the bank two different options for purchasing an ATM. The first option is a brand new machine, and the second option is a reconditioned ATM that was previously used in a larger bank. The supplier has broken the cost down into a daily cost of $500 for the new machine and $350 for the reconditioned machine. The reconditioned ATM is assumed to jam and reset at the same rates as the bank's current ATM machine. Although the new ATM machine is assumed to jam just as often as the current one, it has been enhanced with a faster processor that performs automatic resets in 20 seconds. With this information, the bank management wants your recommendation on whether to purchase an additional ATM machine, and if so, which one.

Use Simio to model this small branch bank as it's described above. In addition to your model, provide responses to the following items that were requested by the bank managers. Your recommendations should be based on what minimizes the bank's operating costs with respect to the cost metrics provided in the problem description ("unhappy-customers" cost, additional ATM cost if applicable, and teller cost). Your responses should be supported by your analysis of the results you obtained through experiments.

1. Should the bank purchase an additional ATM machine? If so, should it be new or reconditioned?

2. Find the number of each classification of tellers that the bank should schedule for each of the four time blocks given in the problem description (i.e., a daily schedule for the bank of tellers).

3. What are the bank's expected daily operating costs with respect to the cost metrics provided in the problem description?

4. What is the expected cost per day associated with "unhappy customers?"

Figure 11.20: Vacation City Airport satellite view.

5. How often does the ATM machine jam? How long does it take to repair it? *(The answers to both of these questions should be accompanied by the method you used to obtain them.)*

6. On average, how many customers is the bank serving per day? This value should include all customers who enter the bank and are involved in some sort of transaction (this should not include customers who leave immediately after entering the bank as a result of a line's being too long.)

7. What proportion of the total arriving customers are considered to be "unhappy customers?"

11.5 Vacation City Airport

The Vacation City Airport (VCA) is a sizeable international airport located near a popular tourist destination, which causes highly variable arrival rates depending on the season. VCA management has surveyed the airline companies about customer service level (CSL) and observed that they were unsatisfied with their total time in system for Terminal 2. Management wants to increase CSL in a cost-conscious way. More specifically, they are looking for objective analysis to help decide between adding new gates or adding a new runway, as well as determining the appropriate staffing. The key parameters are time in system for small and large aircraft, gate utilizations, and utilization of the workers for each operation.

VCA has two international terminals (see Figure 11.20) . Terminal 1 has 12 parking gates and 16 lanes. Terminal 2 has 12 gates and 42 lanes. Both terminals are fed by the same three runways, each with different arrival rates. Once an aircraft has landed through one of these runways, it is immediately directed to connecting taxiways to reach the terminal. The aircraft can be directed to parking lanes and gates according to their density. Once parked, it requires

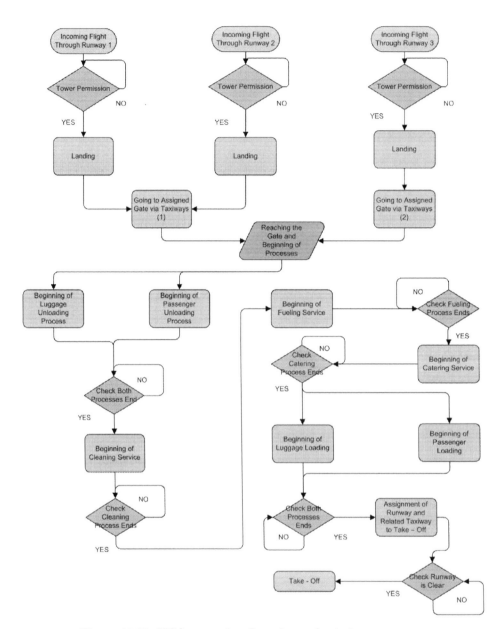

Figure 11.21: VCA operation flow chart of existing system.

several services such as luggage unloading and passenger unloading (which run concurrently), cleaning, fueling, and catering (which run sequentially), and luggage loading and passenger loading (which run concurrently). Each of these service teams has different resources available. Once all services are done, the plane is directed to one of the outbound runways through appropriate taxiways for its take off. The process is summarized in Figure 11.21. Table 11.18 indicates the team sizes and number of teams available in the existing system.

Table 11.18: VCA Terminal 2 existing work force.

Process	Workers per Team	Current Number of Terminal 2 Teams
Luggage	2	6
Passenger (Unloading, Loading)	1	4
Cleaning	2	2
Fueling	1	2
Catering	2	1

Table 11.19: VCA overall plane arrivals for each runway by hour of day.

Time	Runway 1	Runway 2	Runway 3
24 - 1	2	5	3
1 - 2	3	6	2
2 - 3	1	2	2
3 - 4	3	1	1
4 - 5	3	3	2
5 - 6	2	2	3
6 - 7	4	4	4
7 - 8	7	6	6
8 - 9	9	8	8
9 - 10	4	7	5
10 - 11	8	6	7
11 - 12	6	10	8
12 - 13	9	11	11
13 - 14	9	10	10
14 - 15	12	8	9
15 - 16	7	8	7
16 - 17	5	8	6
17 - 18	5	5	3
18 - 19	4	9	7
19 - 20	9	6	6
20 - 21	7	4	5
21 - 22	11	10	9
22 - 23	7	8	8
23 - 24	6	5	4

Plane arrivals depend on the time of day (see Table 11.19). The arrival rate reaches its maximum in the middle of the day. Airport management has categorized aircraft into two groups: large aircraft (at least 120 seats) and small aircraft (fewer than 120 seats). All gates are sized to handle both large and small aircraft. For the planning period in question, the percentage of the above arrivals handled by Terminal 2 are 20% for runway 1, 20% for runway 2 and 30% for runway 3. The balance of the flights are handled by Terminal 1.

Table 11.20: VCA Terminal 2 burdened hourly cost of worker teams.

Process	Rate per Team
Luggage	$142
Passenger (Unloading, Loading)	$40
Cleaning	$108
Fueling	$150
Catering	$145

The cost of a new 11,000 foot runway is $2,200 per linear foot. The cost of a new 21,000 square foot gate is $28 per square foot. The fully-burdened hourly cost (including associated equipment) of each team is indicated in Table 11.20.

VCA management has supplied 100 samples of typical processing times (in minutes) for each of the above operations. These data can be found in the file `AirportCaseStudyInputData.zip` in the student area of the textbook web site as described in the Preface.

Acknowledgments: Vacation City Airport is based on a project by Hazal Karaman and Cagatay Mekiker to solve problems for an actual airport. This project was completed as a major part of a graduate-level introductory simulation course at the University of Pittsburgh.

11.6 Simply The Best Hospital

Simply The Best Hospital (STBH) is a rather large hospital that is adopting a pod configuration of its emergency department (ED). A pod is an enclosed area consisting of seven rooms with staff assigned to that pod. They are planning to have six pods, each staffed by one or two nurses. STBH management wants an evaluation of two things:

- What staffing level is required for each pod to ensure that patients don't have to wait extended periods before being given a room, while not under-utilizing the nurses?

- How much time will it take for a blood test from the time that is ordered until it reaches the lab? The time to carry out the blood test must be kept under fifteen minutes (a policy specific to STBH).

The ED at STBH has two entrances (see Figure 11.22). One entrance is for patients who walk into the ED, and the other entrance is for patients who are brought in by ambulance. Patients who walk in have to go first to the registration desk, where they fill out a form. They then have to go to a triage area where they are diagnosed by a nurse and assigned an acuity. Acuity indicates problem severity (1–5, with 5 being the most serious). When assigned an acuity, the triage nurse enters the patient information into a computer system that assigns every patient to a nurse depending on their availability (if the pod is open, and if they are not busy working with another patient). Then the triage nurse leaves the patient in the corridor for the pod's nurse to pick up and take

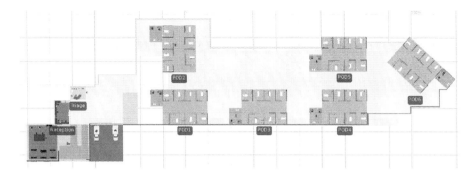

Figure 11.22: STBH facility layout.

Table 11.21: STBH triangular-distribution processing times (minutes), by patient acuity.

Patient Acuity	Registration Time	Triage Time	Time in Room
1	3, 5, 8	4, 5, 7	15, 85, 125
2	3, 5, 8	4, 5, 7	20, 101, 135
3	3, 5, 8	4, 5, 7	35, 150, 210
4	1, 2, 4	1.5, 2, 4	65, 205, 350
5	1, 2, 4	1.5, 2, 4	90, 500, 800

him or her to the entrance of a pod that's open and has a free room. The nurse from the pod then comes to the entrance and takes the patient to his or her room. Here the nurse helps the patient settle into the room, and sets up the necessary equipment. Once the patient has been set up, a doctor comes into the room and may order a blood test for the patient. This happens 75% of the time. When free, the nurse draws blood from the patient, labels it with the patient's name, and then takes it to the hospital unit cleric (HUC). The HUC takes the blood and labels it with the necessary tests to be done, and then sends it to the lab via a chute. The processing time at the HUC's desk has a triangular (3, 4, 5) distribution (in minutes).

The file STBH_ArrivalData.xls in the student area of the textbook web site (see the Preface) contains arrival data by patient acuity level. From past data, it appears that 80% of acuity 4 patients and 95% of acuity 5 patients arrive by ambulance; all other patients arrive as walk-ins. Processing-time data (assumed to be triangularly distributed with the given minima, modes, and maxima, in minutes) are in Table 11.21.

Patients in triage have their blood pressure and temperature taken, and their general complaints are noted. Patients complaining of serious illnesses have expedited triage times, and are taken to the rooms for treatment as soon as possible. Allocation of rooms to patients uses the following strategy: First send incoming patients to pod 1, and when it's full (all seven rooms are occupied), the next patient will be sent to pod 2 until it fills up, and so on. In the meantime, if a patient has been discharged from pod 1, the next patient to enter the system will be taken to pod 1. This is done so that the minimum numbers of pods are open at any given time, which helps the hospital control costs and staffing. Patients leaving the ED are considered to have left the ED, regardless of whether they leave the hospital.

The key measures that STBH management are looking for are:

- ATWA: Average time in waiting area (average across all types of patients entering the system via walk in).

- NBT: Number of blood tests conducted.

- ATB: Average time for blood test.

- A1TS, . . . A5TS: Time in system for each of the 5 acuity levels.

STBH management is particularly interested in six scenarios (other scenarios may also be evaluated):

1. Model a day when the staffing of the hospital is lower than the usual level (10 nurses instead of 12), and there happens to be a mass accident like a train or plane crash so about 54 patients of acuity 4 and 5 are brought to the ED.

2. Model a period when all the nurses are present but we have roughly 72 more patients of acuities 1, 2, and 3 entering the system during a period of two and a half days. This scenario could be one where a minor epidemic occurs and many people have sore throats, headaches, or upset stomachs.

3. Compare a regular week to one where the staffing is low (one nurse per pod) and where the number of patients coming into the hospital is also lower than the norm (200 fewer patients for the week).

4. Compare the normal running of the ED in a week where the service times for all the patients increase by 10% throughout.

5. Model a weekend (after a normal week) with staffing levels of one nurse per pod.

6. Model a period where there is a virus and the ED has 120 patients of acuity 1 and 2 enter over three days. Assume a staffing level that is lower than normal, i.e., there are three pods with two nurses each and the remaining three pods have only one nurse each. Since this involves a virus, assume that 90% of the patients will have blood tests.

Acknowledgments: Simply The Best Hospital is based on a project by Karun Alaganan and Daniel Marquez to solve problems for an actual hospital. This project was completed as a major part of a graduate-level introductory simulation course at the University of Pittsburgh.

Appendix A

Simulation-based Scheduling

As discussed in Chapter 1, and illustrated throughout the book, simulation has been used primarily to enhance the design of a system — to compare alternatives, optimize parameters, and predict performance. Although having a good system design is certainly valuable, there is a parallel set of problems in effectively operating a system.

In the first section of this appendix we will explore some current technology used for planning and scheduling, and some opportunities for improvement. This section is not intended as a thorough treatment, but rather a quick overview of a few concepts and common problems. (For more in-depth coverage see Factory Physics [18].) Then we will discuss Risk-based Planning and Scheduling (RPS), a new technology recently introduced to capitalize on some of the discussed opportunities. We will round out this material with a discussion of the Simio tools available to support RPS and then provide an example using instructions supplied with the Simio software.

A.1 Unsolved Problems in Planning and Scheduling

Planning and scheduling are often discussed together because they are related applications. *Planning* is the "big-picture" analysis – how much can or should be made, when, where, and how, and what materials and resources will be required to make it? *Scheduling* is concerned with the operational details – given the current production situation and work in progress (WIP), what priorities, sequencing, and tactical decisions will result in best meeting the important goals? Where planning is required days, weeks or months ahead of execution, scheduling is often done only minutes, hours, or days ahead.

In many applications, planning and scheduling tasks are done separately. In fact, it is not unusual for only one to be done while the other may be ignored. One simple type of planning is based on lead times. For example, if averages have historically indicated that most parts of a certain type are "normally" shipped 3 weeks after order release, it will be assumed that — regardless of other factors — when we want to produce one, we should allow 3 weeks. This often does not adequately account for resource utilization. If you have more parts in process than "normal," the lead times may be optimistic. Another simple type of planning uses a magnetic board, white board, or a spreadsheet to manually create a Gantt chart to show how parts move through the system and how resources are utilized. This can be a very labor-intensive operation, and the quality of the resulting plans may be highly variable, depending on the complexity of the system and the experience level of the planners.

A third planning option is a purpose-built system - a system that is designed and developed using custom algorithms usually expressed in a programming language. These are highly customized to a particular domain and a particular system. Although they have the potential to perform quite well, they often have a very high cost and implementation time and low opportunity for reuse because of the level of customization.

One of the most popular general techniques is *Advanced Planning and Scheduling* (APS). APS is a process that allocates production capacity, resources, and materials optimally to meet production demand. There are a number of APS products on the market designed to integrate detailed production scheduling into the overall Enterprise Resource Planning (ERP) solution, but these solutions have some widely recognized shortcomings. For the most part the ERP system and day-to-day production remain disconnected largely due to two limitations that impede their success: Complexity and Variation.

Complexity. The first limitation is the inability to effectively deal with indeterminately complex systems. Although purpose-built systems can potentially represent any system, the cost and time required to create a detailed, custom-built system often prevents it from being a practical solution. Techniques such as the others discussed above tend to work well if the system is very close to a standard benchmark implementation, but to the extent the system varies from that benchmark, the tool may lack enough detail to provide an adequate solution.

Variation. A second limitation is the inability to effectively deal with variation within the system. All processing times must be known and all other variability is typically ignored. For example, unpredictable downtimes and machine failures aren't explicitly accounted for; problems with workers and materials never occur, and other negative events don't happen. The resulting plan is by nature overly optimistic. Figure A.1. illustrates a typical scheduling output in the form of a Gantt chart where the green dashed line indicates the slack between the (black) planned completion date and the (gray) due date. Unfortunately it is difficult to determine if the planned slack is enough. It is common that what starts off as a feasible schedule turns infeasible over time as variation and unplanned events degrade performance. It is normal to have large discrepancies between predicted schedules and actual performance. To protect against delays, the scheduler must buffer with some combination of extra time, inventory, or capacity; all these add cost to the system.

The problem of generating a schedule that is feasible given a limited set of capacitated resources (e.g. workers, machines, transportation devices) is typically referred to as Finite Capacity Scheduling. There are two basic approaches to Finite Capacity Scheduling. The first approach is an optimization approach in which the system is defined by a set of mathematical relationships expressed as constraints. An algorithmic Solver is then used to find a solution to the mathematical model that satisfies the constraints while striving to meet an objective such as minimizing the number of tardy jobs. Unfortunately these

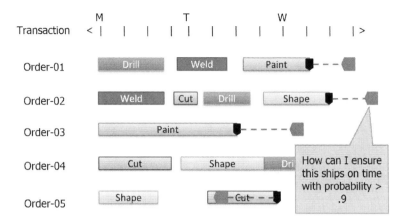

Figure A.1: Typical Gantt chart produced in planning.

mathematical models fall into a class of problems referred to as NP-Hard for which there are no known efficient algorithms for finding an optimal solution. Hence, in practice, heuristic solvers must be used that are intended to find a "good" solution as opposed to an optimal solution to the scheduling problem. Two well-known examples of commercial products that use this approach are the ILOG product family (CPLEX) from IBM, and APO-PP/DS from SAP.

The mathematical approach to scheduling has a number of well-known shortcomings. Representing the system by a set of mathematical constraints is a very complex and expensive process, and the mathematical model is difficult to maintain over time as the system changes. In addition, there may be many important constraints in the real system that cannot be accurately modeled using the mathematical constraints and must be ignored. The resulting schedules may satisfy the mathematical model, but are not feasible in the real system. Finally the solvers used to generate a solution to the mathematical model often take many hours to produce a good candidate schedule. Hence these schedules are often run overnight or over the weekend. The resulting schedules typically have a short useful life because they are quickly outdated as unplanned events occur (e.g. a machine breaks down, material arrives late, workers call in sick).

A.2 Simulation-based Scheduling

The second approach to Finite Capacity Scheduling is based on using a simulation model to capture the limited resources in the system. The concept of using simulation tools as a planning and scheduling aid has been around for decades. One of the authors used simulation to develop a steel-making scheduling system in the early 1980s. In scheduling applications we initialize the simulation model to the current state of the system and simulate the flow of the actual planned work through the model. To generate the schedule, we must eliminate all variation and unplanned events when executing the simulation.

Simulation-based scheduling generates a heuristic solution – but is able to do

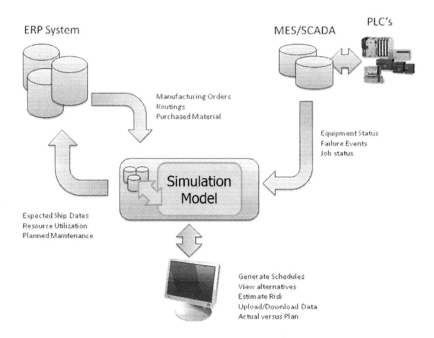

Figure A.2: Architecture of a typical simulation-based scheduling system.

so in a fraction of the time required by the optimization approach. The quality of the simulation-based schedule is determined based on the decision logic that allocates limited resources to activities within the model. For example, when a resource such as a machine goes idle, a rule within the model is used to select the next entity for processing. This rule might be a simple static ranking rule such as the highest priority job. or a more complex dynamic selection rule such as a rule that minimizes a sequence dependent setup time, or a rule that selects the job based on urgency by picking the job with the smallest value of the time remaining until the due date, divided by the work time remaining (critical ratio).

Many of the simulation-based scheduling systems have been developed around a data-driven pre-existing, or "canned," job shop model of the system. For example, the system is viewed as a collection of workstations, where each workstation is broken into a setup, processing, and teardown phase, and each job that moves through the system follows a specific routing from workstation to workstation. The software is configured using data to describe the workstations, materials, and jobs. If the application is a good match for the canned model, it may provide a good solution; if not, there is limited opportunity to customize the model to your needs. You may be forced to ignore critical constraints that exist in the real system but are not included in the canned model.

It is also possible to use a general purpose discrete event simulation (DES) product for Finite Capacity Scheduling. Figure A.2 illustrates a typical architecture for using a DES engine at the core of a planning and scheduling system. The advantages of this approach include:

- It is flexible. A general-purpose tool can model any important aspects of the system. Just like in a model built for system design.

- It is scalable. Again, similar to simulations for design, it can (and should) be done iteratively. You can solve part of the problem and then start using the solution. Iteratively add model breadth and depth as needed until the model provides the schedule accuracy you desire.

- It can leverage previous work. Since the system model required for scheduling is very similar to that which is needed (and hopefully was already used) to fine tune your design, you can extend the use of that design model for planning and scheduling.

- It can operate stochastically. Just as design models use stochastic analysis to evaluate system configuration, a planning model can stochastically evaluate work rules and other operational characteristics of a scheduling system. This can result in a "smarter" scheduling system that makes better decisions from the start.

- It can be deterministic. You can disable the stochastic capabilities while you generate a deterministic schedule. This will still result in an optimistic schedule as discussed above, but because of the high level of detail possible, this will tend to be more accurate than a schedule based on other tools. And you can evaluate how optimistic it is (see next point).

- It can evaluate risk. It can use the built-in stochastic capability to run AFTER the deterministic plan has been generated. By again turning on the variation – all the bad things that are likely to happen - and running multiple replications against that plan, you can evaluate how likely you are to achieve important performance targets. You can use this information to objectively adjust the schedule to manage the risk in the most cost effective way.

- It supports any desired performance measures. The model can collect key information about performance targets at any time during model execution, so you can measure the viability and risk of a schedule in any way that is meaningful to you.

However there are also some unique challenges in trying to use a general purpose DES product for scheduling, since they have not been specifically designed for that purpose. Some of the issues that might occur include the following:

- Scheduling Results: A general purpose DES typically presents summary statistics on key system parameters such as throughput and utilization. Although these are still relevant, the main focus in scheduling applications is on individual jobs (entities) and resources, often presented in the form of a Gantt chart or detailed tracking logs. This level of detail is typically not automatically recorded in a general purpose DES product.

- Model Initialization: In design applications of simulation we often start the model empty and idle and then discard the initial portion of the simulation to eliminate bias. In scheduling applications, it is critical that we are able to initialize the model to the current state of the system – including jobs that are in process and at different points in their routing through the system. This is not easily done with most DES products.

- Controlling Randomness: Our DES model typically contains random times (e.g. processing times) and events (e.g. machine breakdowns). During generation of a plan, we want to be able to use the expected times, and turn off all random events. However once the plan is generated, we would like to include variation in the model to evaluate the risk with the plan. A typical DES product is not designed to support both modes of operation.

- Interfacing to Enterprise Data: The information that is required to drive a planning or scheduling model typically resides in the companys ERP system or databases. In either case, the information typically involves complex data relations between multiple data tables. Most DES products are not designed to interface to or work with relational data sources.

- Updating Status: The planning and scheduling model must continually adjust to changes that take place in the actual system e.g. machine breakdowns. This requires an interactive interface for entering status changes.

- Scheduling User Interface: A typical DES product has a user interface that is designed to support the building and running of design models. In scheduling and planning applications, a specialized user interface is required by the staff that employs an existing model (developed by someone else) to generate plans and evaluate risk across a set of potential operational decisions (e.g. adding overtime or expediting material shipments).

A new approach, Risk-based Planning and Scheduling (RPS), is designed to overcome these shortcomings to fully capitalize on the significant advantages of a simulation approach.

A.3   Risk-based Planning and Scheduling

Risk-based Planning and Scheduling (RPS) is a tool that combines deterministic and stochastic simulation to bring the full power of traditional DES to operational planning and scheduling applications [55]. RPS extends traditional APS to fully account for the variation that is present in nearly every production system, and provides the necessary information to the scheduler to allow the upfront mitigation of risk and uncertainty. RPS makes dual use of the underlying simulation model. The simulation model can be built at any level of detail and can incorporate all of the random variation that is present in the real system.

RPS begins by generating a deterministic schedule by executing the simulation model with randomness disabled (deterministic mode). This is roughly equivalent to the deterministic schedule produced by an APS solution. However,

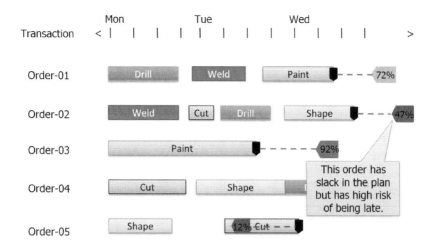

Figure A.3: Gantt chart identifying a high-risk order.

RPS then uses the same simulation model with randomness enabled (stochastic) to replicate the schedule generation multiple times (employing multiple processers when available), and record statistics on the schedule performance across replications. The recorded performance measures include the likelihood of meeting a target (such as a due date), the expected milestone completion date (typically later than the planned date based on the underlying variation in the system), as well as optimistic and pessimistic completion times (percentile estimates, also based on variation). Contrast Figure A.1 with the RPS analysis presented in Figure A.3 . Here the risk analysis has identified that even though Order-02 appears to have adequate slack, there is a relatively low likelihood (47%) that it will complete on time after considering the risk associated with that particular order, and the resources and materials it requires. Having an objective measure of risk while still in the plan development phase provides the opportunity to mitigate risk in the most effective way.

RPS uses a simulation-based approach to scheduling that is built around a purpose-built simulation model of the system. The key advantage of this is that the full modeling power of the simulation software is available to fully capture the constraints in your system. You can model your system using the complete simulation toolkit. You can use custom objects for modeling complex systems (if your simulation software provides that capability). You can include moving material devices, such as forklift trucks or AGVs (along with the congestion that occurs on their travel paths), as well as complex material handling devices such as cranes and conveyors. You can also accurately model complex workstations such as ovens and machining centers with tool changers.

RPS imposes no restrictions on the type and number of constraints included in the model. You no longer have to assume away critical constraints in your production system. You can generate both the deterministic plan and associated risk analysis using a model that fully captures the realities of your complex production and supply chain. You can also use the same model that is developed for evaluating changes to your facility design to drive an RPS installation which means a single model can be used to drive improvements to your facility design as well as to your day-to-day operations.

A.4   Planning and Scheduling With Simio Enterprise

Simio Enterprise Edition is an extended version of Simio that has additional features specifically designed for planning and scheduling applications. This extended capability is not part of every academic package, but if you do not currently have it, instructors can apply for an upgrade for both institution and student software by sending a request to academic@simio.com. The simplest way to determine whether you have Enterprise and RPS capability is to look for a **Planning** tab just above the Facility view — many of the Enterprise features are accessed within the Planning tab.

Although the model used to drive a Simio RPS solution can be built using any of the Simio family of products, the Simio Enterprise Edition is required for preparing the model for RPS deployment. This preparation includes adding output states and scheduling targets to tables, and the customizing the user interface for the scheduler. In traditional Simio applications, the data tables are only used to supply inputs to the model. However in RPS applications, the data tables are also used to record values during the running of the model. This is done by adding state columns to table definitions, in addition to the standard property columns. For example a date-time table state might be used to record the ship date for a job, and a real table state value might be used to record the accumulated cost for the job. Table states can be written to by using state assignments on the objects in the Standard Library, or more flexibly by using the Assign step in any process logic. Output state columns are added to tables using the States ribbon in the Data tab of the Simio Enterprise Edition.

A scheduling *target* is a value that corresponds to a desired outcome in the schedule. A classic example of a target is the ship date for a job (we would like to complete the job before its due date). However, targets are not limited to completion dates. Targets can report on anything in the simulation that can be measured (e.g., material arrivals, proficiency, cost, yield, quality, OEE) and at any level (e.g., overall performance, departmental, sub-assemblies, and other milestones).

Targets are defined by an expression that specifies the value of the target, along with upper and lower bounds on that value, and labels for each range relative to these bounds. For example, a target for a due date might label the range above the upper limit as "Late," and the range below the upper limit as "On Time." Simio will then report statistics using the words Late and On Time. Figure A.4 illustrates the definition of a target (TargetShipDate) based on comparing the table state ManufacturingOrders.ShipDate to the table property ManufacturingOrders.DueDate. It has three possible outcomes, OnTime, Late, or Incomplete. In a similar fashion, a cost target might have its ranges labeled "Cost Overrun" and "On Budget." Targets can be based on date/time, or general values such as the total production cost. Some targets, such as due date or cost, are values that we want to be below their upper bound; others, such as work completed, are values we want to be above their lower bound. A

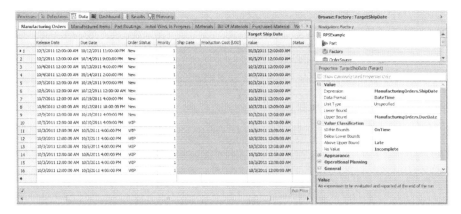

Figure A.4: Defining a target based on table states and properties.

special Target ribbon is provided in the Data tab of Simio Enterprise Edition for adding targets to data tables.

Simio automatically records performance of each target value relative to its defined bounds. In a deterministic plan run, it simply records data on where the target value ended up relative to its defined bounds (On Time, Cost Overrun, etc.). However, when performing risk analysis based on variation and unplanned events, Simio records the performance of each target value across all replications of the model and then uses this information to compute risk measures such as the probability of on time delivery, and the expected/optimistic/pessimistic ship date.

The standard Simio user interface is focused on model building and experimentation. In RPS applications, there is a need for a different user interface that is tailored for the planning and scheduling staff. They do not build models, but instead use models in an operational setting to generate plans and schedules. A separate and dedicated user interface is needed for the planning and scheduling user, which is provided in Simio Scheduling Edition. Simio Enterprise Edition lets you easily tailor the user interface for the planning and scheduling staff who will use the Scheduling Edition. You can fully configure the types of data that the scheduler can view and edit, and fully customize the implementation to specific application areas

A.5 Deployment with Simio Scheduling Edition

Simio Scheduling Edition (SSE) is the deployment software used with Enterprise Edition. Academic installations do not have direct access to SSE, but it is important to understand the looks and capability of the tool that would be provided to a scheduler. You can use your Enterprise software to visualize that tool simply by shifting to the Planning tab and ignoring Simio components outside of the Planning tab.

SSE provides a specialized user interface for the planning and scheduling staff. This user interface is smaller and simpler than the standard Simio user interface, since it provides no support for building new models. SSE executes a

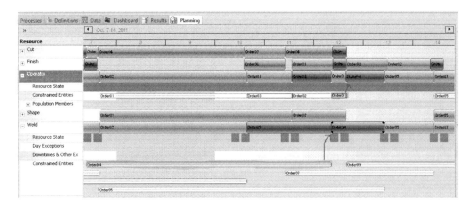

Figure A.5: Examine constraints to investigate why Order-04 is late.

Simio model that was built using one of the standard Simio products, and then prepared for deployment using the Enterprise Edition.

The main purpose of SSE is to generate a plan/schedule by running the planned jobs through the model in a deterministic mode, where all random times are replaced by their expected values, and all random events are suppressed. This deterministic schedule (like all deterministic schedules) is optimistic; however we can then use the features in SSE to analyze random events.

SSE provides a number of both static and dynamic graphic views of the resulting plan/schedule, along with a number of specialized reports. The graphic views include both entity and resource centric Gantt charts, as well as a 3D animation of the schedule generations. The entity Gantt chart displays each entity (e.g., a job) as a row in the Gantt, and rectangles drawn along a horizontal date-time scale depict the time span that each resource was held by that entity. The resource Gantt chart displays each resource (e.g. machine or worker) as a row in the Gantt, and rectangles drawn along a horizontal time scale depict the entities that used this resource. Both of these Gantt charts are part of the Simio example named RPSExample that will be discussed later. Figure A.5 illustrates a resource Gantt chart from RPS Example with the resource constraints partially exploded. It is annotated to identify some contributors to the lateness of Order-04, notably a long time waiting for the weld machine and then additional time waiting for the operator required to run it.

The specialized reports include a work-to list for each resource. The list defines the start and end time for each operation to be performed by that resource during the planning period. This work-to list is typically provided to the operator of each workstation in the system to provide a preview of the expected workload at that workstation. The standard reports also include a resource utilization report to show the expected state over time during the planning period, and a constraint report to summarize the constraints (material not available, workers were busy, etc.) that created non-value added time in the system during the planning period. Figure A.6 shows an example of a resource dispatch list for the operator in RPS Example. More examples of these reports will be covered in the study of RPS Example that follows.

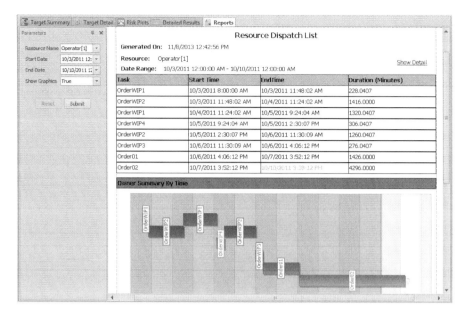

Figure A.6: Resource dispatch list for operator.

SSE can also display custom dashboards that have been designed using the Enterprise Edition to summarize information on a specific job, machine, worker, or material. These dashboards can combine several types of graphical and tabular schedule information into a single summary view of a job or machine to provide additional insights into a plan/schedule.

SSE allows the planners and schedulers to test out alternatives to improve a high risk schedule. Alternatives can be evaluated to see the impact of over time, expediting of materials, or reprioritizing work on meeting specific targets.

A.6 RPS Example

The Simio software includes an example named RPSExample[1] that can be loaded from the Examples button on the Support ribbon. This is a simple job shop example with four workstations, three part types, and a single worker. Its many components illustrate the various RPS features and many of the issues that might commonly arise. A description file of over 30 pages accompanies that example and does a good job describing those features, so we refer you to that file and will not duplicate that information here.

A.7 Summary

In this appendix we have discussed some common approaches to planning and scheduling and their strengths and weaknesses. Then we explored new technol-

---

[1]Simio 5.91.10321 and earlier permit only one target to be defined, so you will need to delete one of the two targets before running.

ogy to use simulation in Risk-based Planning and Scheduling to address many of those problems and provide important new capabilities to schedulers. We briefly introduced some of the Simio features that make RPS possible and concluded with a small example to illustrate the concepts. Although this was only a brief introduction to RPS, we hope that we have sparked your interest. There are many other potential advantages to using RPS, and new application areas have opened up. This could be a rich area for student projects and research.

Bibliography

[1] S.C. Albright, W.L. Winston, and C.J. Zappe. *Data Analysis and Decision Making With Microsoft Excel*. South-Western Cengage Learning, Mason, Ohio, revised third edition, 2009.

[2] R.G. Askin and C.R. Standridge. *Modeling and Analysis of Manufacturing Systems*. Wiley, New York, 1993.

[3] J. Banks, J.S. Carson II, B.L. Nelson, and D.M. Nicol. *Discrete-Event System Simulation*. Pearson Prentice Hall, Upper Saddle River, New Jersey, fourth edition, 2005.

[4] B. Biller and B.L. Nelson. Fitting time series input processes for simulation. *Operations Research*, 53:549–559, 2005.

[5] G.M. Birtwistle, O.-J. Dahl, B. Myhraug, and K. Nygaard. *Simula BEGIN*. Auerbach, Philadelphia, Pennsylvania, 1973.

[6] J. Boesel, B.L. Nelson, and S. Kim. Using ranking and selection to "clean up" after simulation optimization. *Operations Research*, 51:814–825, 2003.

[7] R. Conway, W. Maxwell, J.O. McClain, and L.J. Thomas. The role of work-in-process inventory in serial production lines. *Operations Research*, 35(2):229–241, 1988.

[8] L. Devroye. *Non-Uniform Random Variate Generation*. Springer-Verlag, New York, 1986.

[9] A.K. Erlang. The theory of probabilities and telephone conversation. *Nyt Tidsskrift for Matematik*, 20, 1909.

[10] A.K. Erlang. Solution of some problems in the theory of probabilities of significance in automatic telephone exchanges. *Elektrotkeknikeren*, 13, 1917.

[11] M. Evans, N. Hastings, and B. Peacock. *Statistical Distributions*. Wiley, New York, second edition, 2000.

[12] Gartner. Gartner identifies the top 10 strategic technologies for 2010. *Gartner Symposium/ITxpo*, 2009.

[13] Gartner. Gartner identifies the top 10 strategic technology trends for 2013. *Gartner Symposium/ITxpo*, 2012.

[14] L.J. Gleser. Exact power of goodness-of-fit tests of kolmogorov type for discontinuous distributions. *Journal of the American Statistical Society*, 80:954–958, 1985.

[15] F. Glover and G.A. Kochenberger, editors. *Handbook of Metaheuristics*. Kluwer Academic Publishers, Norwell, Massachusetts, 2003.

[16] D. Gross, J.F. Shortle, J.M. Thompson, and C.M. Harris. *Fundamentals of Queueing Theory*. Wiley, Hoboken, New Jersey, fourth edition, 2008.

[17] S. Harrod and W.D. Kelton. Numerical methods for realizing nonstationary poisson processes with piecewise-constant instantaneous-rate functions. *Simulation: Transactions of The Society for Modeling and Simulation International*, 82:147–157, 2006.

[18] W.J. Hopp and M.L. Spreaman. *Factory Physics*. McGraw Hill, New York, New York, 2008.

[19] J.R. Jackson. Networks of waiting lines. *Operations Research*, 5:518–521, 1957.

[20] N.L. Johnson, A.W. Kemp, and S. Kotz. *Univariate Discrete Distributions*. Wiley, New York, third edition, 2005.

[21] N.L. Johnson, S. Kotz, and N. Balakrishnan. *Continuous Univariate Distributions*, volume 1. Wiley, New York, second edition, 1994.

[22] N.L. Johnson, S. Kotz, and N. Balakrishnan. *Continuous Univariate Distributions*, volume 2. Wiley, New York, second edition, 1995.

[23] W.D. Kelton. Implementing representations of uncertainty. In S.G. Henderson and B.L. Nelson, editors, *Handbook in Operations Research and Management Science, Vol. 13: Simulation*, pages 181–191. Elsevier North-Holland, Amsterdam, The Netherlands, 2006.

[24] W.D. Kelton. Representing and generating uncertainty effectively. *Proceedings of the 2009 Winter Simulation Conference*, pages 40–44, 2009.

[25] W.D. Kelton, R.P. Sadowski, and N.B. Zupick. *Simulation With Arena*. McGraw-Hill Education, New York, sixth edition, 2015.

[26] S. Kim and B.L. Nelson. A fully sequential procedure for indifference-zone selection in simulation. *ACM Transactions on Modeling and Computer Simulation*, 11:251–273, 2001.

[27] P.J. Kiviat, R. Villanueva, and H.M. Markowitz. *The SIMSCRIPT II Programming Language*. Prentice Hall, Englwood Cliffs, New Jersey, 1969.

[28] L. Kleinrock. *Queueing Systems: Volume I - Theory*. Wiley, New York, 1975.

[29] M. Kuhl, S.G. Sumant, and J.R. Wilson. An automated multiresolution procedure for modeling complex arrival processes. *INFORMS Journal on Computing*, 18:3–18, 2006.

[30] A.M. Law. *Simulation Modeling and Analysis*. McGraw-Hill, New York, fourth edition, 2007.

[31] P. L'Ecuyer. Uniform random number generation. In S.G. Henderson and B.L. Nelson, editors, *Handbook in Operations Research and Management Science, Vol. 13: Simulation*, pages 55–81. Elsevier North-Holland, Amsterdam, The Netherlands, 2006.

[32] P. L'Ecuyer and R. Simard. Testu01: A c library for empirical testing of random number generators. *ACM Transactions on Mathematical Software*, 337:Article 22, 2007.

[33] L. Leemis. Nonparametric estimation of the intensity function for a non-homogeneous poisson process. *Management Science*, 37:886–900, 1991.

[34] D.H. Lehmer. Mathematical methods in large-scale computing units. *Annals of the Computing Laboratory of Harvard University*, 26:141–146, 1951.

[35] D.V. Lindley. The theory of queues with a single server. *Proceedings of the Cambridge Philosophical Society*, 48:277–289, 1952.

[36] J.D.C. Little. A proof for the queuing formula $l = \lambda w$. *Operations Research*, 9:383–387, 1961.

[37] J.D.C. Little. Little's law as viewed on its 50th anniversary. *Operations Research*, 59:536–549, 2011.

[38] M. Matsumoto and T. Nishimura. Mersenne twister: A 623-dimensionally equidistributed uniform pseudo-random number generator. *ACM Transactions on Modeling and Computer Simulation*, 8:3–30, 1998.

[39] L. Mueller. Norwood fire-department simulation models: Present and future. Master's thesis, Department of Quantitative Analysis and Operations Management, University of Cincinnati, 2009.

[40] R.E. Nance and R.G. Sargent. Perspectives on the evolution of simulation. *Operations Research*, 50:161–172, 2002.

[41] B.L. Nelson. *Stochastic Modeling, Analysis and Simulation*. McGraw-Hill, New York, first edition, 1995.

[42] B.L Nelson. The more plot: Displaying measures of risk & error from simulation output. *Proceedings of the 2008 Winter Simulation Conference*, pages 413–416, 2008.

[43] B.L. Nelson. Personal communication. 2013.

[44] C.D. Pegden, R.E. Shannon, and R.P. Sadowski. *Introduction to Simulation Using SIMAN*. McGraw-Hill, New York, second edition, 1995.

[45] C.D Pegden and D.T. Sturrock. *Rapid Modeling Solutions: Introduction to Simulation and Simio*. Simio LLC, Pittsburgh, Pennsylvania, 2013.

[46] A.A.B. Pritsker. *The GASP IV Simulation Language*. Wiley, New York, 1974.

[47] A.A.B. Pritsker. *Introduction to Simulation and SLAM II*. Wiley, New York, fourth edition, 1995.

[48] D. Robb. Gartner taps predictive analytics as next big business intelligence trend. *Enterprise Apps Today*, 2012.

[49] S.M. Ross. *Introduction to Probability Models*. Academic Press, Burlington, Massachusetts, tenth edition, 2010.

[50] S.L. Savage. *The Flaw of Averages: Why We Underestimate Risk in the Face of Uncertainty*. Wiley, Hoboken, New Jersey, 2009.

[51] T.J. Schriber. *Simulation Using GPSS*. Wiley, New York, 1974.

[52] T.J. Schriber. *An Introduction to Simulation Using GPSS/H*. Wiley, New York, 1991.

[53] A.F. Seila, V. Ceric, and P. Tadikamalla. *Applied Simulation Modeling*. Thomson Brooks/Cole, Belmont, California, 2003.

[54] W.S. Stidham. A last word on $l = \lambda w$. *Operations Research*, 22:417–421, 1974.

[55] D.T. Sturrock. New solutions for production dilemmas. *Industrial Engineer Magazine*, 44:47–52, 2012.

[56] D.T. Sturrock. Simulationist bill of rights. *Success in Simulation Blog*, 2012.

[57] J.J. Swain. Simulation: A better reality? (simulation software survey). *OR/MS Today*, 40(5):48–59, 2013.

[58] J.W. Tukey. *Exploratory Data Analysis*. Addison-Wesley, Reading, Massachusetts, 1977.

[59] H.M. Wagner. *Principles of Operations Reserach, With Applications to Managerial Decisions*. Prentice Hall, Englwood Cliffs, New Jersey, 1969.

Index